PRIMARY PRODUCTIVITY IN AQUATIC ENVIRONMENTS

Primary Productivity in Aquatic Environments

Proceedings of an I.B.P. PF Symposium
Pallanza, Italy, April 26–May 1, 1965
Edited by C. R. Goldman

UNIVERSITY OF CALIFORNIA PRESS
Berkeley, Los Angeles, London

University of California Press
Berkeley and Los Angeles, California

University of California Press, Ltd.
London, England

Third Printing, 1974
ISBN: 0-520-01425-1

Library of Congress Catalog Card Number: 66-23180

Printed in the United States of America

FOREWORD

Welcome to Pallanza and to the Istituto Italiano di Idrobiologia.

I do not want to take the valuable time of this audience to recall the scope and the philosophy of the International Biological Programme. I am quite sure that all of you know that its principal purpose is the basic study of productivity and that we are all therefore perfectly on the crest of this wave.

It would perhaps be interesting to inform you that this Symposium is the first one to be held in the I.B.P., and it is of some significance that is concerned with primary productivity in water, « mater vitae ». I sincerely believe that your experience and brilliant talents will produce a first class first I.B.P. symposium on the first link of the aquatic food-chain, which will stimulate the future I.B.P. symposia dealing with successive steps.

As you know, a triad, Dr. C. R. Goldman, Dr. R. A. Vollenweider and myself, is responsible for the organization of this meeting, but I must acknowledge that the first two took on their shoulders most of the work for the realization of the scientific scheme. The privilege of having more white hair is the only reason I am speaking to you first.

We organizers are convinced that this convention of marine and fresh water specialists will not produce a meromictic state of affairs, because the eddy diffusion of ideas and the desirable turbulence of discussion will produce a mixed water, flavoured « cum sale sapientiae ».

The first three days will be spent in listening to and discussing individual contributions, which, as you have agreed, should be printed as proceedings in a Supplement volume of the « Memorie » of this Institute. The sessions will be directed by chairmen, who will carefully control the time allotted for each paper.

The last three days, as it has already been announced, will be devoted to discussions, through which decisions will be reached for the preparation of a methodological manual on primary productivity in water. This will greatly help the future work of many scientists in the choice of the most suitable methods of measurement, offering them, whenever possible, intercalibration procedures.

Working groups have been selected for this second half of the week. Outlines of what form the manual may take have already been distributed. This second part of our work requires our deepest attention, because of the large distribution which the Central Office of I.B.P. intends to give to these methodological manuals, and because the financial support we have received for the organization of this Symposium, namely from I.B.P. itself and from the Italian National Research Council, was intended primarily for this purpose.

And now, « buon lavoro »!

Vittorio Tonolli

Verbania Pallanza, 26 April 1965

PREFACE

The International Biological Programme has aptly chosen « The Biological Basis of Productivity and Human Welfare » as the integrating theme for aquatic and terrestrial environments. The section of this program on productivity of freshwater logically selected primary productivity as the subject for the first symposium, leaving higher levels of the food pyramid for subsequent symposia.

In the last decade there have been many meetings on problems of primary productivity in aquatic food chains (*e.g.,* East Lansing 1955; Bergen 1957; Honolulu 1961; Princeton 1961, 1962; Edinburgh 1964). The great research effort that has been and will continue to be expended on investigations of primary productivity in both marine and freshwaters is good reason for repeated re-evaluation of both methods and principles. The symposium here reported constitutes a conscious effort to extend rather than duplicate its predecessors at a time when both theory and methods of measurement deserve particular re-evaluation.

To prevent a « meromixis of ideas » as Professor Tonolli so aptly describes it, the organizers assembled an international group with both marine and freshwater experience in the general areas of Limnology, Oceanography, Plant Physiology, and Biochemistry. The charge to these specialists from twelve countries was to review the achievements in primary productivity research, as well as to discuss and formulate ideas for further exploration of the most promising areas of research.

A subject as broad as primary productivity in aquatic environments does not lend itself to rigid organization in this volume. The general categories under which the papers are listed have been selected with full recognition of their not inconsiderable interrelationship; they are intended only as a general guide to the reader. Although the often exciting and sometimes heated discussions that followed each paper have not been reported *per se,* they have in a real sense been incorporated, in many cases, as useful modification of the papers presented here.

A particular effort has been made for rapid publication of this symposium which regrettably may result in some errors. I am particularly indebted to Professor Vittorio Tonolli, Convener of the Productivity of Freshwater section of the IBP, for arranging for expeditious publication and for maintaining the flow of manuscripts among the editor, contributors, and printer. I should also like to particularly acknowledge the conscientious and imaginative assistance of Mrs. Kathleen C. Green in the task of editing. The editor was a John Simon Guggenheim Fellow during the preparation of these proceedings.

Charles R. Goldman

Davis, California
November 1965

CITATION OF PAPERS FROM THIS VOLUME

For consistency and to prevent confusion, workers citing papers from this volume are urged to adopt the following form:

Author. 1965. Title of paper, p. 000–000. *In* C. R. Goldman [ed.], *Primary Productivity in Aquatic Environments*. Mem. Ist. Ital. Idrobiol., 18 Suppl., University of California Press, Berkeley.

LIST OF PARTICIPANTS

ROBERT J. BEYERS. *Laboratory of Radiation Ecology, Building 772-6 SRP, Aiken, South Carolina (U.S.A.)*

E. K. DUURSMA. *I.A.E.A. - Musée Océanographique, Monaco (Monaco Principauté)*

HANS-JOACHIM ELSTER. *Limnologisches Institut der Universität Freiburg, Falkau (Schwarzwald) (Germany)*

INGO FINDENEGG. *Biologische Station Lunz, Lunz-am-See (Niederösterreich, Austria)*

G. E. FOGG. *Westfield College, Botany Department, Hampstead, London N.W. 3 (Great Britain)*

GIORGIO FORTI. *Laboratorio di Fisiologia Vegetale, Istituto di Scienze Botaniche dell'Università, Milano (Italy)*

MARCO GERLETTI. *Istituto Italiano di Idrobiologia, Pallanza (Novara) (Italy)*

CHARLES R. GOLDMAN. *Department of Zoology, University of California, Davis, California (U.S.A.)*

HANS E. GOLTERMAN. *Hydrobiological Institute, Nieuwersluis (Netherland)*

JOHN E. HOBBIE. *Department of Zoology, North Carolina State University, Raleigh, N.C. (U.S.A.)*

RAMON MARGALEF. *Instituto de Investigaciones Pesqueras, Paseo Nacional, Barcelona 3 (Spain)*

W. OHLE. *Hydrobiologische Anstalt der Max-Planck-Gesellschaft, 232 Plön, Holstein (Germany)*

M. OWENS. *Water Pollution Research Laboratory, Elder Way, Stevenage, Herts (Great Britain)*

T. V. R. PILLAY. *F.A.O. Biology Branch, Roma (Italy)*

DOMENICO POVOLEDO. *Istituto Italiano di Idrobiologia, Pallanza (Novara) (Italy)*

WILHELM RODHE. *Limnologiska Institutionen, Uppsala Universitet, Uppsala (Sweden)*

C. S. SOEDER. *Limnologisches Institut der Universität Freiburg, Falkau (Schwarzwald) (Germany)*

IU. I. SOROKIN. *Institute of Freshwater Biology, USSR Academy of Sciences, Borok, Jaroslav, Nekouz (U.R.S.S.)*

JOHN H. STEELE. *Marine Laboratory, P.O. Box 101, Victoria Road, Aberdeen, Scotland (Great Britain)*

E. STEEMANN NIELSEN. *Royal Danish School of Pharmacy, Copenhagen (Denmark)*

MIROSLAV STEPANEK. *Rosy Luxemburkové 20, Praha-Smichov (Czechoslovakia)*

J. F. TALLING. *Freshwater Biological Association, The Ferry House, Ambleside, Westmorland (Great Britain)*

VITTORIO TONOLLI. *Istituto Italiano di Idrobiologia, Pallanza (Novara) (Italy)*

JOHN R. VALLENTYNE. *Department of Zoology, Cornell University, Ithaca, New York (U.S.A.)*

RICHARD A. VOLLENWEIDER. *Istituto Italiano di Idrobiologia, Pallanza (Novara) (Italy)*

D. F. WESTLAKE. *Freshwater Biological Association, The River Laboratory, East Stoke, Wareham (Dorset, Great Britain)*

ROBERT G. WETZEL. *Department of Botany and Plant Pathology, Kellogg Gull Lake Biological Station, Michigan State University, Hickory Corners, Michigan (U.S.A.)*

CHARLES S. YENTSCH. *Woods Hole Oceanographic Institution, Woods Hole, Massachusetts (U.S.A.)*

TABLE OF CONTENTS

I

THE PHOTOSYNTHESIS AND ADAPTATION OF ALGAE

LIGHT ENERGY UTILIZATION IN PHOTOSYNTHESIS

GIORGIO FORTI

Laboratorio di Fisiologia Vegetale
Istituto di Scienze Botaniche dell'Università, Milano

For bibliographic citation of this paper, see page 10.

Abstract

The current views and experimental approaches to the mechanism of photosynthesis are discussed with regard to the biochemical processes involved in the metabolic utilization of light energy. Light energy absorbed by the molecules of photosynthetic pigments is utilized to generate ATP and TPNH, which are then utilized for the reduction of CO_2 to the carbohydrate level. Other metabolic utilizations of photosynthetically-generated ATP and TPNH are also discussed, as are the possible regulatory mechanisms by which photosynthesis and other metabolic processes might influence each other.

Primary productivity of aquatic environments is basically dependent on the photosynthetic activity of the autotrophic organisms, namely the photosynthetic organisms capable of transforming CO_2 into organic matter. For this reason, the estimation of the primary productivity of natural waters is based upon the measure of their photosynthetic activity, which is due primarily to algae, but possibly also to photosynthetic bacteria.

Photosynthesis was discovered by Priestley as a process by which green plants « purify » air. Soon after, Ingenhousz recognized that the photosynthetic process is of fundamental physiological importance for plant nutrition, and that green plants can « feed on light ». In more recent times, it was recognized that light energy is utilized in photosynthesis to move electrons against the thermochemical gradient. The most general equation that can be written for photosynthesis of green plants

$$6\,CO_2 + 6\,H_2O \xrightarrow{\text{light}} C_6H_{12}O_6 + 6\,CO_2 \qquad (1)$$

shows clearly that the overall process is a redox reaction: CO_2 is reduced, and water is oxidized. Van Niel (1941) first emphasized this concept, and extended it to interpret also bacterial photosynthesis, where no oxygen is evolved. Equation (1) can thus be written more generally as

$$2\,AH_2 + CO_2 \xrightarrow{h\nu} (HCOH) + 2\,A + H_2O \qquad (2)$$

where AH_2 stands for a reductant, which may be water, H_2S, $Na_2S_2O_3$, or an organic carbon compound.

The fact that, experimentally, the same amount of light energy is required per mole of CO_2 assimilated regardless of the nature and redox potential of the reductant utilized, suggested to Van Niel the idea that the primary reaction is the same in all organisms, and that it consists of the splitting of a molecule of H_2O to yield a reducing agent (H) and an oxidant (OH).

$$H_2O \xrightarrow{h\nu} (H) + (OH) \tag{3}$$

(H) and (OH) here indicate in very general terms a reductant and an oxidant).

Such a general presentation of photosynthesis as a light-driven redox reaction involving H_2O proved to be extremely productive and was the basis of the recent most successful experimental work.

According to Van Niel's hypothesis, the oxidant (OH) can be employed in green plants to oxidize water with production of O_2, or, for instance in sulfur bacteria, to form sulfur:

$$\text{green plants:} \quad 2\,H_2O + 4\,(OH) \longrightarrow O_2 + 4\,H_2O \tag{4}$$

$$\text{purple bacteria:} \quad 2\,H_2S + 4\,(OH) \longrightarrow 2\,S + 4\,H_2O \tag{5}$$

An interesting implication of these reactions is that the O_2 evolved in photosynthesis of green plants must be derived from water and not from the CO_2. This was indeed experimentally demonstrated by Ruben *et al.* (1941) in a classical experiment utilizing O^{18} labelled H_2O. These experiments were questioned by Warburg (1958), who pointed out that the exchange of the oxygen atoms of CO_2 with those of water accounts for these observations.

More recently Brown (1953) confirmed Ruben and Kamen's results, under conditions where Warburg's objections cannot be relevant.

THE OVERALL REACTION OF PHOTOSYNTHESIS: ENERGETIC ASPECTS

In the reaction

$$CO_2 + H_2O \xrightarrow{118{,}000\text{ cal}} (HCOH) + O_2 \tag{6}$$

the free energy gain is of 118 K cal/mole. Quite obviously, this energy has to be provided by light. In order to be effective in the biochemical transformations involved in photosynthesis, light has to be absorbed by the molecules of pigments, raising the energetic level of these molecules well above the ground state. A pigment molecule in an excited state is then responsible for the reduction of a redox substance, X, which must be present in the oxidized state in order to be able to accept it. As a result, the pigment will be left with a positive charge, and the oxidant generated in this way is responsible for the oxidation of an intermediate, and ultimately of AH_2. The reductant generated in the light reaction, XH, will be used eventually for the reduction of CO_2. This sequence of events can be represented as follows:

$$CO_2 \leftarrow \leftarrow \leftarrow XH \overset{X}{\curvearrowleft} \overset{h\nu \downarrow}{Chl} \leftarrow \leftarrow \leftarrow \leftarrow Y \leftarrow H_2O \rightarrow O_2 \qquad (7)$$

where the arrows point out the direction of electron flow. Each arrow indicates one step; however, it should be clearly understood that the number of steps is far from being ascertained at our days. The scheme, as presented, indicates only that one (or more than one, as we shall see later) light reaction provides the reagents for a series of « dark » enzymatic reactions. It follows from this that the dark reactions should be limiting the rate at low light intensities. This is indeed the case: the rate versus light-intensity curve, shortly called the intensity curve, starts with a constant slope, and then bends and becomes parallel to the abscissa. At high light intensities, the maximal rate attainable is called the « light saturation » rate. The saturation rate can be lowered by lowering the concentration of CO_2 or the temperature. These treatments do not change the slopes of the light intensity curves at low light intensity. These observations are taken as a demonstration that photosynthesis involves temperature-dependent dark reactions and temperature-independent light reactions (photochemical reactions). As stated above, CO_2 can be made a limiting factor for photosynthesis, and this can certainly occur under natural conditions. Photosynthesis of green plants is probably seldom limited by the concentration of H_2O and certainly not in aquatic organisms. However, it might be pertinent to point out that the hydrogen donor concentration might well be a limiting factor, for instance, in purple bacteria photosynthesis utilizing H_2S (or other reducing substances from the environment) for the reduction of CO_2. This fact should be taken into account in all instances where bacterial photosynthesis contributes appreciably to the productivity of an aquatic environment.

The evidence for the occurrence of light and dark reactions in photosynthesis was obtained also with a different approach. Emerson and Arnold (1932) measured photosynthesis (as oxygen evolution) of green algae with very short flashes of saturating light, separated by short dark periods. Using flashes of 10^{-5} seconds (of saturating intensity) they observed that the amount of oxygen produced *per flash* (flash yield) increased with the length of the dark period separating two successive flashes up to dark periods of a few hundreths of a second. For longer dark periods, there was no further increase in rate even if more intense flashes were used. It was also found that the maximum yield was of one O_2 molecule per 2000 - 2500 chlorophyll molecules. The presence of inhibitors of the dark reactions, such as CN^- or phenylurethanes, did not change the maximum flash yield, but caused the dark periods needed to be of longer duration.

These experiments led to the hypothesis, proposed by Gaffron and Wohl (1936), of the « photosynthetic unit ». The history of the photosynthetic unit concept is beyond the scope of this work, and we refer to Rabinowitch (1956) for a more complete discussion. Only the most fundamental observations and theoretical discussions will be briefly summarized here, freely borrowing from Duysens (1952, 1964). The model proposed by Duysens for the photosynthetic unit implies that light energy absorbed by a large number of pigment molecules is transferred by the physical mechanisms of induced resonance to pigment molecules, called the reaction centers, present in a small concentration as compared to the bulk of photosynthetic pigments. In algae, each reaction center receives energy from about 200 chlorophyll molecules. The assembly of these molecules with the reaction center is the « photosynthetic unit ». The molecule at the reaction center initiates the chemical reactions (consisting of redox reactions) of photosynthesis. The transfer of the energy of a quantum by induced resonance from the pigment molecules of the unit to the reaction center occurs within 10^{-8} second after the absorption of the quantum. Upon excitation, the reaction center molecule reacts photochemically with a molecule, X, bound to it, and this reaction produces a reductant, XH; the « oxydized » reaction center molecule is rapidly restored into the initial state by another molecule, YH, which donates an electron to it. The time required for this dark reaction (the regeneration of the reaction center in its initial state) is at least 10^{-5} second. A quantum absorbed by the unit cannot be utilized unless the reaction center molecule is ready to accept it, i.e. in its initial state. Tamiya and Chiba (1949) and Tamiya (1949) used flashes of the duration of 10^{-2} seconds, and under these conditions found a maximum flash yield higher than Emerson's (on a chlorophyll basis), but requiring much longer dark intervals to be attained. The dark period needed between the flashes was longer at lower temperatures. During the flash of 10 milliseconds the dark reaction evidently could restore the reaction center several times while a longer dark interval was needed for the subsequent enzymatic reactions to utilize the larger amounts of XH and of Y generated.

The results of Emerson and of Tamiya suggested that the primary dark reaction which regenerates the reaction center takes place in a time of 10^{-3} second or shorter, but longer that 10^{-5} second. We shall see later that the concept of a « photosynthetic unit » has received a larger body of evidence by the recent studies on the sequence of the biochemical reactions of photosynthesis with *in vitro* systems, as well as by the *in vivo* studies of fluorescence emission of photosynthetic pigments.

HILL REACTION AND PHOTOSYNTHETIC PHOSPHORYLATION

The understanding of the mechanisms involved in photosynthetic electron transport made rapid progress after Hill's (1937, 1939) success in demonstrating part of the photosynthetic process in a cell-free system from higher plants. Working with chloroplasts isolated from higher plants, Hill was able to obtain O_2 evolution equivalent to the reduction of an electron acceptor, according to the equation

$$A + H_2O \xrightarrow{\text{light}} AH_2 + \frac{1}{2} O_2 \qquad (8)$$

The isolated chloroplasts were unable to assimilate CO_2, and the electron acceptors used by Hill were non-physiological compounds, such as ferric oxalate, benzoquinone, ferricyanide. Later, it was demonstrated (Vishniac and Ochoa 1951, Arnon 1951, Tolmach 1951) that isolated chloroplasts can utilize NADP, the well known electron transport coenzyme, as a « Hill oxidant ». Work with chloroplasts isolated from higher plants led to the important discovery of photosynthetic phosphorylation by Arnon, Allen and Whatley (1954); at the same time, Frenkel (1954) discovered this process in chromatophores from photosynthetic bacteria. In this process, photosynthetic particles synthesize ATP from ADP and inorganic phosphate (Pi) at the expense of light energy, according to the reaction

$$ADP + Pi \xrightarrow{\text{light}} ATP \qquad (9)$$

without any evolution or uptake of O_2. The only requirement is the addition of catalytic amounts of a redox substance. A number of substances have proven to be active, the most active being the non-physiological compound phenazonium methosulfate (PMS) (Avron and Jagendorf 1958). As the result of the work of a number of laboratories (for a review see Jagendorf 1962), three types of photosynthetic phosphorylation were distinguished:

a) A « cyclic » photophosphorylation, in which the electrons activated by light reduce an acceptor, X, which is then reoxidized by the oxidant produced in the light reaction, via an as yet scarcely known chain of electron carriers. No oxygen evolution or uptake is involved, and the only reaction product is ATP. The reaction runs under anaerobic as well as aerobic conditions, and needs the addition of a catalytic cofactor. This can be PMS, or pyocyanine. The rates are up to 1000 or more micromoles of ATP per hour per mg of chlorophyll. During the reaction, no exchange of oxygen between H_2O and free O_2 occurs. The reaction, under anaerobic conditions at least, is unaffected by orthophenantroline and substituted ureas like p-chlorophenyldimethylurea (CMU) and the dichloro-analog (DCMU). These substances are strong inhibitors of the other two types of photosynthetic phosphorylation.

b) The second type of photophosphorylation occurs in the presence of cofactors like vitamin K, FMN, FAD, and a number of other redox substances. No net gas exchanges are observed; however, it has been shown that during the reaction an exchange of O between H_2O and O^2 occurs. The reaction is inhibited by CMU, DCMU, and orthophenantroline. The mechanism is interpreted as follows: XH is formed by light, and at the same time an equivalent amount of oxygen is evolved. The oxygen is then used up for the oxidation of XH.

$$\text{a)} \quad H_2O + 2\,X \xrightarrow{\text{light}} 2\,XH + \tfrac{1}{2}\,O_2$$
$$ADP + Pi \rightarrow ATP$$

$$\text{b)} \quad 2\,XH + \tfrac{1}{2}\,O_2 \xrightarrow{\text{(cofactor)}} 2\,X + H_2O \qquad (10)$$

$$\text{sum:} \quad ADP + Pi \xrightarrow{\text{light}} ATP$$

c) The third type of photophosphorylation is the ATP synthesis coupled to the Hill reaction. As stated before, NADP can be utilized, under the appropriate conditions, as an electron acceptor in the Hill reaction. It has been shown (Arnon, Whatley and Allen 1957) that ATP formation is coupled to this Hill reaction, and the stoichiometry of the reaction has been established:

$$H_2O + NADP + ADP + Pi \xrightarrow{\text{light}} \tfrac{1}{2}\,O_2 + NADPH + ATP \qquad (11)$$

The observations in a number of laboratories indicate that one ATP molecule is formed per couple of electrons transferred from H_2O to NADP or ferricyanide. Quite obviously NADP is the natural electron acceptor, as has been confirmed also with *in vivo* experiments (to be discussed further on).

The Hill reaction is strongly inhibited by CMU, DCMU, and orthophenantroline.

In photosynthetic phosphorylation as in oxidative phosphorylation, the energy conservation in the form of the pyrophosphate bonds of ATP is coupled to the transfer of electrons from a more electronegative to a more electropositive substance. The outstanding difference between the two processes consists in the fact that while in oxidative phosphorylation the reduced intermediates are generated upon oxidation of organic matter, in photophosphorylation the reductants are generated at the expense of light energy. Therefore, photosynthetic phosphorylation represents a net transformation of light energy into (Gibbs) free energy of chemical bonds, readily utilizable for biosynthetic purposes in the cells.

The photosynthetic bacteria are unable to evolve O_2, and the chromatophores isolated from these organisms do not catalyse the Hill reaction. They do, however, catalyse cyclic photophosphorylation (with the addition of PMS) and phosphorylation coupled to the photooxidation of a number of substrates (for instance, succinate), pyridine nucleotides being reduced in the light.

To summarize, chloroplasts and bacterial chromatophores are capable of utilizing light to form ATP and reduced pyridine nucleotide; water is the final electron donor in green plants' chloroplasts, a reduced substrate in bacterial chromatophores. The by-products are, respectively, O_2 and oxidized substrate (sulfur, organic acids).

The classical work of Calvin and his associates has demonstrated that ATP and reduced pyridine nucleotide are all that are needed to fix CO_2 and reduce it to the carbohydrate level, through the well known photosynthetic carbon cycle.

COMPONENTS OF THE PHOTOSYNTHETIC ELECTRON TRANSPORT SYSTEM

The work of biochemists in the last two decades has brought about the identification of a number of the intermediates in electron transport in photosynthesis. It is then possible to substitute, in part at least, the X's and Y's of scheme (7) with known enzymes and cofactors.

San Pietro and his group (San Pietro 1958) isolated in the pure state from spinach leaves a reddish protein, readily soluble in water, required for NADP photoreduction by washed chloroplasts. This protein, originally named Photosynthetic Pyridine Nucleotide Reductase (PPNR), has recently been renamed ferredoxin, after it had been shown (San Pietro and Keister 1962; Shin, Tagawa and Arnon 1963) that it does not react directly with NADP, but only by the mediation of another enzyme, a flavoprotein. The flavoprotein was discovered independently by Avron and Jagendorf (1956) and by Marrè and Servettaz (1958) as a NADPH diaphorase and was later demonstrated to be identical with the NADP reductase. So the pathway of NADP reduction includes a compound X, directly reduced by the «reaction center» pigment molecule, ferredoxin, and the chloroplast flavoprotein. It has been proposed, though not demonstrated, that ferredoxin itself is identical with substance X. Ferredoxin has a very negative redox potential (—0.43 volts at pH 7) and is present in photosynthetic bacteria as well as in green plants. These properties make of it a suitable candidate as substance X.

On the water-oxidation side of the light reaction which brings about the reduction of NADP, a number of components have been identified and characterized in part at least.

Back in the early 1950's Hill and his associates discovered a cytochrome peculiar to photosynthetic tissues of higher plants (Hill and Scarisbrick 1951, Davenport and Hill 1952) which they denominated cytochrome *f*. The redox potential of cytochrome *f* is + 0.36 volts (at pH 7); its direct participation in photosynthetic electron transport was first demonstrated by Duysens and his associates (Duysens 1955). These authors have shown that in intact algae cytochrome *f*, in the reduced state in the dark, is rapidly oxidized upon illumination, and again reduced upon darkening.

More recently, Chance and Bonner (1963) have demonstrated that cytochrome *f* photo-oxidation, in higher plants as well as in algae, is temperature independent (the reaction occurs at high rate at 77 °K). This indicates that the cytochrome reacts directly with the primary oxidant generated in the light reaction. Kok and his collaborators (Kok 1963) have shown that a long-wave length absorbing component of the chlorophyll system (called « P 700 ») is the primary oxidant generated in the light reaction. The redox potential of P 700 has been found to be + 0.45 volts.

Cytochrome *f* photo-oxidation has been demonstrated to occur with chloroplast extracts (Kok, Rurainski and Harmon 1964) at rates and quantum efficiency high enough to account for the rate of photosynthesis under physiological conditions. Cytochromes have been shown to be involved also in bacterial photosynthesis in the span from the electron donor to the light reaction (for a review, see Duysens 1964).

The electron transport chain of chloroplasts in the span from cytochrome *f* to H_2O includes a copper protein, called plastocyanin by its discoverer, Katoh (1960), the redox potential of which is approximately + 0.37 v (Katoh, Shiratori and Takamiya 1962), and a quinone of redox potential approximately 0 v. This compound, plastoquinone, was isolated (Bishop 1959) and found to be similar in its properties (though having a definitely different chemical formula) to the ubiquinone compounds of the respiratory chain of mitochondria. The participation of plastocyanine and plastoquinone was demonstrated both *in vivo* and *in vitro* (De Kouchkovsky and Fork 1964, Amesz 1964). Chance, Schleyer and Legallais (1963) reported the photooxidation in green algae of a *b*-type cytochrome, which could be identical with the cytochrome b_6 discovered by Hill (1954). However, the direct participation of this cytochrome in photosynthetic electron transport is still controversial (Duysens 1964).

THE TWO LIGHT REACTIONS OF PHOTOSYNTHESIS

a) - The « Emerson effect ».

Emerson and his collaborators (Emerson 1958) discovered that the quantum yield of photosynthesis in green algae (i.e. the moles of O_2 evolved per quantum absorbed) drops rapidly on the long wave length side of the action spectrum, even at wave lengths (i.e. 690 mμ) where chlorophyll *a* absorption is still high. The study of this anomaly led Emerson and his associates to the discovery of the « enhancement effect », called thereafter Emerson effect. This phenomenon can be shortly described as follows: a light beam of wave length 1 and another beam of properly chosen wave length 2, when presented together, elicit a rate of photosynthesis greater that the sum of the rates when presented separately.

It is not intended here to review the literature pertinent to the Emerson enhancement effect; a number of reviews are available (Emerson and Rabinowitch 1960, Myers 1963).

Though the Emerson effect in itself might have been explained also in a different way, the more recent studies have led to the hypothesis that two light reactions are required for green plants' photosynthesis. These two light reactions are working in series: one of them (light reaction 1) reduces NADP (via the reactions described above) and produces an oxidant (oxidized P_{700}), which is required to oxidize a reduced substance formed in the second light reaction (light reaction 2) upon oxidation of water. O_2 evolution results from the oxidation of H_2O by light reaction 2.

b) - The two light reactions hypothesis.

A large body of evidence has accumulated in favor of this hypothesis, both from studies with intact cells and from the study of *in vitro* reactions of chloroplasts and chloroplast fragments. Duysens (1964) and his associates have demonstrated that in red and blue-green algae two different pigment systems exist, both containing chlorophyll *a*. The first one is excited by light absorbed at the long wave length absorption peak of chlorophyll *a* and gives little fluorescence (« pigment system 1 »). The second pigment system (« pigment system 2 ») is excited by light absorbed by the « accessory » pigments, the phycobilins, at lower wave length. (In higher plants, chlorophyll *b* and the carotenoids replace the phycobilins.) The light absorbed by the phycobilins in algae, and by chlorophyll *b* and the carotenoids in higher plants, is much more efficient in exciting chlorophyll *a* fluorescence than the light absorbed by chlorophyll *a* itself. This proved that part of the chlorophyll *a*, associated with pigment system 1, is weakly fluorescent (« chlorophyll a_1 ») while another part of chlorophyll *a* is strongly fluorescent and associated with system 2 (« chlorophyll a_2 »). Quite obviously, the light absorbed at lower wave length by the accessory pigments can be transferred with high efficiency to the long wave length absorbing chlorophyll *a*, most probably through the induced resonance mechanism.

Such a mechanism is theoretically effective when, as in this case, the fluorescence emission spectrum of the energy-transferring pigment overlaps at least partially the absorption spectrum of the energy-accepting pigment.

On the other hand, the action spectrum of photosynthesis (i.e. the photosynthetic rate versus the wave length of light at constant incident energy) is parallel to the action spectrum of chlorophyll *a* fluorescence, and not to chlorophyll *a* absorption. This does not mean that the light absorbed by « system 1 », that is by chlorophyll *a* at the long wave length peak, is inactive in photosynthesis. Duysens and Amesz (1962) have shown that system 1 absorption activates cytochrome *f* oxidation (also, Katoh and Takamiya 1963) and pyridine nucleotide reduction (Amesz and Duysens 1959). Furthermore, they demonstrated that light absorbed by system 2 induces the reduction of cytochrome *f* (Duysens and Amesz 1962) if this compound is in the oxidized state as a result of system 1 activation.

The hypothesis of two light reactions in series implies also that the product of one photoreaction reacts with the product of the second photoreaction in stoichiometric proportion. If, under certain conditions of illumination, the rate of one of the two photoreactions is lower than the other, then the rate of photosynthesis is equal to the rate of the slower photoreaction.

On the basis of this rationale, Duysens and Amesz (1962) have demonstrated that if photosynthesis is measured in non-saturating light intensities, the addition of light absorbed by system 2 to a background of light absorbed by system 1 gives a 7-fold higher photosynthetic rate than the addition of system 1 light of the same intensity. Vice versa, on a background of system 2 light, the addition of system 1 light is more effective.

Evidence in favor of the two light reactions hypothesis came also, more or less simultaneously, from the studies on isolated chloroplast reactions. As has been mentioned, the inibitors of the Hill reaction and the associated phosphorylation, such as CMU and DCMU, do not affect cyclic photophosphorylation (see Jagendorf 1962), but they do inhibit the « O_2 exchange type » (« b type » as indicated above) phosphorylation. Indeed, only the cyclic type of photophosphorylation does not involve the water splitting reaction of photosynthesis, and this last reaction, promoted by light absorption on system 2, is the step inhibited by CMU. The photo-reduction of NADP by isolated chloroplasts can be made insensitive to (D)CMU if the physiological electron donor, water, is substituted with an artificial reducing system (Losada, Whatley and Arnon 1961) like ascorbate and the redox dye dichlorophenolindophenol. In these last conditions, indeed, photoreaction 2 is by-passed. Also, the photooxidation of cytochrome *f* by chloroplast extracts occurs under conditions where photoreaction 2 is suppressed. Furthermore, the quantum efficiency of cytochrome photooxidation is higher at high wave lengths (over 700 mμ), where chlorophyll *a* still absorbs while the accessory pigments do not (Kok *et al.* 1964). On the other hand, it has also been demonstrated (Forti, Bertolè and Parisi 1963) that cytochrome *f* added in substrate amounts to chloroplast preparations in the oxidized form is photoreduced, and that this reaction is coupled to ATP formation. This process is completely inhibited by CMU. The role of the other components in the photosynthetic electron transport chain has been identified: plastocyanin (Katoh and Takamiya 1963, De Kouchkovsky and Fork 1964) and plastoquinone (Bishop 1959, Amesz 1964) are both reduced by system 2 and oxidized by system 1, as indicated by *in vitro* as well as *in vivo* studies. The recent studies of Duysens (1964) and his associates on fluorescence of algae have led those authors to postulate the existence of a substance, Q, as yet unidentified chemically, that quenches the fluorescence of chlorophyll a_2 when in the oxidized state, while it does not quench when reduced (QH). Strong reductants such as hydrosulfite were indeed found to enhance the fluorescence of chlorophyll a_2 in intact cells. In general, it was found that illumination of system 1 quenches the fluorescent emission activated by system 2 absorption. In the presence of DCMU, however, the fluorescence increases; this is taken as an indication that substance Q oxidation is the step inhibited by DCMU. To summarize, the scheme 7 previously indicated to represent photosynthetic electron transport can now be written as in Fig. 1 on the basis of the above reported evidence.

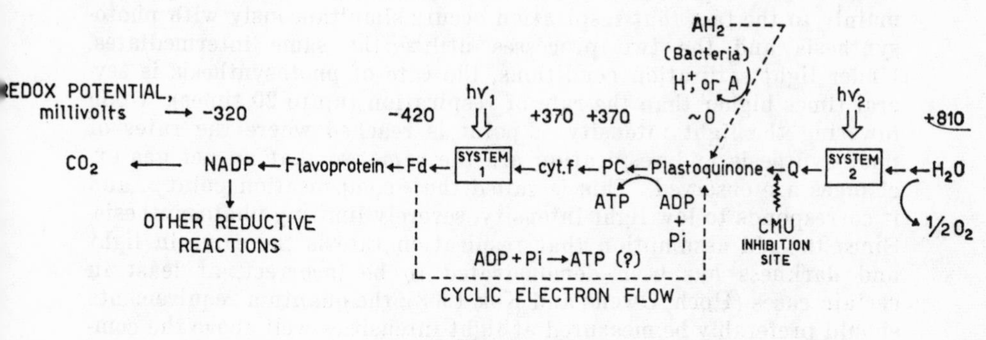

Fig. 1.

As previously stated, cyclic photophosphorylation by isolated photosynthetic particles requires the addition of artificial electron carriers such as PMS. This might indicate that part of the physiological machinery has been lost or inactivated during the preparation, or that cyclic phosphorylation is an artifact caused by PMS. This latter possibility has been excluded by the demonstration that higher plant leaves poisoned with CMU (causing complete inhibition of O_2 evolution) are capable of rapid, light-dependent synthesis of ATP (Forti and Parisi 1962, 1963). Results of other laboratories supported this finding (Simonis and Urbach 1963, Urbach and Simonis 1964), showing that the uptake of phosphate and its incorporation into organic fractions is stimulated by light, also in the presence of CMU.

Interestingly enough, all the above mentioned evidence in favor of two light reactions operating in series does not apply to photosynthetic bacteria, which are unable to evolve O_2. Therefore, no Emerson effect can be observed in photosynthetic bacteria (Blinks and Van Niel 1963), and no (D)CMU inhibition of photosynthesis. Similarly, Gaffron (1960) has shown that hydrogen-adapted algae, which have « learned » to utilize molecular hydrogen (in substitution for H_2O) as a photoreductant for CO_2 assimilation, are insensitive to (D)CMU and do not exhibit the enhancement effect.

QUANTUM REQUIREMENT FOR PHOTOSYNTHESIS AND PHOTOSYNTHETIC REACTIONS

The overall efficiency of photosynthesis, from the energetic point of view, is still controversial. The number of quanta required for the evolution of one molecule of O_2 by green plants (quantum requirement) has been measured in several laboratories, and the data obtained are scattered from 3 quanta/O_2 to 8-10 quanta (see Kok 1960 for a critical discussion). The difficulties in the measurement are related not only to the technical experimental methodology, but mainly to the fact that respiration occurs simultaneously with photosynthesis and the two processes utilize the same intermediates. Under light saturation conditions, the rate of photosynthesis is several times higher than the rate of respiration (up to 20 times). Upon lowering the light intensity, a point is reached where the rates of photosynthesis and respiration are the same, so that no net gas exchanges are observed. This is called the « compensation point », and it corresponds to low light intensity, severely limiting photosynthesis. Since the old assumption that respiration rate is the same in light and darkness has been demonstrated to be incorrect, at least in certain cases (Hoch, Owens and Kok 1963), the quantum requirements should preferably be measured at light intensities well above the compensation point in order to minimize the uncertainties introduced by the correction for respiration. Most of the reported values range from 6-7 to 8-10 quanta per O_2 molecule (see Kok 1960). The one remarkable exception is the values of 3-4 reported by Warburg and his associates on the basis of manometric measurement at light intensities below the compensation point. Under these conditions it is conceivable that respiratory metabolism contributes energy, possibly in the form of ATP and/or TPNH, to CO_2 fixation (and then to O_2 evolution).

According to the pathway proposed originally by Calvin for CO_2 assimilation, 2 molecules of NADPH and 3 molecules of ATP are needed to reduce one CO_2 to the carbohydrate level. On the other hand, it is well known that only one ATP is produced by the photochemical apparatus per molecule of NADP (see above). The additional ATP could be provided by cyclic photophosphorylation, using up additional quanta. The observed quantum requirement for the photoreduction of NADP in intact cells as well as in isolated chloroplasts is of 4-6 quanta per molecule (see Duysens 1964). On the other hand, no data are yet available for the quantum requirement of cyclic photophosphorylation *in vivo*, so that it is not possible to compute the quantum requirement of the overall process from the requirements of partial reactions.

Two very recent findings could account for an overall quantum requirement of not more than 8. Baltscheffsky and de Kiewiet (1964) have provided evidence for the existence of two phosphorylating sites in cyclic photophosphorylation of chloroplasts as well as bacterial chromatophores (Baltscheffsky and Arwidsson 1962). This could conceivably indicate a high quantum efficiency for ATP formation in the cyclic electron transport. On the other hand, the CO_2 fixation experiments of Gibbs and Kandler (1957) indicated that part of the assimilated CO_2 may by-pass 3-phosphoglyceric acid (PGA). More recently, Bassham and Kirk (1960) concluded from the time course of incorporation of radioactive CO_2 into PGA and ribulose-diphosphate during steady state photosynthesis in *Chlorella* that only one molecule of PGA appears in the main pool of PGA per molecule of ribulose-diphosphate carboxylated.

The other 3-carbon compound (not identified) could be a triose molecule. This hypothesis could explain the results of Gibbs and Kandler (1957). Such a finding could be explained by assuming that the molecule of carboxylated ribulose-diphosphate undergoes reductive splitting, yielding one molecule of PGA and one molecule of triose; a reductant with a redox potential of —0.43 volts, like ferredoxin, would be adequate for this reductive splitting. In this way, one ATP molecule per CO_2 assimilated would be saved, and an overall quantum requirement as low as 7-8 could be calculated.

PHOTOSYNTHESIS AS RELATED TO OTHER CELL ACTIVITIES

An understanding of photosynthesis as a physiological process of primary importance must take into consideration the effect of photosynthetic activity on the fine regulation of the complicated network of metabolic reactions on which cell activity and growth are dependent. Indeed, if our knowledge of the basic mechanisms of photosynthesis is far from being satisfactory, our understanding of the influence of photosynthesis on the other metabolic processes and vice versa is even less. It seems a reasonable prediction that future research will have to look more deeply into the possible mechanisms by which photosynthetic cells might regulate photosynthesis at the different stages of the process.

Kandler (1964) has observed that in *Chlorella* illumination causes the inhibition of triosephosphate oxidation; the effect is attributed to phosphate competition between the triosephosphate dehydrogenase reaction and photophosphorylation. It has also been demonstrated (Hoch *et al.* 1963) that algal respiration (O_2 uptake) is stimulated by light, and that the light induced respiration, in higher plants, has a relatively low affinity for O_2 (Krotkov 1963) as compared to the high O_2 affinity of cytochrome oxidase. Though little is known of this « light respiration » process, the above mentioned results would indicate that the O_2 uptake is not due to mitochondrial respiration, but to a different respiratory pathway. This deduction would also be in agreement with the fact that a large part of green tissue respiration is insensitive to carbon monoxide, the well known inhibitor of cytochrome oxidase.

These findings indicate that photosynthetic activity has a relevant regulatory effect on the pathways of cell respiration, both in the quantitative and the qualitative aspect.

The more recent studies have already indicated that assimilation of the CO_2 carbon is not the only function of photosynthesis. It has indeed been demonstrated that ATP and NADPH produced at the expense of light energy can be utilized for a number of energy-requiring metabolic processes, including fatty acid activation and reduction (Stumpf, Bovè and Goffeau 1963), the uptake of ions and molecules and their translocation inside the cell and tissues, nitrite reduction (Marrè *et al.* 1963, Santarius *et al.* 1964), and so on. Particularly interesting is the finding (Marrè *et al.* 1963) that the synthesis of cellulose from glucose is stimulated by light in *Wolffia arrhiza* poisoned with CMU. It is well known that glucose incorporation into cellulose is a biosynthetic process requiring ATP for the activation of glucose; furthermore, this process does not occur inside the chloroplast. The above mentioned effect of light seems then to indicate that the ATP formed inside the chloroplast by cyclic photophosphorylation can be utilized in other cellular structures for biosynthetic reactions.

These still very preliminary studies point out the importance of considering photosynthesis as a very complex metabolic process strictly connected with all the metabolic reactions of the cell, and not merely as a mechanism which accumulates carbohydrates at low cost.

REFERENCES

Amesz, J. 1964. Spectrophotometric evidence for the participation of a quinone in photosynthesis of intact blue-green algae. - Biochim. Biophys. Acta *79*: 257.

— and L.N.M. Duysens. 1959. Spectrophotometric studies on pyridine nucleotide in photosynthetic cells and cellular material. - Disc. Faraday Soc. *27*: 173.

Arnon, D.I. 1951. Extracellular photosynthetic reactions. - Nature *617*: 1008.

— M. B. Allen and F. R. Whatley 1954. Photosynthesis by isolated chloroplasts. Nature *174*: 394.

— F. R. Whatley and M. B. Allen. 1957. Triphosphopyridine nucleotide as a catalyst of photosynthetic phosphorylation. - Nature *180*: 182.

Avron, M. and A. T. Jagendorf. 1956. A TPNH Diaphorase from chloroplasts. Arch. Biochem. Biophys. *65*: 475.

— and A. T. Jagendorf. 1958. Cofactors and Rates of Photosynthetic Phosphorylation by Spinach Chloroplasts. - J. Biol. Chem. *231*, No. 1: 277.

Baltscheffsky, H. and D. Y. de Kiewiet. 1964. Existence and localisation of two phosphorylation sites in photophosphorylation of isolated spinach chloroplasts. - Acta Chem. Scand. *18*, No. 10, 2406.

— and B. Arwidsson. 1962. Evidence for two phosphorylation sites in bacterial cyclic photophosphorylation. - Biochim. Biophys. Acta *65*: 425.

Bassham, J. A. and M. Kirk, 1960. Dynamics of the photosynthesis of carbon compounds. I. Carboxylation reactions. - Biochim. Biophys. Acta *43*: 447.

Bishop, N. I. 1959. The reactivity of a naturally occuring quinone (Q 225) in photochemical reactions of isolated chloroplasts. - Proc. Nat. Acad. Sci. (U.S.) *45* (12): 1696.

Blinks, L. R. and C. B. Van Niel. 1963. The absence of enhancement (Emerson effect) in the photosynthesis of ***Rhodospirillum rubrum***, p. 297. ***In*** Microalgae and Photosynthetic Bacteria. - Univ. of Tokyo Press.

Brown, A. H. 1953. The effects of light on respiration using isotopically enriched oxygen. - Am. J. Botany *40*: 719.

Chance, B. and W. D. Bonner. 1963. The temperature insensitive oxidation of cytochrome *f* in green leaves. A primary biochemical event of photosynthesis, p. 66. *In* Photosynthetic Mechanisms of Green Plants. - N.A.S. - N.R.C. publ. 1145.

— H. Schleyer and Legallais. 1963. Activation of electron transfer in a *Chlamydomonas* mutant by light pulses from an optical maser, p. 337. *In* Studies on Microalgae and Photosynthetic Bacteria. - Univ. of Tokyo Press.

Davenport, H. E. and R. Hill. 1952. Preparation and some properties of cytochrome f. - Proc. Roy. Soc. London, B, *139*: 327.

De Kouchkovsky, Y. and D. C. Fork. 1964. A possible functioning *in vivo* of plastocyanin in photosynthesis as revealed by a light-induced absorbance change. - Proc. Natl. Acad. Sci. *52* (2): 232.

Duysens, L. N. M. 1952. Transfer of Excitation Energy in Photosynthesis. Thesis, Utrecht.

— 1955. Role of cytochrome and pyridine nucleotide in Algae photosynthesis. Science *121*: 210.

— 1961. Cytochrome oxidation by a second photochemical system in the red alga *Porphyridium cruentum*, p. 135. *In* Progress Photobiology, Proc. 3rd. Int. Congr. Photobiology, Elsevier, Amsterdam.

— 1964. Photosynthesis, p. 1. *In* Progress in Biophysics, vol. 14 - Pergamon Press, London.

— and J. Amesz, 1962. Function and identification of two photochemical systems in photosynthesis. - Biochim. Biophys. Acta *64*: 243.

Emerson, R. 1958. The quantum yield of photosynthesis. - Ann. Rev. Plant. Physiol. *9*: 1.

— and W. Arnold. 1932. A separation of the reactions in photosynthesis by means of intermittent light. - J. Gen. Physiol. *15*: 391.

— and E. Rabinowitch. 1960. Red drop and role of auxiliary pigments in photosynthesis. - Plant Physiol. *35* : 47.
Forti, G., M. L. Bertolè and B. Parisi. 1963. On the function of cytochrome f in photosynthetic electron transport. - Biochem. Biophys. Res. Comm. *10* (5) : 384.
— and B. Parisi. 1962. Sur l'éxistence de la photophosphorylation cyclique chez les plantes supérieures. - Bull. Soc. Franc. Physiol. Veg. *8* : 41.
— and B. Parisi. 1963. Evidence for the occurrence of cyclic photophosphorylation *in vivo*. - Biochim. Biophys. Acta *71* : 1.
Frenkel, A. W. 1954. Light induced phosphorylation by cell-free preparations of photosynthetic bacteria. - J. Am. Chem. Soc. *76* : 5568.
Gaffron, H. 1960. Energy storage: photosynthesis. - Plant Physiology vol. I B, p. 3 - Acad. Press N. Y.
— and K. Wohl. 1936. Zur theorie der assimilation. - Naturwissenschaften, *24* :81.
Gibbs, M. and O. Kandler. 1957. Asymmetric distribution of C^{14} in sugars formed during photosynthesis. - Proc. Natl. Acad. Sci. U.S., *43* : 446.
Hill, R., 1937. Oxygen evolved by isolated chloroplasts. - Nature, *139* : 881.
— 1939. Oxygen produced by isolated chloroplasts. - Proc. Roy. Soc. (London), *127* B : 192.
— 1954. The cytochrome *b* component of chloroplasts. - Nature, *174* : 501.
— and R. Scarisbrick, 1951. The haematin compounds of leaves. - New Phytol. *50* : 98.
Hoch, G. E., O. H. Owens, and B. Kok, 1963. Photosynthesis and respiration. Arch. Biochem. Biophys. *101* : 171.
Jagendorf, A. T., 1962. Biochemistry of energy transformations during photosynthesis, p. 181. *In* Survey of Biological Progress, V. IV. - Acad. Press., N.Y.
— and M. Avron, 1958. Cofactors and rates of phosphorylation by spinach chloroplasts. - J. Biol. Chem. *231* : 277.
Kandler, O. 1964. Reported at the Xth Int. Bot. Congress, Edimburgh.
Katoh, S. 1960. A new copper protein from *Chlorella ellipsoidea*. - Nature *186* : 533.
— and A. Takamiya. 1963. Photochemical reactions of plastocyanin in chloroplasts, p. 262. *In* Photosynthesis Mechanisms of Green Plants, N.A.S. - N.R.C. publ. 1145.
— J. Shiratori and A. Takamiya. 1962. Purification and some properties of spinach plastocyanin. - J. Biochem. *51* (1) : 32.
Kok, B. 1960. Efficiency of photosynthesis, p. 566. - Handbuch der Pflanzenphysiologie, Band V, Springer Verlag, Berlin.
— 1963. Photosynthetic electron transport, p. 35. *In* Photosynthetic Mechanisms of Green Plants, N.A.S. - N.R.C., publ. 1145.
— H. J. Rurainski and A. E. Harmon. 1964. Photooxidation of cytochrome c, f, and plastocyanin by detergent treated chloroplast. Plant. Physiol. *39* : 4, 513.
Krotkov, G. 1963. Effect of light on respiration, p. 452. *In* Photosynthetic Mechanisms of Green Plants - N.A.S. - N.R.C. publ. 1145.
Losada, M., F. R. Whatley and D. I. Arnon. 1961. Separation of two light reactions in noncyclic photo-phosphorylation of green plants. - Nature *190* : 606.
Marrè, E. and O. Servettaz. 1958. TPNH-cytochrome *c* reductase of chloroplasts and its role in photosynthetic phosphorylation. - Arch. Biochem. Biophys. *75* : 309.
— G. Forti, R. Bianchetti and B. Parisi. 1963. Utilization of photosynthetic chemical energy for metabolic processes different from CO_2 fixation, p. 557. *In* La Photosynthèse - Centre National de la Recherche Scientifique, Paris.
Myers, F. 1963. Enhancement, p. 301. *In* Photosynthetic Mechanisms of Green Plants - N.A.S. - N.R.C. publ. 1145.

Rabinowithch, E. I. 1956. Photosynthesis and Related Processes. - Interscience, New York.
Ruben, S., M. Randall, M. Kamen and J. L. Hyde. 1941. - J. Am. Chem. Soc. *63* : 877.
San Pietro, A. 1958. Photochemical reduction of triphosphopyridine nucleotide by chloroplasts. *In* The Photochemical Apparatus, Brookhaven Symposia in Biology, N. 11.
— and D. L. Keister. 1962. Pyridine nucleotide transhydrogenase from spinach. II. Requirement of enzyme for photochemical accumulation of reduced pyridine nucleotides. - Arch. Biochem. Biophys. *98* : 235.
Santarius, K. A., U. Heber, W. Ulbrich and W. Urbach, 1964. Intracellular translocation of ATP, ADP and inorganic phosphate in leaf cells of *Elodea densa*. - Biochem. Biophys. Res. Comm. *15* : 2, 139.
Shin, M., K. Tagawa and D. I. Arnon, 1963. Crystallization of ferredoxin - TPN reductase and its role in the photosynthetic apparatus of chloroplasts. Biochem. Zeits, *38* : 84.
Simonis, W. and W. Urbach. 1963. Untersuchungen zur lichtabhängigen Phosphorylierung bei *Ankistrodesmus Braunii* IX. Beeinflussung durch phosphatkonzentrationen, temperatur, hemmstoffe, Na-ionen und vorbelichtung p. 597. *In* Microalgae and Photosynthetic Bacteria - Univ. of Tokyo Press.
Stumpf, P. K., J. M. Bovè and A. Goffeau. 1963. Fat metabolism in higher plants. XX. Relation of fatty acid synthesis and photophosphorylation in lettuce chloroplasts. - Biochim. Biophys. Acta. *70* : 260.
Tamiya, H. 1949. Studies from the Tokugawa Institute *6* (2) : 43.
— and Y. Chiba. 1949. Studies from the Tokugawa Institute, *6* (2) : 7.
Tolmach, L. J. 1951. Effects of triphosphopyridine nucleotide upon oxygen evolution and carbon dioxide fixation by illuminated chloroplasts. - Nature *167* : 946.
Urbach, W. and W. Simonis. 1964. Inhibitor studies on the photophosphorylation *in vivo* by unicellular algae (*Ankistrodesmus*) with antimycin A. HOQNO, salicylaldoxine and DCMU. - Biochem. Biophys. Res. Comm. *17*, (1) : 39.
Van Niel, C. B. 1941. The bacterial photosyntheses and their importance for the general problem of photosynthesis. - Adv. in Enzymol. *1* : 263.
Vishniac, W. and S. Ochoa. 1951. Photochemical reduction of pyridine nucleotides by spinach extracts and coupled carbon dioxide fixation. - Nature, *167* : 768.
Warburg, O. Photosynthesis - Science *128* : 68.

ADAPTATION IN PLANKTON ALGAE

ERIK G. JORGENSEN and E. STEEMANN NIELSEN
Royal Danish School of Pharmacy
Botanical Department, Copenhagen

For bibliographic citation of this paper, see page 10.

Abstract

Plankton algae adapt themselves to factors such as temperature and light intensity. The adaptation takes place primarily by varying the concentration of pigments, photosynthetic enzymes, and other enzymes. Due to the increase in the amount of all enzymes at low temperatures, the total amount of organic matter per cell increases. Therefore, more organic matter must be produced at low temperatures for doubling the number of cells.

INTRODUCTION

Plankton algae are able to adapt themselves more or less to the conditions found in their habitat. The physiological behavior of a species will thus vary, and often to a very considerable degree. However, the adaptation does not take place immediately when an organism is transferred to new conditions. In many cases it does not take place at all. If, for example, the changes in conditions are too pronounced, these changes will damage or even kill the organisms. A gradual adaptation to the new conditions must take place.

It is therefore imperative to take the adaptation of the algae into consideration when using laboratory experiments with plankton algae as means for understanding their ecology. Otherwise the results may be completely misleading. Plankton algae adapt themselves to factors such as temperature, light intensity, light color, CO_2-concentration, and salinity. We shall concentrate here on light intensity and temperature, and mainly discuss the influence of variations in these factors on the rates of photosynthesis and growth.

It is apposite to mention that the units used to present the rate of photosynthesis are of great importance. Curves presenting the rate of photosynthesis as a function of light intensity may differ greatly, according to whether the rate is given per number of cells, per concentration of pigments, or per unit of dry weight (Steemann Nielsen *et al.* 1962). Under stable conditions growth rates are the same, whether given by increase in cell number or by such measures as increase in dry weight or pigments. However, during the period of adaptation to new conditions this is by no means the case. Algal cells adapted to different conditions may have highly varying contents of organic matter and pigments.

Photosynthesis requires at least a certain degree of equilibrium between the photochemical and the enzymatic parts of the mechanism. If the enzymatic part of the process is unable to keep pace with the photochemical, damage may take place in the cells (Steemann Nielsen 1962). The slope of the initial part of the light intensity-photosynthesis curve (Fig. 2, p. 44) is a function of the photochemical part of photosynthesis. On the other hand, the light saturated rate represents the maximum rate of the enzymatic processes. Therefore the light intensity at which the initial slope and the horizontal part of the light intensity-photosynthesis curve intersect describes to a degree the ratio between the two kinds of processes. This light intensity, introduced as I_K by Talling (1957), and illustrated in Fig. 2 at the 2°C curve, is an important means of describing the physiological state of an alga. It presents in fact, at a certain temperature, an expression for the ratio between the concentration of enzymes active in photosynthesis and the concentration of pigments active in photosynthesis. Due to adaptation, I_K seems always to be higher in algae grown at high light intensities than in algae grown at low intensities. However, different species vary quantitatively in this respect. In *Chlorella vulgaris* grown at 20°C at either 3 or 30 klux, I_K is 4 and 12 klux respectively. In *Skeletonema costatum,* the diatom used for the experiments presented in this article, I_K under the same conditions is 9 and 12 klux respectively. Roughly speaking, we might expect that *Skeletonema* grown at a low light intensity is much better fit to be transferred suddenly to a high light intensity than *Chlorella,* a fact which has been shown in this laboratory (unpublished). This, on the other hand, means that *Skeletonema* grown at low light intensities uses a higher proportion of the cell material for photosynthetic enzymes than *Chlorella,* which is by no means unimportant for the rate of growth as noted below.

Chlorella varies at a certain temperature the ratio

$$\frac{\text{concentration of photosynthetic enzymes}}{\text{concentration of photosynthetic pigments}}$$

by varying practically only the concentration of pigments (Steemann Nielsen *et al.,* 1962). The diatom *Cyclotella Meneghiniana* does it by varying the concentration of the enzymes (Jorgensen 1964). Finally *Skeletonema* varies the concentration of both pigments and enzymes.

We have for the time being considered only the photosynthesis apparatus in the algae. For the understanding of growth it is necessary, however, to take all parts of the cells into consideration. Photosynthesis is only one of the processes necessary for growth.

We may schematically divide a plankton alga into the following four parts:

a) photosynthetic pigments
b) photosynthetic enzymes
c) all other enzymes
d) the rest (cell wall, DNA, etc.)

It is important to remember that the processes controlled by (b) and (c) are temperature dependent, whereas temperature has very little influence on (a) and (d).

A cell is not able to increase the concentration of the different parts *ad libidum.* There is thus a limit for the increase of enzymes. A considerable decrease in temperature can partially, but generally not completely, be counteracted by increasing the concentration of all enzymes. We must consider that the enzymes represent a major part of the organic matter of the cell, and by far the largest part of the nitrogen-containing organic matter.

A lowering of the temperature by 20°C, a situation which *Skeletonema* is exposed to in nature, decreases the rates of the enzymatic processes by a factor of about 5. If this should be counteracted by increasing the concentration of the enzymes, this increase should also be by a factor of about 5.

Only a smaller part of this increase is possible by substituting a corresponding amount of enzymes for a part of the pigments. For completely counteracting the lowering of the temperature by 20°C the total content of organic matter in the cells has to be increased many fold.

In the cases where the effect of lowering the temperature is fully counteracted by increasing the content of enzymes, the rate of light-saturated photosynthesis per number of cells may not be influenced. The growth rate on the other hand is seriously influenced, as doubling of the cell number now calls for a much larger production of organic matter. This has been obvious during the present investigation; see p. 42.

If a plankton alga is to grow well, not only the processes involved in photosynthesis, but all processes — as for example respiration — must match each other. The conditions for the algae are thus not the same in a culture illuminated continuously as in a culture alternately illuminated and in the dark. As processes like respiration in the latter case take place for relatively much longer periods than photosynthesis, the rates of processes like respiration per time unit may be lower. The algae are thus able to place a relatively larger part of the total enzymatic organic matter in the chloroplasts. This was clearly demonstrated in the *Skeletonema* experiments to be shown in the following section.

EXPERIMENTS

The diatom *Skeletonema costatum* was used in these experiments. One special reason for using this particular plankton alga for studies in adaptation to different light and temperature conditions was the appearance of a work by Jitts *et al.* (1964). According to them, *Skeletonema* should be unable to grow at all when the temperature decreases to 6°C. A cross-grading incubator was used in this study. As *Skeletonema* in nature grows even at negative temperatures, the results could hardly be considered ecologically significant. Preliminary experiments showed that *Skeletonema costatum* was unable to stand a direct transfer from 20°-6°C if illuminated. A gradual adaptation was necessary.

An experimental series with *Skeletonema* for studying the adaptation to light and temperature was therefore started November 1st, 1964. The series has been running continuously since that time. As many aspects of light will be included, such as intensity, quality. and duration of periods of light and dark, in May 1965 we are still far from having finished the experimental series. All changes in light and temperature are made separately and in small steps, allowing for gradual adaptation.

It is thus premature to present the complete observations obtained until now. Only a few characteristic photosynthesis curves shall be given here. They should be sufficient to show the complexity and the ecological importance of the adaptation. All rates of photosynthesis are given per unit of cell number. The C^{14} technique has been used. The measurements are corrected to indicate the rate of real photosynthesis.

Fig. 1 shows the rate of photosynthesis at 20, 14, 8, and 2°C, at the light intensities (continuous illumination) at which the algae have been cultured, 3, 10 or 20 klux. The highest rate is found at 10 klux, 8°C. However, this does not mean that 10 klux and 8°C present the optimum conditions for the species. Growth rate, which must be considered the only meaningful criterion in this respect, does not have its optimum under these conditions. The daily growth rate at 10 klux was 3.7 for the cells adapted to 8°C against 5.4 for the cells adapted to 20°C. The daily growth rates express how many times the number of cells have increased during 24 hours.

Fig. 2 shows the rate of photosynthesis as a function of light intensity of algae grown at 3 klux, continuous illumination, at either 20, 14, 8, or 2°C. It is interesting to note that the curves for the three higher temperatures are practically the same.

The content of pigments per cell number must be the same as the initial part of the curves are identical. Pigment analyses showed this to be correct. On the other hand, the concentration of photosynthetic enzymes must increase with decreasing temperature.

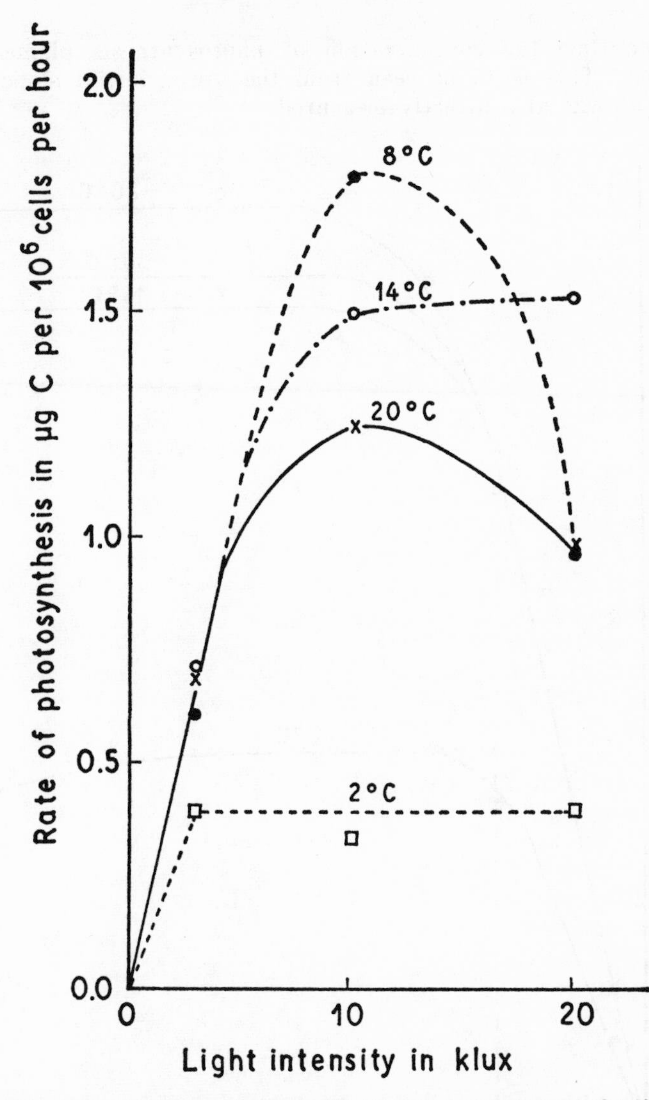

Fig. 1. - The rate of photosynthesis in *Skeletonema costatum* at the light intensity and temperature at which the alga has been cultured.

If the algae grown at 20°C are transferred directly to 8°C, the immediate light photosynthesis curve has a saturation value even lower than the saturation value for the 2°C-curve shown in Fig. 2.

Finally, the 2°C-curve shows that the alga at this low temperature has not been able to increase the concentration of photosynthesis enzymes sufficiently to counteract the low temperature. However, at the same time the concentration of photosynthesis pigments has decreased. This is to be seen from the lower initial slope of the curve, but was also directly measured.

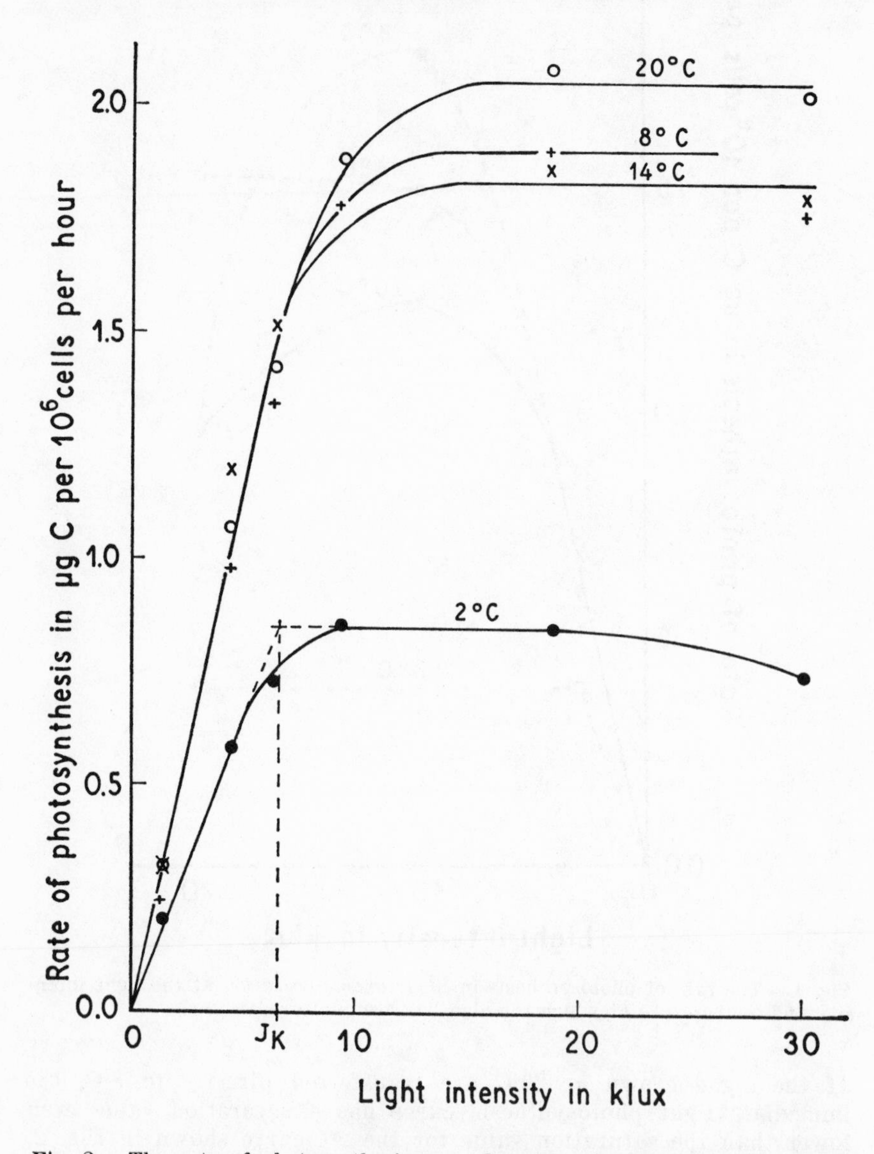

Fig. 2. - The rate of photosynthesis as a function of light intensity. *Skeletonema costatum* grown at 3 klux, continuous illumination, 20, 14, 8, or 2°C. Temperature during experiment the same as during growth. I_K shown for the 2°C curve.

Fig. 3 shows the rate of photosynthesis of algae grown at 3 klux and 2°C. The illumination was either continuous, in periods of 15 hours light and 9 hours dark, or 9 hours light and 15 hours dark. The three curves show that the concentrations of both photosynthetic pigments and photosynthetic enzymes are influenced by the periodicity. The algae having the relatively longest dark period present by far the highest rate of photosynthesis at any light intensity. As in this case all processes other than photosynthesis have a much longer period for work, the algae are able to place a relatively larger part of the organic cell substance at the disposal of the photosynthetic process.

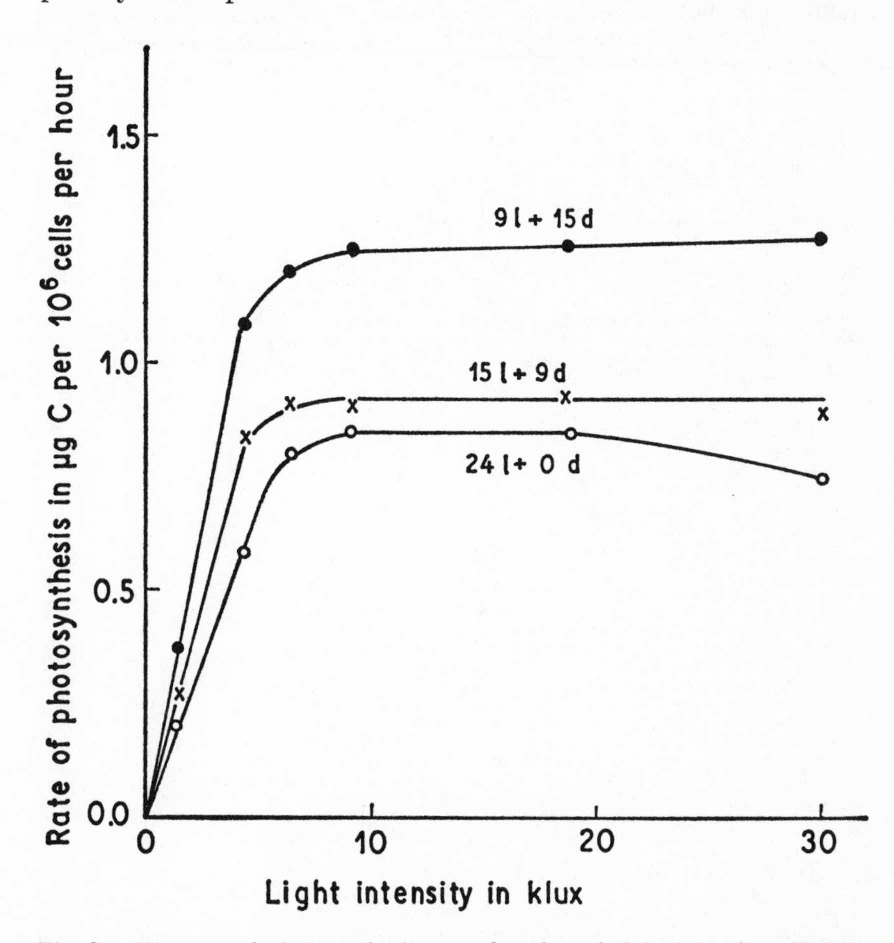

Fig. 3. - The rate of photosynthesis as a function of light intensity. *Skeletonema costatum* grown at 3 klux and 2°C. The light was given either continuously or in periods of 15 hours light and 9 hours dark, or 9 hours light and 15 hours dark.

REFERENCES

Jitts, H. R., C. D. McAllister, K. Stephens, and J. D. H. Strickland. 1964. The cell division rates of some marine phytoplankters as a function of light and temperature. - J. Fish. Res. Bd. Canada *21* : 139-157.

Jorgensen, E. G. 1964. Adaptation to different light intensities in the diatom *Cyclotella Meneghiniana Kütz.* - Physiol. Plant. *17* : 136-145.

Steemann Nielsen, E. 1962. Inactivation of the photochemical mechanism in photosynthesis as a means to protect the cells against too high light intensity. - Physiol. Plant. *15* : 161-171.

Steemann Nielsen E., V. K. Hansen, and E. G. Jorgensen. 1962. The adaptation to different light intensities in *Chlorella vulgaris* and the time dependence on transfer to a new light intensity. - Physiol. Plant. *15* : 505-517.

Talling, J. F. 1957. Photosynthetic characteristics of some freshwater plankton diatoms in relation to underwater radiation. - New Phytol. *56* : 1-50.

SOME ASPECTS OF PHYTOPLANKTON GROWTH AND ACTIVITY

CARL J. SOEDER
Limnologisches Institut
der Universität Freiburg, Falkau

For bibliographic citation of this paper, see page 10.

Abstract

The possible role of endogenous factors in metabolic activity of phytoplankton is discussed and illustrated by some experiments with synchronous cultures of *Chlorella.* It is demonstrated that *Chlorella* suspensions can be completely synchronized by the natural light and dark regimen (day and night).

From the data in the literature we can estimate that environmental factors and the availability of inorganic nutrients are more important for the intensity of primary production than are the fluctuations of photosynthetic activity which may occur in synchronized populations of algae. However, the effects of synchrony in natural phytoplankton deserve special attention and should not be neglected.

Phytoplanktonic ecosystems are considered to be composed of photosynthetically active, neutral, and inactive organisms. The proportions of these « activity groups » are thought to be decisive for the actual photosynthetic capacity of the ecosystem as a whole.

Ecological determinations of photosynthetic activity in phytoplankton are sometimes difficult to interpret (Talling 1962, Yentsch 1962). Among others, we are confronted with the following problems:

1. Within certain limits, there is no general correlation between the trophic level of a lake and the photosynthetic capacity of its phytoplankton at a given time (Gessner 1959, Elster 1965).

2. Although the « Assimilation Number » permits coarse estimates of productivity (Gessner 1959), the assimilation rates per chlorophyll unit may vary over wide ranges (Ryther and Yentsch 1958, Bursche 1961). On a cell volume basis, photosynthetic rates of natural phytoplankton are scattered over nearly two orders of magnitude (Rodhe 1958, Nauwerck 1963, Elster 1965).

3. At similar or equal light intensities, temperatures, and phytoplankton densities, relative photosynthetic activities can differ by more than one order of magnitude (Elster 1965).

Doubtless the productivity of phytoplankton depends on a complicated and multidimensional network of external and intrinsic factors besides the availability of inorganic carbon, light, or temperature. Many of the discrepancies mentioned above point to the importance of the physiological status of the algae. Some of these endogenous factors controlling productivity shall be considered here.

Vigorous growth of planktonic algae can lead to an exponential increase of population density (Lund 1949, 1950, Verduin 1950, 1952). Having reached a climax, population density either remains constant for a while or breaks down. This series of events is reminiscent of the general sigmoid growth pattern which is typical for many types of undiluted laboratory cultures of microorganisms. The growth curve of algal cultures has been reviewed by Myers (1953, 1962).

Exponential growth of an algal species will cease as soon as a limiting factor such as a mineral deficiency begins to block the unrestricted growth. A phase of linear growth may follow. More and more, division activity and dry matter production slow down, and may come to a complete standstill. In case of various mineral deficiencies, the once actively growing populations switch over to the exclusive production of storage material (von Witsch 1948, Winokur 1949, Spoehr and Milner 1949, Krauss 1958, Pirson 1958). The cells may finally degenerate and decay. All these events are accompanied by an increasing loss of photosynthetic activity (Pirson 1960, Yentsch 1962). Different mineral deficiencies cause more or less specific symptoms including the different ability to recover if viability is not lost entirely (Pirson 1958, 1960). It is, therefore, extremely important to know whether we deal with a « juvenile » or a « senile » phytoplankton.

The rhythmic alternation of day and night causes another complication of the physiological status, namely the synchrony of cell division and growth. Judging from the laboratory experiments, we may expect the algae to be synchronized at least under conditions which enable the organisms in question to grow exponentially. The term « synchrony » describes the phenomenon that many or all members of a certain species perform the same steps of their life cycle at an almost identical time.

There are three types of synchrony to be distinguished: partial synchrony, synchrony in alternating groups, and complete synchrony (Senger 1961).

Laboratory cultures of various unicellular or colonial algae, both marine and freshwater, have been synchronized. Reviews of these studies were published by Pirson (1962), Hoogenhout (1963), and the synchronizing techniques by Kuhl and Lorenzen (1964). Far less is known about synchrony in natural phytoplankton. The observations of Overbeck (1962) and of Nauwerck (1963) should be mentioned here.

Fig. 1 describes the series of events that can be observed microscopically in a completely synchronous culture of *Chlorella* (strain 211/8 b of the culture collection at Göttingen). The synchronizing regimen consisted of a 16 hrs light period and an 8 hrs dark period.

Starting from the autospore, the life cycle is concluded by the synchronous liberation of several or many new autospores, showing up by a sudden steep increase of cell number and a concomitant disappearance of autospore mother cells.

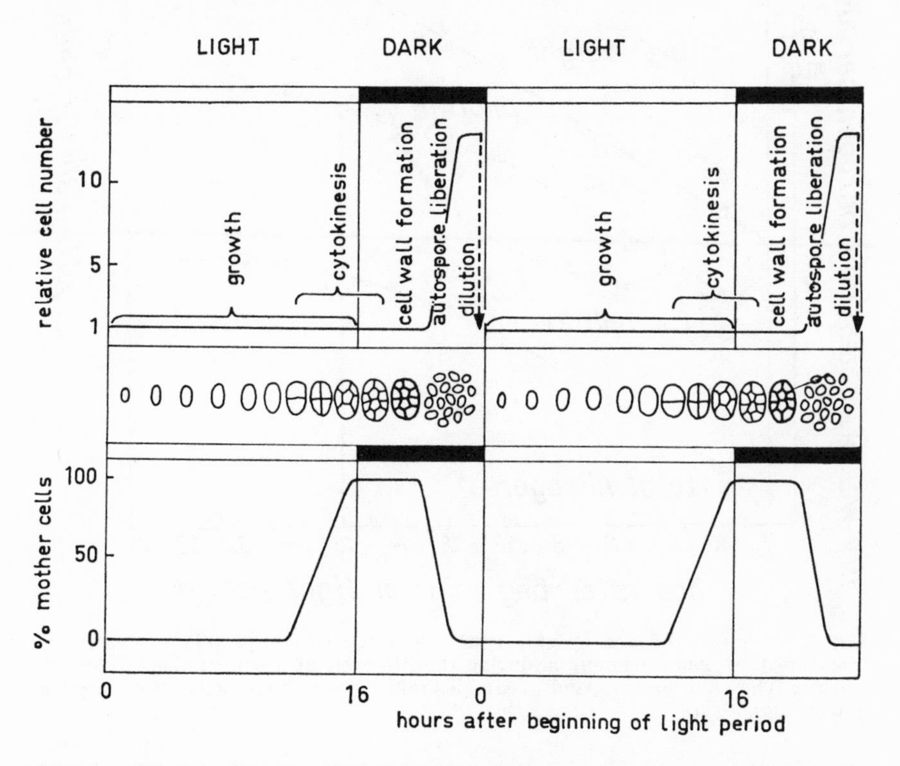

Fig. 1. - Scheme of events during 2 cycles of synchronous growth of *Chlorella* in the type of culture that was studied in the experiments presented in Figs. 2-5 and 6a. Culture conditions were essentially as described by Ried, Soeder and Müller (1963). Changes of relative cell number (top; 1 unit corresponds to about 10^6 cells/ml), cytological events (middle), and changes of the percentage of autospore mother cells, i.e. divided mature cells (bottom). As in the following figures, dark periods are indicated by black bars.

Metabolic activities and chemical composition of the cells vary drastically in the course of such a cycle (Fig. 2). The increase of dry weight proceeds in an exponential fashion for 14 hrs of the 16 hrs light period. Protein synthesis ceases earlier to increase exponentially and continues during the dark period. The relation dry weight: protein is, therefore, regularly unstable (Fig. 3). The problems involved in taking these characters as the reference bases of metabolic activities are obvious and do not deserve special comment.

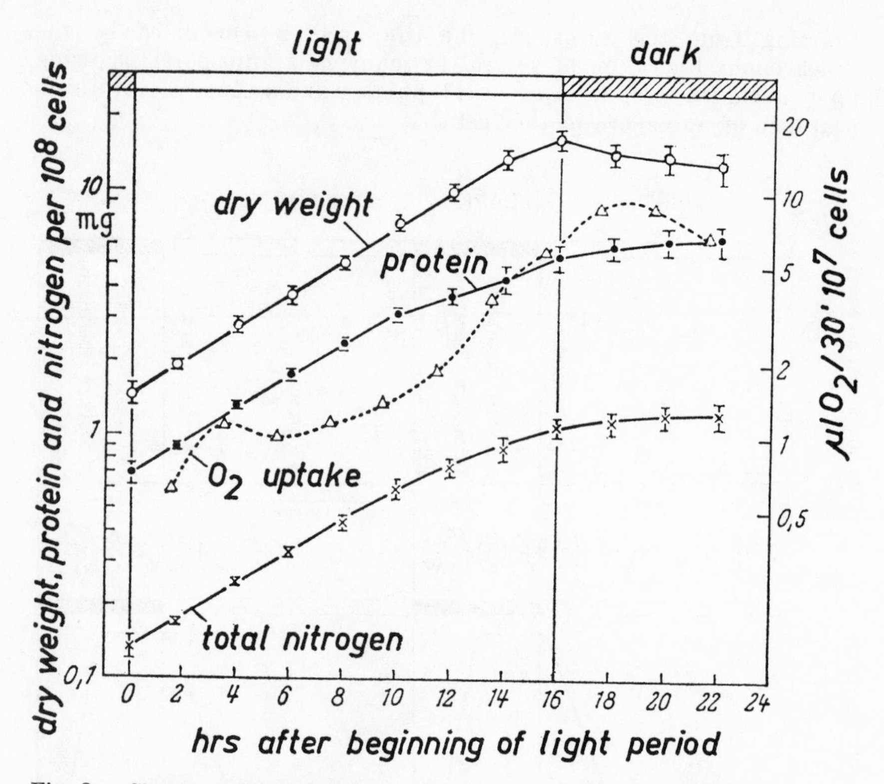

Fig. 2. - Changes of dry weight, total nitrogen content, protein content, and respiratory oxygen consumption during the life cycle of a synchronous *Chlorella* culture. The values correspond to samples taken every 2 hrs (after Ried, Soeder and Müller 1963).

During the life cycle of the same *Chlorella* strain (211/8 b) Lorenzen (1959) and Metzner and Lorenzen (1960) found regular changes in photosynthetic activity (Fig. 4). Since the chlorophyll content does not remain constant during the cycle (Ruppel 1962), the activity changes per unit of chlorophyll are not likely to be parallel to the curve of oxygen evolution per unit of dry weight. The time course of photosynthetic rate on a dry weight basis is in good agreement with the findings of Sorokin (1957) for the so called « high temperature strain » of *Chlorella pyrenoidosa*, though this alga had been cultured and synchronized under other conditions. We are, however, not definitely sure yet whether these changes in photosynthetic activity depend directly on cellular development.

It is not easy to prove that certain shifts of metabolic activity are necessarily linked to the life cycle of an organism. It might as well be that we observe mere side effects of the synchronizing

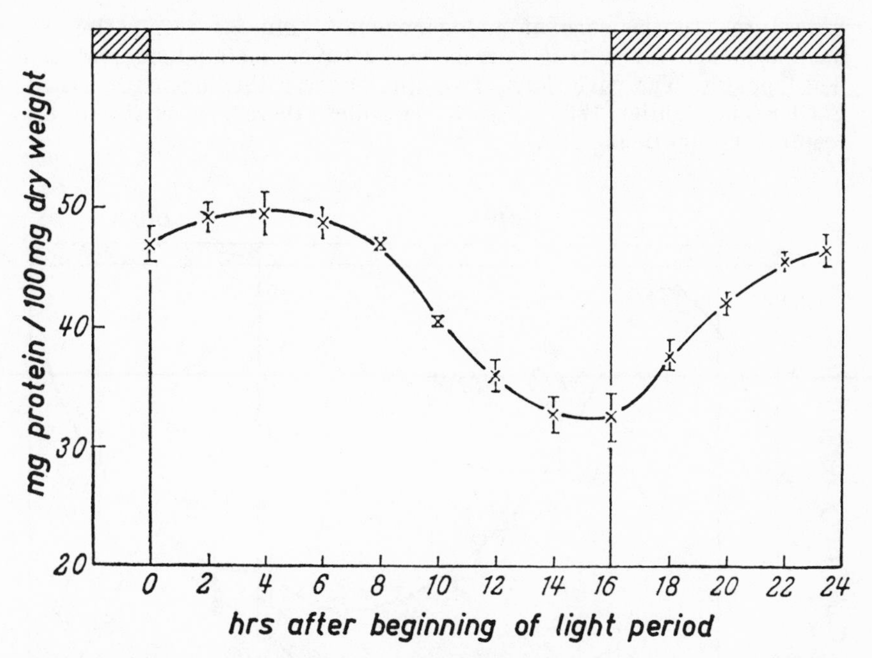

Fig. 3. - Changes of protein/dry weight during the life cycle of a synchronous *Chlorella* culture (after Ried, Soeder and Müller 1963).

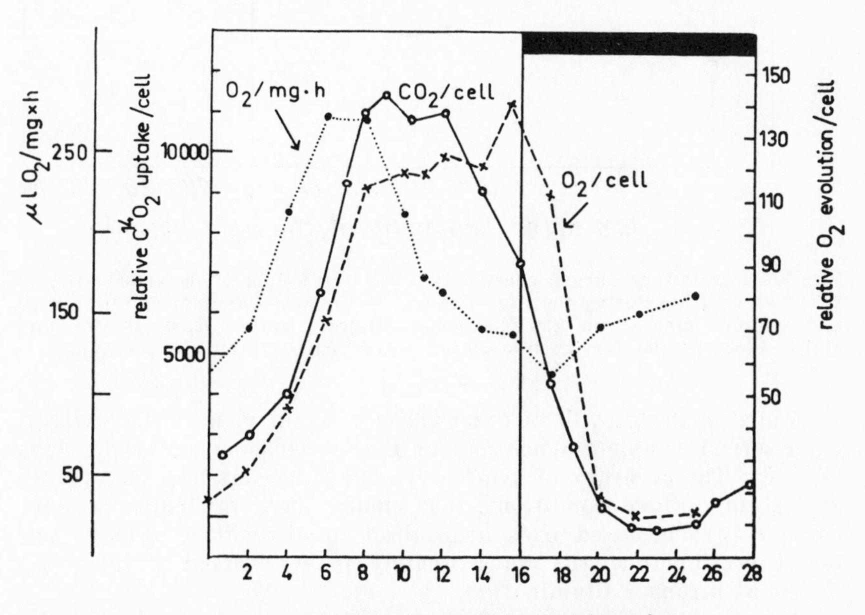

Fig. 4. - Photosynthetic gas exchange during the life cycle of synchronous *Chlorella* cultures. Oxygen evolution per mg × h after Lorenzen (1959), oxygen evolution and CO_2 uptake per cell after Metzner and Lorenzen (1960).

procedure. In the case of endogenous and glucose respiration we have evidence for a true correlation between metabolism and cell development. The variations of O_2-uptake are rather dramatic (Ried, Soeder and Müller 1963; Fig. 5). Besides, there are shifts of the respiratory quotient from 1.2 to 1.9.

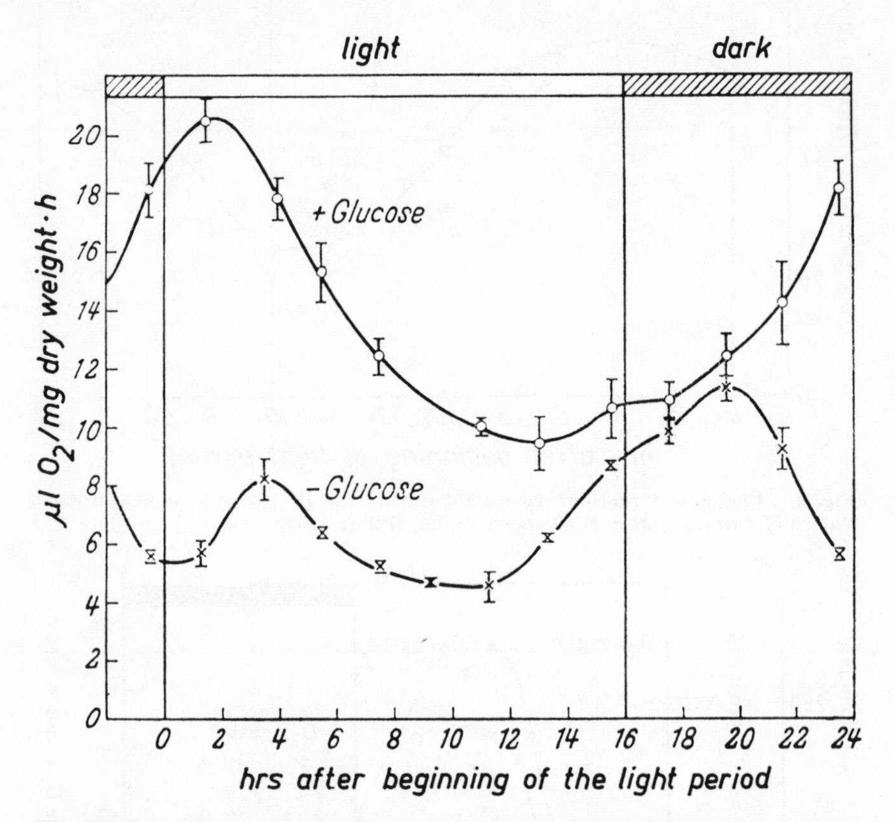

Fig. 5. - **Respiratory oxygen consumption per dry weight of darkened samples taken every 2 hrs during the life cycle of synchronous *Chlorella* cultures, with and without glucose as an exogenous substrate (after Ried, Soeder and Müller 1963). Values for the first 30 minutes of each Warburg experiment.**

Studying the growth of our *Chlorella* strain in natural daylight, we observed complete synchrony of the suspensions on bright days (Fig. 6). The accuracy of synchrony was somewhat less than under normal laboratory conditions. On cloudy days the cultures were partially synchronized or synchronized in alternating groups, i.e., the generation time was approximately 46 hrs instead of the usual 22 hrs at stronger illumination.

According to Pirson and Lorenzen (1958), the beginning of the light period of the light-and-dark cycle acts as the timer of cell development in *Chlorella*. This fits well into the general pattern of light-induced circadian rhythms in plants and animals (Bünning 1964). Our experiments show that the development of *Chlorella* can be timed sufficiently by the natural day-and-night alternation. This is certainly a general behavior of algae (Rieth 1939, Overbeck 1962, Nauwerck 1963).

The less accurate periodicity of our outdoor cultures both at saturating and underoptimal light intensities might best be explained by the assumption that during dawn a light intensity is reached which induces cell division and thereby the start of the life cycle. The response of the cells to the minimum timing light intensity

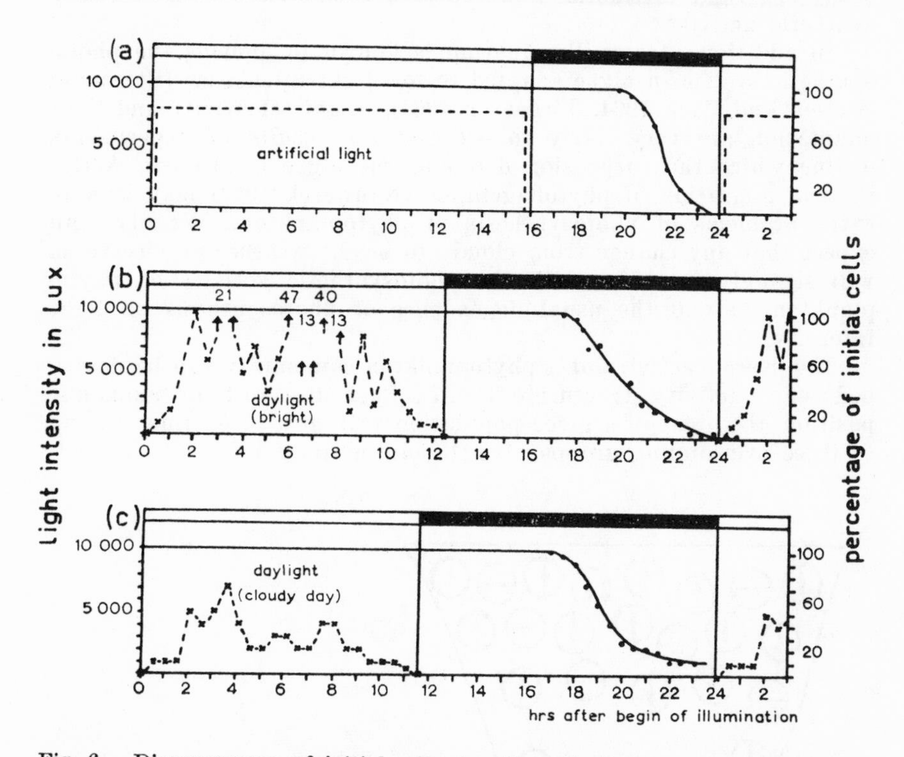

Fig. 6. - Disappearance of initial cells, i.e. the cells that were present at the beginning of the light-and-dark cycle in artificial (a) and natural (b and c) light regimens. Light intensity (broken lines), percentage of initial cells in the suspension (solid lines). 30°C, nutrient solution N5. A 1% CO_2-in-air mixture was bubbled through the cultures. At the beginning of the dark periods, the suspensions consisted of ripe autospore mother cells (not quantitatively in (c)). Since we did not follow the time course of cytokinesis (as in Fig. 1), we have to speak of « initial cells » in this case. They disappear because of the autospore liberation and are replaced by a new generation. **Figures in the upper portion of (b) refer to high light intensities in 1000 Lux.**

apparently varies to such an extent that the reaction of the culture as a whole is less exact than after the abrupt and saturating beginning of illumination in the laboratory.

Though deserving special attention, the changes in metabolic activity during the life cycle of an algal cell are minor in comparison to the discrepancies between biomass and productivity that have been reported by Rodhe (1958), Elster (1965) and others. Another aspect of synchrony might be more important in algal ecology: the resistance of *Chlorella* to high light intensity depends on the developmental stage of the cells (Sorokin 1963, Pirson and Ruppel 1962, Soeder 1964). « Light injury » at the sensitive stage already occurs at comparatively low light intensities and causes a considerable loss of photosynthetic activity.

In addition, strong illumination is known to damage the photosynthetic system in algae adapted to low light intensities (Steemann Nielsen and Park 1964, Jörgensen 1964). Light sensitivity and light adaptation are very likely to overcast the results of experiments during which the suspension depth of the algae is altered. Active vertical migration of phytoflagellates (Nauwerck 1963) may lead to extra problems of photosynthesis in phytoplankton. We also can expect that any change from cloudy to bright weather or vice versa will strongly interfere with the photosynthetic activity of phytoplankton, beyond the usual dependence of photosynthesis on light intensity.

The overall activity of a phytoplankton community can be described as an « activity structure » (Fig. 7). Regardless of their taxonomic position, the algae of a given population will, at a given time, belong to three overlapping groups of metabolic activity.

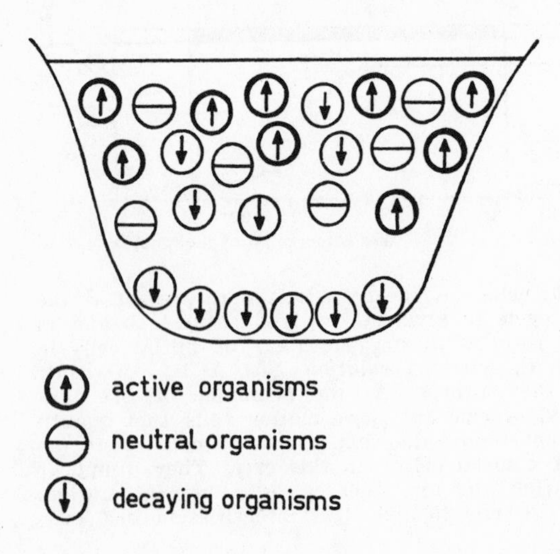

Fig. 7. - Scheme of the « activity structure » of a phytoplankton community. The overall activity in the upper layers of the water body is supposed to be anabolic. Catabolic activity prevails at the bottom.

a) *Active organisms.* Cell numbers increasing (sometimes exponentially). Division rates outruling the losses by sinking and grazing. Activity strongly anabolic.
In some cases an increase of cell number may by recognized only if we consider the cells present in the samples in addition to the number of individuals which have disappeared from the trophogenic zone by sinking (or by the grazing activity of zooplankton).

b) *Neutral organisms.* Cell numbers more or less constant over a given period of time. Activity slightly anabolic, not leading to significant positive or negative changes of total planktonic biomass.

c) *Inactive organisms.* Resting stages, degenerating or decaying algae. Cell numbers mostly decreasing. Activity more or less catabolic.

If we were dealing with laboratory cultures, such a classification would be a platitude. However, we hope that the proposed classification will help to describe and to analyze the status of phytoplanktonic communities. The literature provides many examples for the mentioned differences in population dynamics of different algal groups or species (e.g. Willen 1961 a, b, Nauwerck 1963). The correlation between division rate and metabolic activity has still to be proven.

Primary production of a whole phytoplanktonic ecosystem will be affected by the respective proportions of the biomasses of active, neutral and inactive algae. In lakes as well as in the sea, different « activity groups » are likely to dominate in different strata of the water body.

Looking at the finer structure of activity distribution, we have to regard phytoplankton as a mixture of many pure cultures, each of which consists of one algal species holding a momentary position between metabolic « youth » and « senescence ». This concept has something in common with the point of view of Margalef (1962) who describes phytoplankton activity in terms of « maturity ».

The endogenous factors governing photosynthetic activity of phytoplankton cannot be evaluated unless we know enough about the past history of the algal community under investigation. Knowledge of the actual physiological status of the algae seems to be an essential prerequisite in the process of finding out the true meaning of even serial measurements of primary production in phytoplankton.

Acknowledgments

I thank Miss Gudrun Schulze and Miss Dorothee Thiele for their valuable help with some of the experiments, and for the drawing of the figures.

REFERENCES

Bünning, E. 1964. The physiological clock. - Springer (Berlin).

Bursche, E. M. 1961. Änderungen im Chlorophyllgehalt und im Zellvolumen bei Planktonalgen, hervorgerufen durch unterschiedliche Lebensbedingungen. - Intern. Rev. ges. Hydrobiol. *46* : 610-652.

Elster, H.-J. 1965. Absolute and relative assimilation rates in relation to phytoplankton populations. - Mem. Ist. Ital. Idrobiol. *18* Suppl.: 77-103.

Gessner, F. 1959. Hydrobotanik. II. - Deutscher Verlag der Wissenschaften (Berlin).

Hoogenhout, H. 1963. Synchronous cultures of algae. - Phycologia *2* : 135-147.

Jörgensen, E. G. 1964. Adaptation to different light intensities in the diaton *Cyclotella Meneghiniana* Kütz. - Physiol. Plant. *17* : 136-145.

Krauss, R. W. 1958. Physiology of the fresh-water algae. - Ann. Rev. Plant. Physiol. *9* : 207-244.

Kuhl, A. and H. Lorenzen, 1964. Handling and culturing of *Chlorella*. In: Synchrony in Cell Division and Growth, ed. by E. Zeuthen. Vol. I. of « Methods in Cell Physiology ». - Academic Press (New York and London).

Lorenzen, H. 1959. Die photosynthetische Sauerstoffproduktion wachsender *Chlorella* bei langfristig intermittierender Belichtung. - Flora *147* : 382-404.

Lund, J. W. G. 1949. Studies on *Asterionella formosa Hass*. I. The origin and nature of the cells producing seasonal maxima. - J. Ecol. *37* : 389-419.

— 1950. Studies on *Asterionella formosa Hass*. II. Nutrient depletion and the spring maximum. - J. Ecol. *38* : 1-35.

Margalef, R. 1962. Organisation spatiale et temporelle des populations dans un secteur du littoral méditerranéen espagnol. - Pubbl. Staz. Zool. Napoli *32* (Suppl.) : 336-348.

Metzner, H. and H. Lorenzen. 1960. Untersuchungen über den Photosynthese-Gaswechsel an vollsynchronen *Chlorella*-Kulturen. - Ber. deutsch. Bot. Ges. *73* : 410-417.

Myers, J. 1953. Growth characteristics of algae in relation to the problems of mass cultures. In « Algal Culture From Laboratory to Pilot Plant » (J. S. Burlew, ed.). - Carnegie Inst. Wash. Publ. No. *600* : 37-54.

— 1962. Laboratory cultures. In « Physiology and Biochemistry of Algae » (R. A. Lewin, ed.), 603-615. - Academic Press (New York and London).

Nauwerck, A. 1963. Die Beziehungen zwischen Zooplankton und Phytoplankton im See Erken. - Symbol. Bot. Upsal. *17*, tome 5.

Overbeck, J. 1962. Untersuchungen zum Phosphathaushalt von Grünalgen. III. Das Verhalten der Zellfraktionen von *Scenedesmus quadricauda* (Turp.) Bréb. im Tagescyclus unter verschiedenen Belichtungsbedingungen und bei verschiedenen Phosphatverbindungen. - Arch. Mikrobiol. *41* : 11-26.

Pirson, A. 1958. Functional aspects in mineral nutrition of green plants. - Ann. Rev. Plant Physiol. *6* : 71-114.

— 1960. Photosynthese und mineralische Faktoren. In: Handb. Pflanzenphysiol., Bd. *V/2*: 123-151. - Springer (Berlin).

— 1962. Synchronisierung durch Licht-Dunkel-Wechsel. - Vortr. Gesamtgeb. Bot., N. F. *1* : 178-186.

— and H. Lorenzen. 1958. Ein endogener Zeitfaktor bei der Zellteilung von *Chlorella*. - Z. Bot. *46* : 53-66.

— and H. G. Ruppel. 1962. Über die Induktion einer Teilungshemmung in synchronen Kulturen von *Chorella*. - Arch. Mikrobiol. *42* : 299-309.

Ried, A., C. J. Soeder and I. Müller. 1963. Über die Atmung synchron kultivierter *Chlorella*. I. Veränderungen des respiratorischen Gaswechsels im Laufe des Entwicklungscyclus. - Arch. Mikrobiol. *45* : 343-358.

Rieth, A. 1939. Photoperiodizität bei zentrischen Diatomeen. - Planta *30* : 294-296.

Rodhe, W. 1958. Primärproduktion und Seentypen. - Verh. Int. Ver. Limnol. *13* : 121-141.

Ruppel H. G. 1962. Untersuchungen über die Zusammensetzung *Chlorella* bei Synchronisation im Licht-Dunkel-Wechsel. - Flora (Jena) *152*: 113-138.
Ryther, J. H. and C. S. Yentsch. 1958. Primary production of continental shelf waters off New York. - Limnol. and Oceanog. *3*: 327-335.
Senger, H. 1961. Untersuchungen zur Synchronisierung von *Chlorella*-Kulturen. - Arch. Mikrobiol. *40*: 47-72.
Soeder, C. J. 1964. Some characteristics of photosynthetic O_2 evolution by synchronized *Chlorella* cells. - Yearbook Carnegie Inst. Wash. *63*: 477-480.
Sorokin, C. 1957. Changes in photosynthetic activity in the course of cell development in *Chlorella*. - Physiol. Plant. *10*: 659-666.
— 1963. Injury and recovery of photosynthesis in cells of successive developmental stages: temperature effects. In « Studies on Microalgae and Photosynthetic Bacteria », special issue of Plant and Cell Physiology (pp. 99-110). - Tokyo.
Spoehr, H. A. and H. W. Milner. 1949. The chemical composition of *Chlorella*; effect of environmental conditions. - Plant Physiol. *24*: 120-149.
Steemann Nielsen, E., V. K. Hansen, and E. G. Jörgensen. 1962. The adaptation to different light intensities in *Chlorella vulgaris* and the time dependence on transfer to a new light intensity. - Physiol. Plant. *15*: 505-517.
— and T. S. Park. 1964. On the time course of adaptation to low light intensities in marine phytoplankton. - J. Cons. Esplor. Mer *29*: 19-33.
Talling, J. F. 1962. Freshwater algae. In « Physiology and Biochemistry of Algae » (R.A. Lewin, ed.), pp. 743-753. - Academic Press (New York and London).
Verduin, J. 1950. Comparison of spring diatom crops of western Lake Erie in 1949 and 1950. - Ecology *32*: 662-668.
— 1952. Photosynthesis and growth rates of two diatom communities in western Lake Erie. - Ecology *33*: 163-68.
Willen, T. 1961 a. The phytoplankton of Ösbysjön, Djursholm. - Oikos *12*: 36-69.
— 1961. b. The phytoplankton of Ösbysjön, Djursholm. II. Ecological aspects. - Oikos *12*: 195-224.
Winokur, M. 1949. Ageing effects in *Chlorella* cultures. - Amer. J. Bot. *36*: 287-291.
von Witsch, H. 1948. Physiologischer Zustand und Wachstumsintensität bei *Chlorella*. - Arch. Mikrobiol. *14*: 128-141.

THE PATTERN OF PHOTOSYNTHESIS AND RESPIRATION IN LABORATORY MICROECOSYSTEMS

ROBERT J. BEYERS [1]

Dept. of Zoology and Institute of Radiation Ecology
University of Georgia, Athens, Georgia, U.S.A.

[1] Mailing address: Savannah River Ecology Laboratory, Building 772-G, SROO, P. O. Box A, Aiken, South Carolina, 29802, U.S.A.

For bibliographic citation of this paper, see page 10.

Abstract

Data are presented to show the similarity in the course of diurnal metabolism in several different types of aquatic laboratory microecosystems. Both net photosynthesis and nighttime respiration are maximal in the first half of the light or dark periods. This pattern also occurs on electrical analogue circuits set up as a theoretical model relating metabolic rates to storage accumulations. Effects of lengthening the photoperiod of experimental microcosms are discussed.

INTRODUCTION

Natural aquatic ecosystems are difficult to study under control conditions, due to their large size and the inability of the investigator to dictate the various environmental parameters which affect most of the phenomena taking place in the system. Since no two bodies of natural water can be studied under exactly the same conditions, the validity of comparison of any factor between two such bodies is necessarily limited. In situations where ecosystems small enough to be conveniently subjected to experimental conditions and replicable enough to give some measure of control to the experimental technique were desired, many researchers have resorted to the use of laboratory microecosystems or microcosms (Odum and Hoskin 1957, Whittaker 1961. Beyers 1963 a, b, McConnell 1962, Butler 1964, McIntyre, *et al.* 1964).

The purpose of this report is to present data gathered by myself and others which relate to the pattern of photosynthesis and respiration in the laboratory ecosystems, and to some of the possible reasons for the existence of this pattern.

MATERIALS AND METHODS

The systems to be discussed are the following: 1) Twelve benthic fresh water systems whose primary producers were *Vallisneria* and *Oedogonium*. The major animal component was an oligochaete (*Su-*

Table 1. - Resumé of conditions and total rates of net photosynthesis and nighttime respiration for the microecosystems.

Type of Microecosystem	System Number	Light in Foot Candles	Temperature °C	Partial Community Composition	Salinity in ppt.	Volume in Liters	Net Photosynthesis $mMCO_2/L/12$ hr.	Nighttime Respiration $mMCO_2/L/12$ hr.	$\frac{P\ net}{R\ night}$
Mean of 100 Benthic Systems Curves	1	1000	23	*Vallisneria*, *Oedogonium*, *Sutroa*, *Physa*	0	3.000	0.20	0.17	1.18
Algal Mat Microcosm	2	1000	14	*Desulfovibrio*, *Lyngbya*, *Oscillatoria*, Purple Sulfer Bacteria	34	0.166	0.80	0.69	1.16
51° Microcosm	3	784	51	*Phormidium*, *Oscillatoria*, *Anabaena*	0	4.020	0.28	0.28	1.00
Temporary Pond Microcosm	4	850	22	Cladocera, Ostracoda, Small Green Flagellates	0	4.000	0.17	0.11	1.47
Brine Microcosm	5	467	23	*Artemia*, *Dunaliella*	188	2.030	0.29	0.26	1.12

troa) and the snail *Physa*. This community taken from the San Marcos River in Texas is more fully described by Beyers (1963a). 2) An algal mat microcosm containing *Desulfovibrio, Lyngbya, Oscillatoria,* and purple sulfer bacteria. This is the system of Armstrong and Odum (1964). 3) A hot spring system from Mimbres, New Mexico, held at 51°C and containing *Phormidium, Oscillatoria,* and *Anabaena.* 4) A brine microecosystem from the La Parguera Salt Works, Puerto Rico, containing *Dunaliella* and the brine shrimp *Artemia.* 5) A temporary pond type system from Enchanted Rock, Texas, containing cladocera, ostracods, and small green flagellates. Systems 3, 4, and 5 were more fully described by Beyers (1963b). 6) An ecological analog computer of the type discussed by Odum (1960, 1962). All of the microecosystems were held in a constant temperature room under artificial light with a 12 hour photoperiod. In one series of experiments with the benthic systems (number 1), the photoperiod was lengthened to 24 hours of light followed by 24 hours of dark. Table 1 gives a summary of the temperature and salinity conditions under which the various experiments were conducted. Except for the analogue computer, all systems were studied using the carbon dioxide diurnal rate of change curve method (Beyers 1963a, Beyers *et al.* 1963). The advantage of this method is that the data are presented in such a way that hour by hour changes in the rate of community photosynthesis or respiration are readily apparent. This method involves the use of recording pH meters to determine the diurnal variation in the pH of the microecosystem water. A sample of water from the system is bubbled with nitrogen or any other inert, carbon dioxide-free gas to raise the pH. The sample is then titrated with distilled water saturated with carbon dioxide under approximately one atmosphere pressure using a special burette. This procedure is essentially a titration with gaseous carbon dioxide using distilled water as a carrier. From this titration the relationship between pH and total dissolved carbon dioxide is established. Once the relationship is known, any change in pH can be translated into changes in concentration of carbon dioxide.

RESULTS AND DISCUSSION

In considering the overall metabolism of a closed ecosystem, two similar but opposite processes are of importance. They are total respiration and gross photosynthesis. Total respiration is the entire heterotrophic activity of all the organisms, both animals and plants, in the system, while gross photosynthesis is the entire autotrophic effort of all the primary producers in the community. Neither of these quantities is measurable in the light because they are essentially the same net chemical change proceeding simultaneously in both directions. During the day only the excess of gross photosynthesis over daytime respiration can be measured. This excess is termed net photosynthesis (P_n).

In the dark only respiration takes place. Therefore, it is possible to measure it. However, on a twenty-four hour basis the nighttime respiration (R_{ni}) is not the total respiration of the system. The respiration which takes place during the light period does not figure in the nighttime respiration. Hence we are left with two measurable quantities, net photosynthesis and nighttime respiration. The values for these two quantities for five different types of microecosystems are shown in Table 1. The figures are expressed as millimoles carbon dioxide absorbed or liberated per liter of microcosm water per twelve hour light or dark period. The ratios of these two quantities (P_n/R_{ni}) are also shown in Table 1.

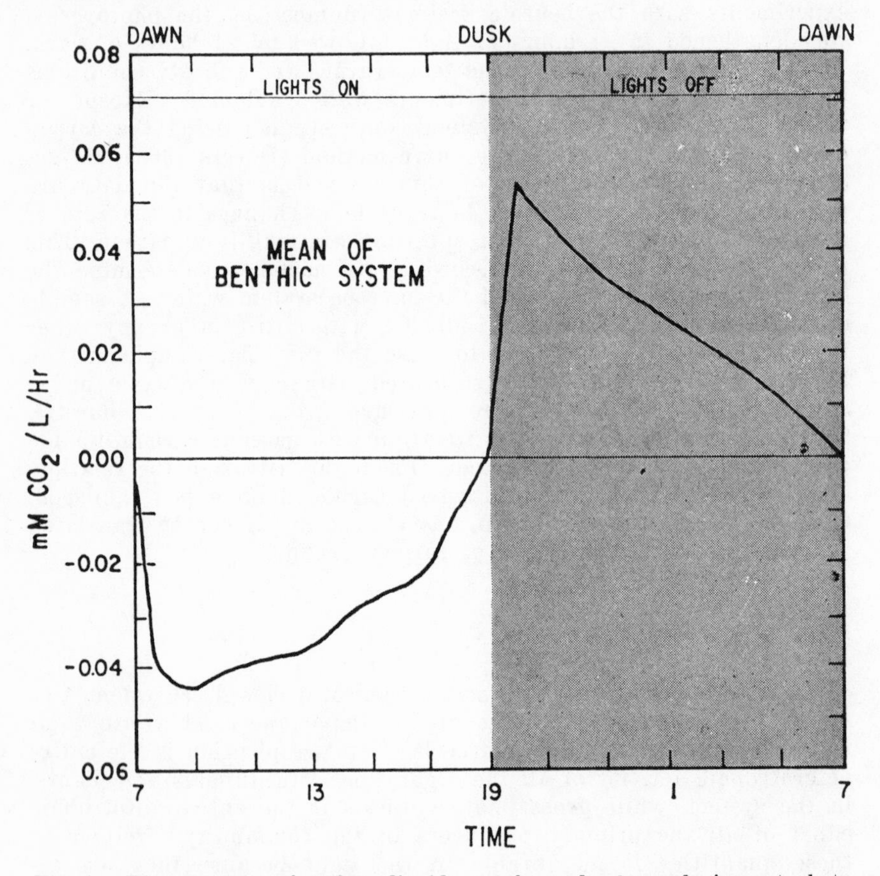

Fig. 1. - Diurnal rates of carbon dioxide uptake and release during net photosynthesis and nighttime respiration in twelve benthic freshwater microecosystems. This curve is the mean of one hundred curves.

The Metabolic Pattern in Laboratory Systems. In an early study (Beyers 1963a), I noted that the carbon dioxide rate of change curves from twelve freshwater benthic microcosms showed a fairly regular pattern. The nighttime respiration was at a maximum in the early evening after the lights were extinguished. The respiratory rate then dropped until the lights came on again the next morning. The net photosynthesis showed a similar pattern, with production at a maximum immediately after the lights were turned on and with decreasing photosynthesis during the rest of the day. Fig. 1 shows an average of 100 carbon dioxide diurnal rate of change curves in the benthic systems.

In an attempt to determine whether the metabolic pattern shown by the benthic systems was of a general nature, several other microcosms of various types were measured under similar conditions. They were the algal mat community, hot spring community, the temporary pond community, and the brine community. Representative curves of these four microcosms are given in Fig. 2. The same type of metabolism is shown by all four systems. The most extreme cases are in the hot spring (51°C.) and brine microcosms, which are also the simplest communities. However, the algal mat and the temporary pond also show decreasing metabolism as either the day or night progresses. These latter two microcosms are much more complex systems in terms of species variety than the hot spring or brine microcosms.

The photosynthetic pattern under constant illumination corresponds closely to the results of Doty and Oguri (1957), Yentsch and Ryther (1957), and McAllister (1963), for marine planktonic systems. The decreasing nighttime respiratory rate in natural aquatic systems is demonstrated by Jackson and McFadden (1954) for Sanctuary Lake and by Park, Hood and Odum (1958) in four out of seven of their carbon dioxide diurnal rate of change curves for several Texas bays.

Analogue Circuits and the Metabolic Pattern. It must be remembered when considering the photosynthetic pattern that these curves (Figs. 1 and 2) represent net photosynthesis. They are graphs of the difference between gross photosynthesis and daytime respiration. Hence, without other evidence the shape may be explained in three ways. First, gross photosynthesis may be constant under constant illumination while daytime respiration increases during the day. In fact, daytime respiration could increase to a level almost equalling gross photosynthesis and thus account for curves of the shape of the brine microcosm or the 51° microcosm (Fig. 2). Second, the gross photosynthesis may decrease during the day while the daytime respiration remains constant, or third, the daytime respiration may increase while the gross photosynthesis decreases during the course of the light period. Due to the fact that gross photo-

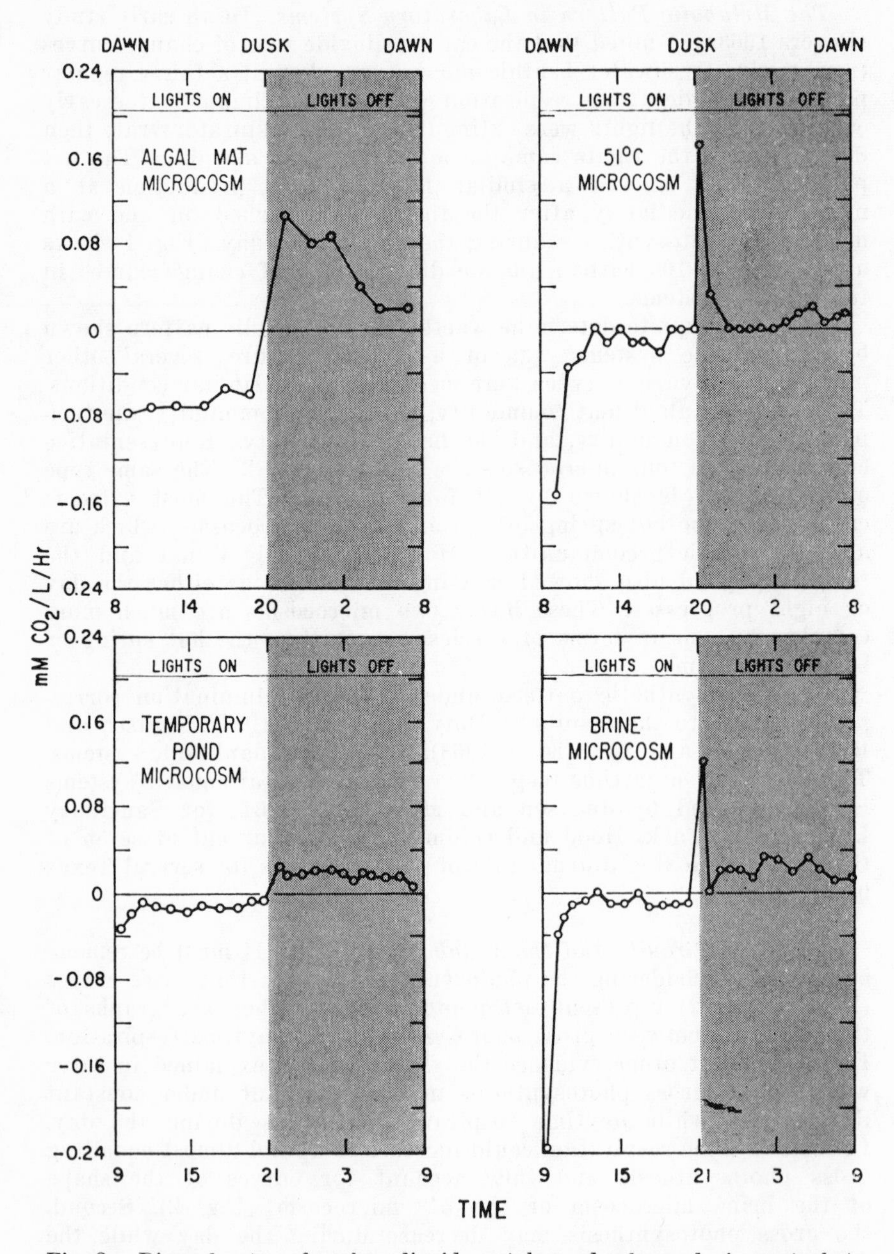

Fig. 2. - Diurnal rates of carbon dioxide uptake and release during net photosynthesis and nighttime respiration in a marine algal mat microecosystem, a hot spring microecosystem, a temporary pond microecosystem, and a brine microecosystem.

synthesis and daytime respiration occur simultaneously, it is not possible at present to test these three hypotheses to determine which is correct. However, it can be demonstrated that nighttime respiration does not remain constant (Figs. 1 and 2), and since nighttime respiration is not static, there is no reason to assume that daytime respiration stays constant. This situation casts doubt on the validity of the second hypothesis mentioned above.

If one assumes the first hypothesis to be correct and assumes a constant illumination, an analogue circuit (Fig. 3) similar to the one proposed by Odum (1960, 1962) can be constructed to simulate the metabolism in a closed ecosystem. This is a simple electrical analogue model where gross photosynthesis is represented by a battery (B) connected to simulate photosynthesis during daylight hours. The gross production (P) of labile organic matter, represented by electron flow, goes first to a storage capacitor (S) representing the organisms of the community and their capacity to take on the high energy compounds and organic matter just produced. From this storage and with a rate proportional to the concentration of the labile organic matter stored, a circuit resistance (R) goes to ground. This resistance circuit represents daytime respiration of the community and returns the organic carbon to the inorganic state of the ground reservoir. The difference between the rate of charging of the capacitor by the battery and the rate of loss of charge through the resistance is analogous to net production. When the battery is disconnected, the electrons stored in the condenser (S) bleed off through the resistance (R). This electron flow represents nighttime respiration.

During either the simulated day or night, voltage on the condenser may be recorded against time with a recorder placed between E and ground. Differentiating the voltage curve produces a rate of change curve analogous to a carbon dioxide diurnal rate of change curve. Such a curve is reproduced in Fig. 3 (Odum, Beyers, and Armstrong 1963). The data are expressed in units of simulated carbon dioxide metabolism.

The curve in Fig. 3 is virtually indistinguishable from a typical microcosm curve of carbon dioxide metabolism. Varying the capacitance and resistance involved in the circuit (time constant) changes the shape of the curve from the more usual type (Fig. 1) to the more extreme classes shown by the brine microcosm in Fig. 2. The remarkable similarity between the computer curves and the microcosm curves tends to lend credence to the hypothesis that during the day gross photosynthesis remains constant while daytime respiration increases in an analogous, but opposite manner to the decrease of respiration during the night.

Lengthened Photoperiod and the Metabolic Pattern. Four series of experiments were performed with the benthic systems to investi-

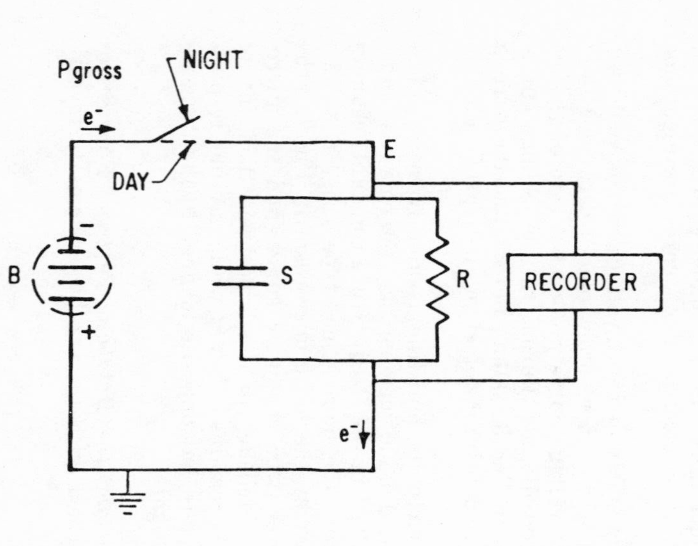

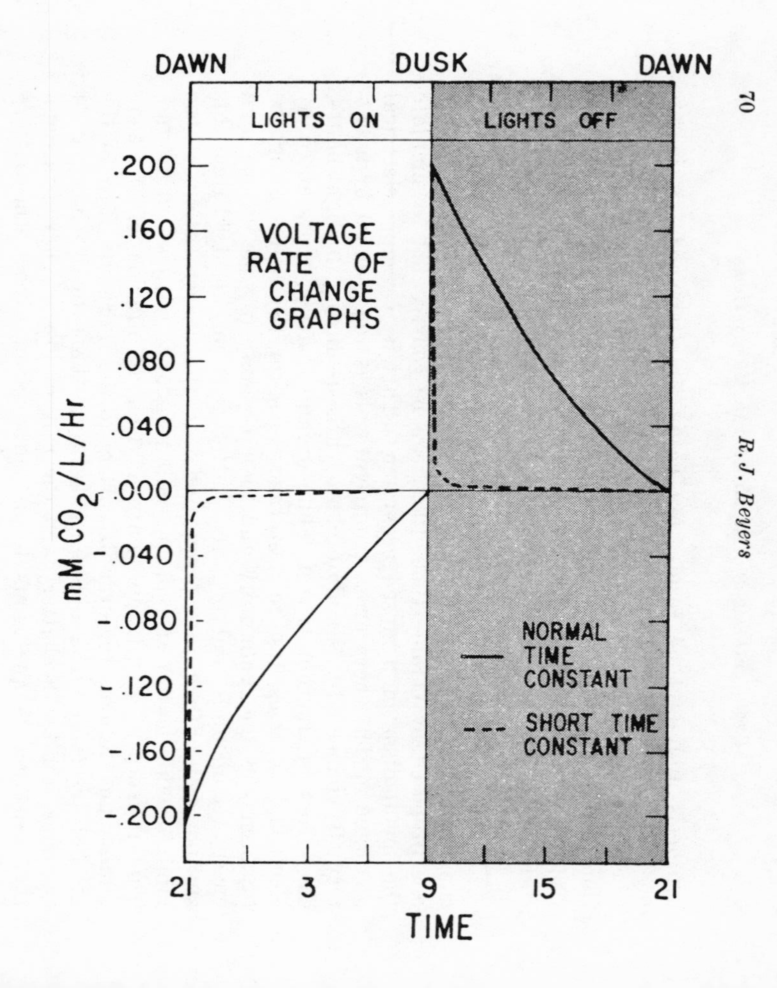

Fig. 3. - Electrical analogue circuit model for diurnal metabolism of balanced artificial microcosms without net growth, export, import, or reproduction, and diurnal rate of change graph of voltage simulating the time course of storage (net photosynthesis) and nighttime respiratory metabolism in the circuit. Results are expressed in units of simulated carbon dioxide metabolism for comparison purposes with other figures. System with short time constant has high respiratory conductivity and small storage capacity with respect to normal time constant system.

gate the effect of lengthened photoperiod on the metabolic pattern. Each series began and ended with a normal carbon dioxide diurnal rate of change curve. The experimental curve lasted 48 hours (24 hours light and 24 hours dark) instead of the normal 24 hours. Since the light and dark periods ran consecutively, the first 12 hour period of both the 24 hours of light and the 24 hours of dark corresponded to the normal day of the microecosystems, while the second 12 hour period corresponded to the normal night.

A diurnal rate of change curve for a typical series is shown in Fig. 4. The control curves are plotted on top and the experimental curves are plotted on the bottom. In the first 12 hours of the light period there was a mean carbon dioxide uptake of 0.87 grams per square meter per 12 hours, and in the second 12 hour period, a mean carbon dioxide output of 0.34 grams. The first half of the dark period showed a mean output of 0.99 grams and the second half a mean output of 0.10 grams.

In interpreting these data at least two factors must be considered: first, the concentration of the various gases and other metabolites, and second, the possible existence of a respiratory or photosynthetic rhythm. The most unexpected facet of these data is the absence of any net photosynthesis in the second 12 hour period of the continuous light. Possible causes are exhaustion of carbon dioxide, exhaustion of other nutrient materials, and oxygen supersaturation in the microecosystem water after 12 hours of photosynthesis. At the end of the first 12 hour light period the oxygen concentration was in the range of 12 to 17 parts per million.

Fig. 4 suggests that between 1900 and 0100 of the first day of the experimental curves a respiratory peak occurs, even though there is light falling on the microecosystems. The time immediately after 1900 is the period of the normal respiratory burst and since the burst seems to persist even in the presence of light it is possible that some inherent respiratory rhythm may be in operation.

In the 48 hour runs the lights were turned off at 0700. This hour is normally dawn and often the start of a period of maximum net production. However, since no light energy was available for photosynthesis, any photosynthetic rhythm was masked and the usual burst of respiration which accompanies the onset of darkness appeared. In the second 12 hour dark period the respiration was almost zero. No second burst of respiration between 1900 and 0100 was possible since all of the oxygen and probably the labile organic matter had been used up in the first 12 hours of respiration in the dark.

The almost complete cessation of respiration in the second half of the dark period allows speculation on the effect of extremely cloudy or dark days on natural, balanced, aquatic ecosystems with small quantities of stored organic material. It may be postulated that if a balanced system lives on a hand to mouth basis, making each day only enough labile organic matter for the ensuing night's needs, a succession of low light intensity days might be a real catastrophe. A diminution of light energy requires the oxidation of the respiratory and photosynthetic machinery of the community to fulfill the usual energy requirements of its component organisms. This destruction of vital metabolic tissue depresses both the pro-

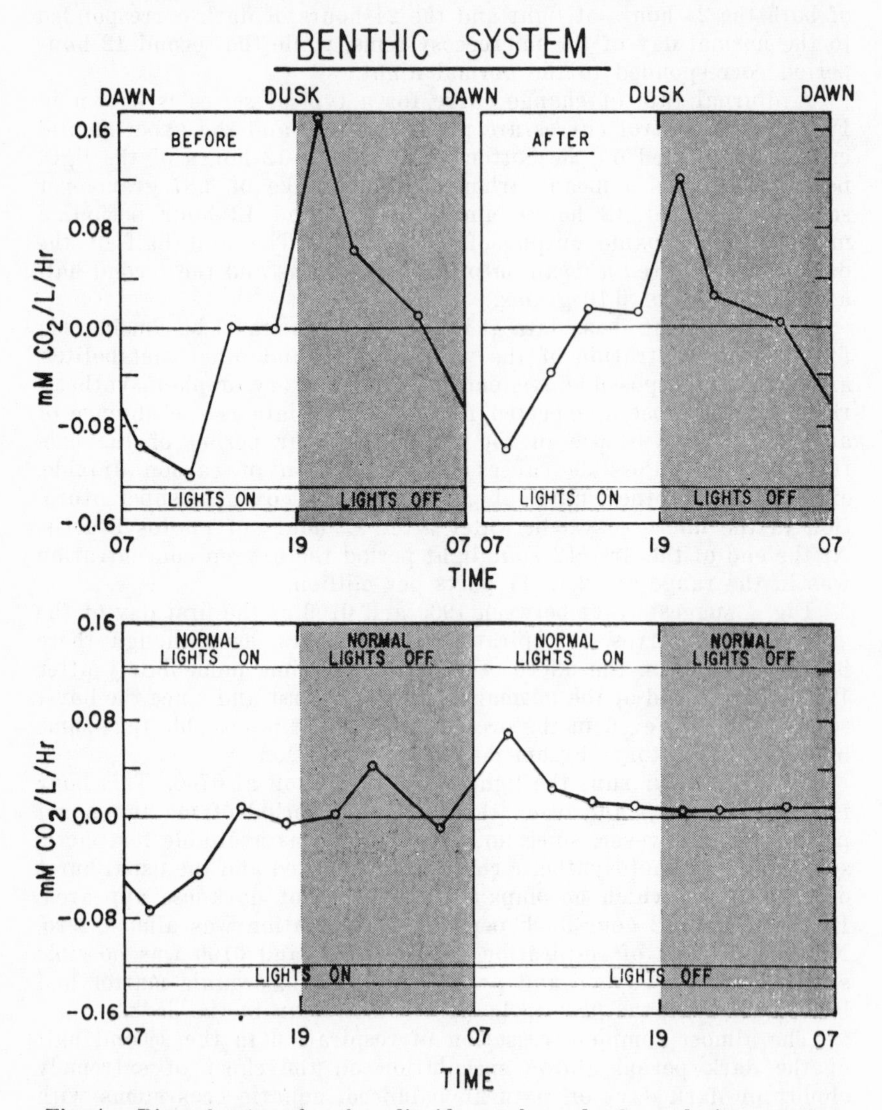

Fig. 4. - Diurnal rates of carbon dioxide uptake and release during net photosynthesis and nighttime respiration in one of the benthic microecosystems. The bottom curve shows a 24 hour light and 24 hour dark experiment. The top curves are normal curves taken one day before and four days after the experiment.

duction and the respiration of the community until such time as the machinery can be replaced from what little net photosynthesis might be available over and above each night's requirements. However, if a balanced ecosystem compensates for prolonged darkness by reducing its respiration before metabolizing its structure, as these microecosystems seem to have done, then we may have a mechanism to avoid the catastrophic effects of a period of bad weather and which would permit the system to survive these adverse conditions.

An ecosystem may react to reduced light by running the respiratory part of its metabolism at minimum rate in order to conserve its tissue until the return of normal light intensities. Such a mechanism would have survival value for an ecosystem in that the system possessing it would be prepared to return to its usual metabolic rate as soon as light levels were again normal, without having to gradually rebuild its photosynthetic and respiratory structure. As an additional survival value, large quantities of energy need not be stored for use during cloudy days, and even more energy would not be required for the maintenance of storage facilities. The energy thus freed would then be expended in other ways to promote the survival of the system.

Metabolic Pattern and Natural Light. The data presented here agree very well with measurements made in the field. Verduin (1957), Yentsch and Ryther (1957), Doty and Oguri (1957), Ohle (1958), Vollenweider and Nauwerck (1961), Copeland, Butler and Shelton (1961), Copeland and Dorris (1962), and Wetzel (1965), have all found photosynthetic maximum under natural light conditions to occur in midmorning. The photosynthetic pattern becomes particularly striking when it is realized that the input energy in nature does not reach a maximum until noon, as opposed to microecosystem experiments, where maximum light becomes available at dawn and continues until dusk. Thus one finds a compromise between the system's tendency toward maximum photosynthesis at dawn and the maximum energy for driving photosynthesis at noon, with an actual photosynthetic maximum sometime in midmorning. This lack of uniformity in the rate of photosynthesis should be kept in mind by those who wish to extrapolate to diurnal or annual primary productivity measurements from light and dark bottle or C^{14} experiments six hours or less in length.

Acknowledgements

Preparation of this manuscript was supported by contract At (38-1)-310 between the United States Atomic Energy Commission and the University of Georgia and grant NSG-706/11-003-001 from the U.S. National Aeronautics and Space Administration to the University of Georgia. Computer work was done by Dr. H. T. Odum and Mr. Neal Armstrong. Figures were prepared by Mr. M. E. Chavis.

REFERENCES

Armstrong, N. E. and H. T. Odum. 1964. Photoelectric ecosystem. Science *143*: 256-258.

Beyers, R. J. 1963 a. The metabolism of twelve aquatic laboratory microecosystems. - Ecol. Monogr. *33*: 281-306.

— 1963b. A characteristic diurnal metabolic pattern in balanced microcosms. Publ. Inst. Mar. Sci. Univ. Texas. *9*: 19-27.

— J. Larimer, H. T. Odum, R. B. Parker, and N. E. Armstrong. 1963. Directions for the determination of changes in carbon dioxide concentration from changes in pH. - Publ. Inst. Mar. Sci, Univ. Texas. *9*: 454-489.

Butler, J. L. 1964. Interaction of effects by environmental factors on primary productivity in ponds and microecosystems. - Ph. D. Thesis. Oklahoma State University.

Copeland, B. J., J. L. Butler, and W. L. Shelton. 1961. Photosynthetic productivity in a small pond. - Proc. Okla. Acad. Sci. *42*: 22-26.

— and T. C. Dorris. 1962. Photosynthetic productivity in oil refinery effluent holding-ponds. - J. Water Pollution Contr. Fed. *34*: 1104-1111.

Doty, M. S. and M. Oguri. 1957. Evidence for a photosyntetic daily periodicity. Limnol. & Oceanog. *2*: 37-40.

Jackson, D. F., and J. McFadden. 1954. Phytoplankton photosynthesis in Sanctuary Lake, Pymatuning Reservoir. - Ecology *35*: 1-4.

McAllister, C. D. 1963. Measurements of diurnal variation in productivity at ocean station « P. ». - Limnol Oceanog. *8*: 280-291.

McConnel, W. J. 1962. Productivity relations in carboy microcosms. - Limnol. Oceanog. *7*: 335-343.

McIntyre, C. D., R. L. Garrison, H. K. Phinney, and C. E. Warren. 1964. Primary production in laboratory streams. - Limnol. Oceanog. *9*: 92-102.

Odum, H. T. 1960. Ecological potential and analogue circuits for the ecosystem. Amer. Sci., *48*: 1-8.

— 1962. The use of a network energy simulator to synthesize systems and develop analogous theory: the ecosystem example. - Proc. Conference on Biomathematics, Department of Statistics, N. C. State College; pp. 291-297.

— and C. M. Hoskin. 1957. Metabolism of a laboratory stream microcosm. Publ. Inst. Mar. Sci. Univ. Texas *4*: 115-133.

— R. J. Beyers, and N. E. Armstrong. 1963. Consequences of small storage capacity in nannoplankton pertinent to measurement of primary production in tropical waters. - J. Mar. Res. *21*: 191-198.

Ohle, W. 1958. Diurnal production and destruction rates of phytoplankton in lakes. - Rapp. Proc. Verb. *144*: 129-131.

Park, K., D. W. Hood and H. T. Odum. 1958. Diurnal pH variation in Texas Bays, and its application to primary production estimation. - Publ. Inst. Mar. Sci. Univ. Texas *5*: 47-64.

Verduin, J. 1957. Daytime variations in phytoplankton photosynthesis. - Limnol. Oceanog. *2*: 333-336.

Vollenweider, R. A., and A. Nauwerck. 1961. Some observations on the C^{14} method for measuring primary production. - Verh. Int. Ver. Limnol. *14*: 134-139.

Wetzel, R. G. 1965. Techniques and problems of primary productivity measurements in higher aquatic plants and periphyton. - Mem. Ist. Ital. Idrobiol. 18 Suppl.: 249-267.

Whittaker, R. H. 1961. Experiments with radiophosphorus tracer in aquarium microcosms. - Ecol. Monogr. *31*: 157-188.

Yentsch. C. S. and J. H. Ryther. 1957. Short term variations in phytoplankton chlorophyll and their significance. - Limnol. Oceanog. *2*: 140-142.

II

FACTORS LIMITING THE PRODUCTIVITY OF NATURAL PHYTOPLANKTON POPULATIONS

ABSOLUTE AND RELATIVE ASSIMILATION RATES IN RELATION TO PHYTOPLANKTON POPULATIONS

HANS JOACHIM ELSTER

Limnologisches Institut (Walter-Schlienz-Institut)
der Universität Freiburg at Falkau, Schwarzwald, Germany

For bibliographic citation of this paper, see page 10.

Abstract

Absolute and relative assimilation rates have been determined in several lakes of the Black Forest, in Lake Constance, and in the Schleinsee. The values differ by about one order of magnitude. Relative assimilation rates (on a cell volume basis) are not strongly correlated to light intensity, water temperature, and the trophic level of the lakes. The importance of the actual physiological status of phytoplankton populations for the observed photosynthetic rates is pointed out. A combination of more detailed field investigations, including hydrophysics, with laboratory experiments seems to be necessary in order to elucidate the external and intrinsic factors conditioning primary production in lakes of different types.

In the following report, the absolute assimilation rate is defined as mg C/time per 1 m^2 of lake surface or per m^3. The reference base of relative assimilation rate is the unit volume of the algal cells which are present in the sample of lake water. The estimation of algal volume was based upon the values of A. Nauwerk (1963; cf. also Grim 1939). If necessary, these were corrected after microscopical determination of true cell dimensions in the respective lakes.

The author is conscious of the fact that algal volume is not an ideal base of reference for the evaluation of assimilatory efficiency in a population of algae, but at present it is the only practical means of estimating the relative biomasses of the principal organisms in the standing crop. Samples were taken with a lucite Ruttner sampler and darkened immediately until the beginning of exposure. The bottles were exposed at noon in a horizontal position. Exposure time was 4 hrs.

If possible the following tests were carried out simultaneously (see also Sorokin 1956).

1) Dark bottles in sampling depths (D)
2) Light bottles in sampling depths (T-T)
3) Surface samples in different depths (O-T)
4) Samples from different depths at the lake surface (T-O)
5) Samples from different depths in a water cooled clinostat at 8000 Lux (fluorescent light: day light type), 1 r.p.m. (K1).

The following lakes were investigated:

a) In the Black Forest: Feldsee, Titisee and Schluchsee

b) Lake Constance: Obersee (Überlingersee included) and Untersee, and

c) in the Lake Constance region (near Langenargen-Nonnenhorn) the Schleinsee.

The Black Forest lakes differ from Lake Constance and the Schleinsee mainly by higher altitude, lower alkalinity, and oligo- or meso-dystrophy (Table 1). Only the artificially dammed up Schluchsee (30 m above the original level) has gained greater alkalinity — which can amount to 1.4 (ml n/10 HCl per 100 ml, methyl orange method) — by the pumped up Rhine water (Elster and Schmolinsky 1952/3, Elster 1962, Eckstein 1963/4).

The general tendencies of phytoplankton distribution and photosynthetic activity in the Black Forest lakes and Lake Constance shall be considered first (Fig. 1). The average algal volume has its maximum between 0.5 and 3 m and decreases slightly towards the surface. Below 3 m it declines almost linearly down to 20 m.

Absolute assimilation rates decrease much more towards the lake surface and culminate between 0.5 and 2 m depth. From here on they steeply decline to 1/3 maximum photosynthetic activity at 5 m and 1/10 at 8.5 m depth.

The respective relative assimilation rates also show a decrease of activity towards the lake surface, and a maximum (13.5 - 15 μg C fixed/h $\times 10^9$ μ^3) at 0.5 - 2 m. Between 2 and 5 m the relative assimilation rate drops also, to 1/3 of its maximum. With increasing depth the gradient is less steep, reaching zero values at about 12.5 m. The curve for surface samples exposed in different depths runs at a higher level. The decline of relative assimilation rates is therefore steeper. This points to a better adaption to lower light intensities of samples taken from the depth.

By contrast, samples taken from the depth and exposed at the lake surface exhibit a less marked decline with increasing depth. Most of the algae do not begin dying unless they have sunk below the trophogenic zone. This is of great importance for the evaluation of the decomposition processes in lakes. The major part of the settling phytoplankters can photosynthesize again if they are carried upward from the hypolimnion by fall turnover or turbulence.

In the different lakes, average absolute assimilation rates under 1 m^2 lake surface and the respective average algal volumes vary over a range of 1 : 10 (Fig. 2), and enable us to classify the lakes into 3 groups: the lowest productivity is found in the Black Forest lakes, Lake Constance stations hold a medium position, and the maximal rates were obtained in the Schleinsee. Although the relation « assimilation rate : algal volume » has a tendency to increase with higher

Table 1.

	BLACK FOREST			LAKE OF CONSTANCE and its vicinity		
	Feldsee	Titisee	Schluchsee	Obersee	Untersee	Schleinsee
HEIGHT above sea level	1109 m	846 m	930 m	395 m	395 m	475 m
AREA	9.2 ha	110 ha	513 ha	476 km²	63 km²	15 ha
MAX. DEPTH	34.5 m	39.5 m	63 m	253 m	46 m	11.6 m
MEAN DEPTH	17.6 m	20.4 m	21.8 m	100 m	13.2 m	6.4 m
ALKALINITY	0.1-0.2	0.2	(0.2)-1.4	2-2.5	2-2.5	2.5

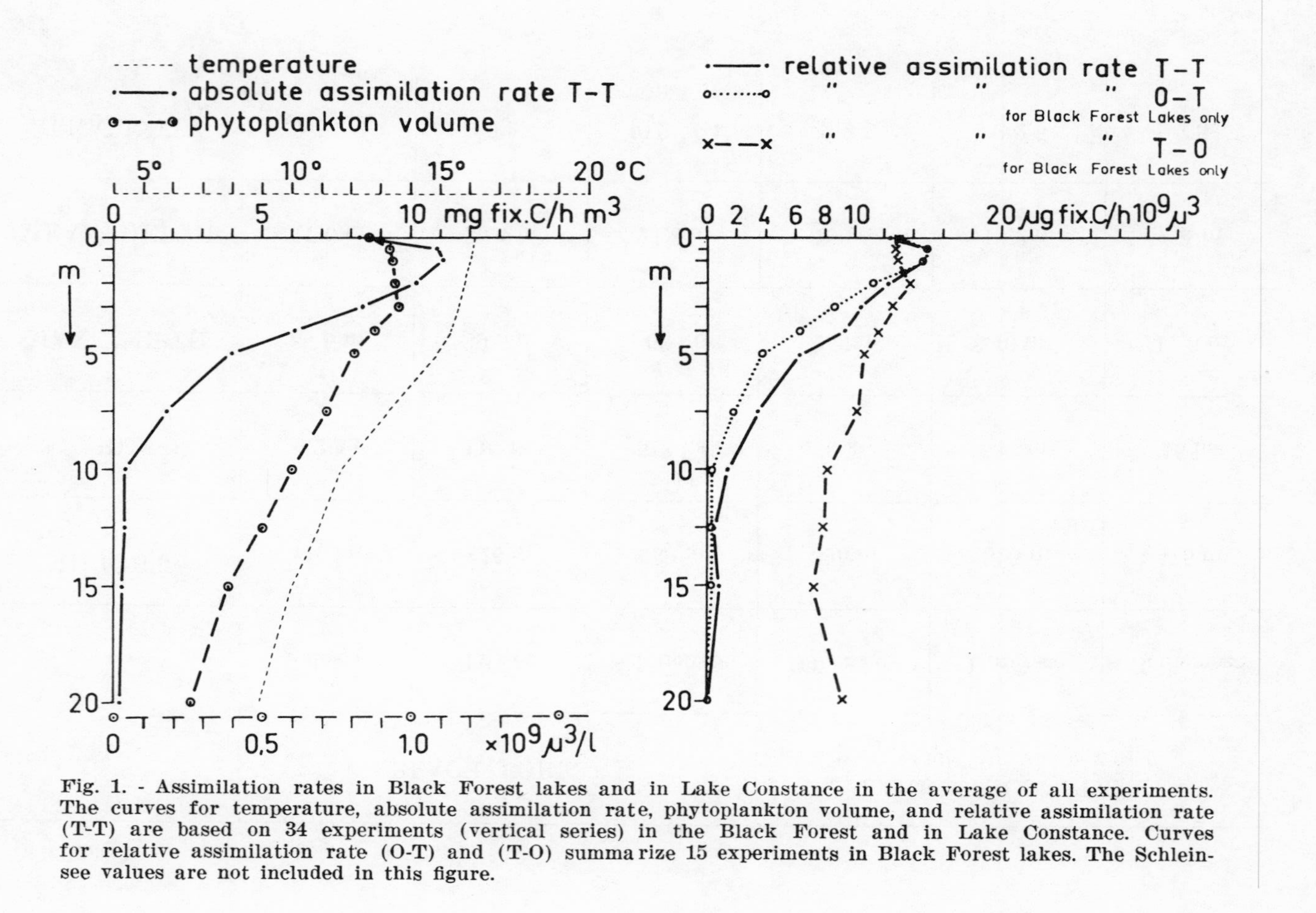

Fig. 1. - Assimilation rates in Black Forest lakes and in Lake Constance in the average of all experiments. The curves for temperature, absolute assimilation rate, phytoplankton volume, and relative assimilation rate (T-T) are based on 34 experiments (vertical series) in the Black Forest and in Lake Constance. Curves for relative assimilation rate (O-T) and (T-O) summarize 15 experiments in Black Forest lakes. The Schleinsee values are not included in this figure.

productivity, it is not clearly correlated to the trophic level, as is especially demonstrated by the Schleinsee values. It has to be considered that Fig. 2 gives only the average of our measurements but not the true annual average, since the experiments were less frequent in winter.

The average spectral transparency is given as %/m (Fig. 3). The two extremes are among the Black Forest lakes: the greenish Feldsee is the most transparent, and the yellowish to brownish Titisee the

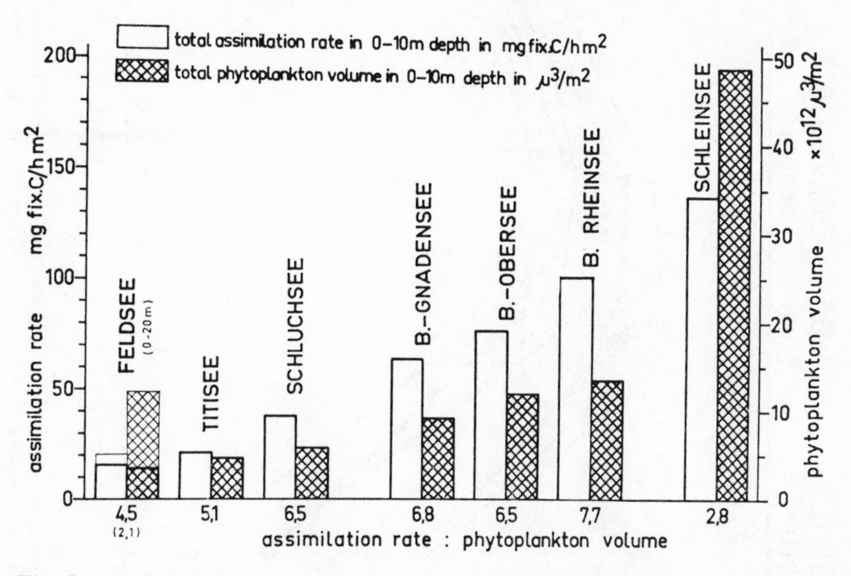

Fig. 2. - Assimilation rate and phytoplankton volume under 1 m² of lake surface in the different lakes. Thin-lined columns for the Feldsee indicate the values between 0 and 20 m depth.

most opaque. The biggest differences in transparency are found in the range of short wave lengths. They are even greater in the ultra violet range. At 375 mµ Schmolinsky (1954) found for the Feldsee a transparency of 36 %/m and for the Titisee 1.4 %/m. We can therefore not detect a good correlation between the level of relative assimilatory rates and the light quality in the lakes.

Before dealing with some problems of more general interest, some special characteristics of our lakes should be mentioned. The Feldsee is remarkable because of a maximum of algal volume in 10 - 12 m depth (Fig. 4), which is especially pronounced in summer, and does not significantly contribute to the lake's total production. This maximum essentially consists of algae being abundant in the surface layers also. At first we thought the trophogenic zone of the highly transparent Feldsee would extend down to the upper hypolimnion

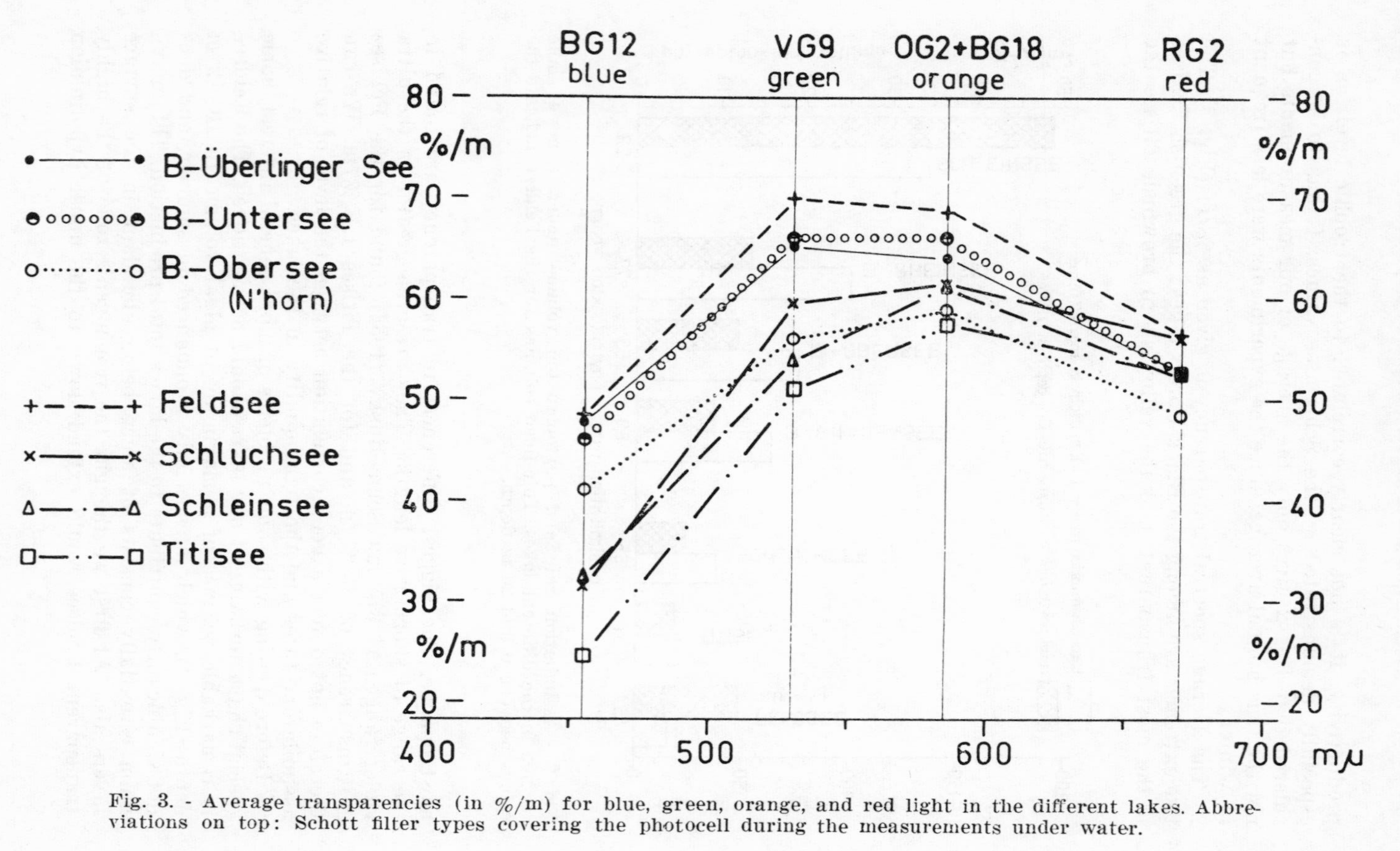

Fig. 3. - Average transparencies (in %/m) for blue, green, orange, and red light in the different lakes. Abbreviations on top: Schott filter types covering the photocell during the measurements under water.

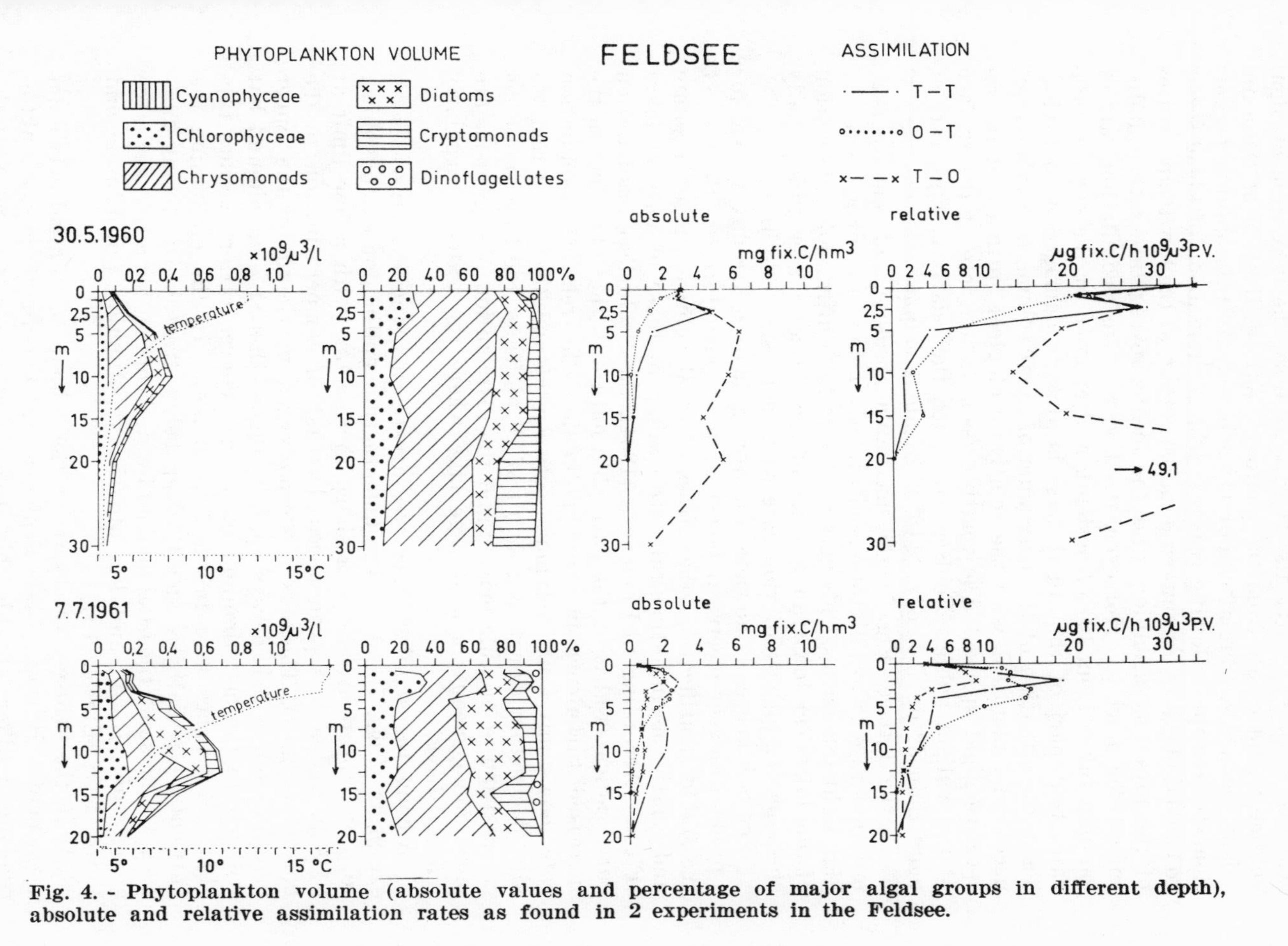

Fig. 4. - Phytoplankton volume (absolute values and percentage of major algal groups in different depth), absolute and relative assimilation rates as found in 2 experiments in the Feldsee.

which is richer in nutrients. If this were so, the maximum of algal volume would result from the photosynthetic utilization of these nutrient resources. But after our C^{14} experiments this opinion became doubtful for the following reasons: The accumulation of algae seems to consist of passively sunken organisms that, at the low temperatures of the lower metalimnion, find the means to vegetate rather undisturbed for a while. Moreover, the low relative assimilation rate of samples taken from 10 - 12 m depth and exposed at the surface indicates that part of this algal mass is already damaged or decaying. Thus, the situation of the maximum of algal volume in a lake is not necessarily identical with the relative ecological optimum of the respective algal species. The relative assimilation rate at the surface of the « oligotrophic » Feldsee can reach the relatively high rate of more than 32 μg C fixed/h $\times 10^9$ μ^3 of algal volume.

In the Titisee (Fig. 5), the maximum absolute and relative assimilation rate is usually observed directly below the surface, even at high light intensities. This may be due to high ultra violet absorption of this lake. The low transparency of the Titisee causes an especially steep decline of assimilation rate with increasing depth.

The Schluchsee is nowadays a storage lake in a reflux system for « production » of electrical power at peak output times. Because of the stable stratification, the curves of algal volume and of absolute and relative assimilation rate are similar to those of natural lakes (Fig. 6). According to Findenegg (1964), the Schluchsee assimilation curve indicates already the eutrophic nature of the lake. Due to the increased turbulence in the hypolimnion, the relative assimilation rate shows only a flat maximum between 0.5 and 3.0 m, and remains almost constant down to a depth of 20 m. Thus, the hypolimnic decomposition of organic matter and the distribution of organisms are more homogeneous in a storage lake where the water is regularly pumped back.

Among the smaller lakes we investigated, the Schleinsee is an exception (Figs. 7 and 8). The homothermic epilimnion extends down to 4 or 5 m, then follows a sharp thermocline with a low limit at 8 - 9 m, and a less pronounced decrease of temperature on to the bottom (Fig. 7). There is a steep oxygen maximum in the upper metalimnion, which decays in fall. Below this O_2 - maximum, but still within the metalimnion, the oxygen concentration declines rapidly, and reaches zero in the hypolimnion. In the epilimnion the nutrients, especially N and P, drop below traceability during summer. At the same time the mud-and-water contact zone releases large amouts of NH_4, P, and Fe into the hypolimnion (cf. Einsele and Vetter 1938, Einsele 1941).

As in the Feldsee, we observe in the Schleinsee an algal maximum in the depth. However, in the Schleinsee the hypolimnic phytoplankton community differs considerably from the epilimnic (Fig. 8). Near the surface we find a phytoplankton typical of slightly eutrophic lakes.

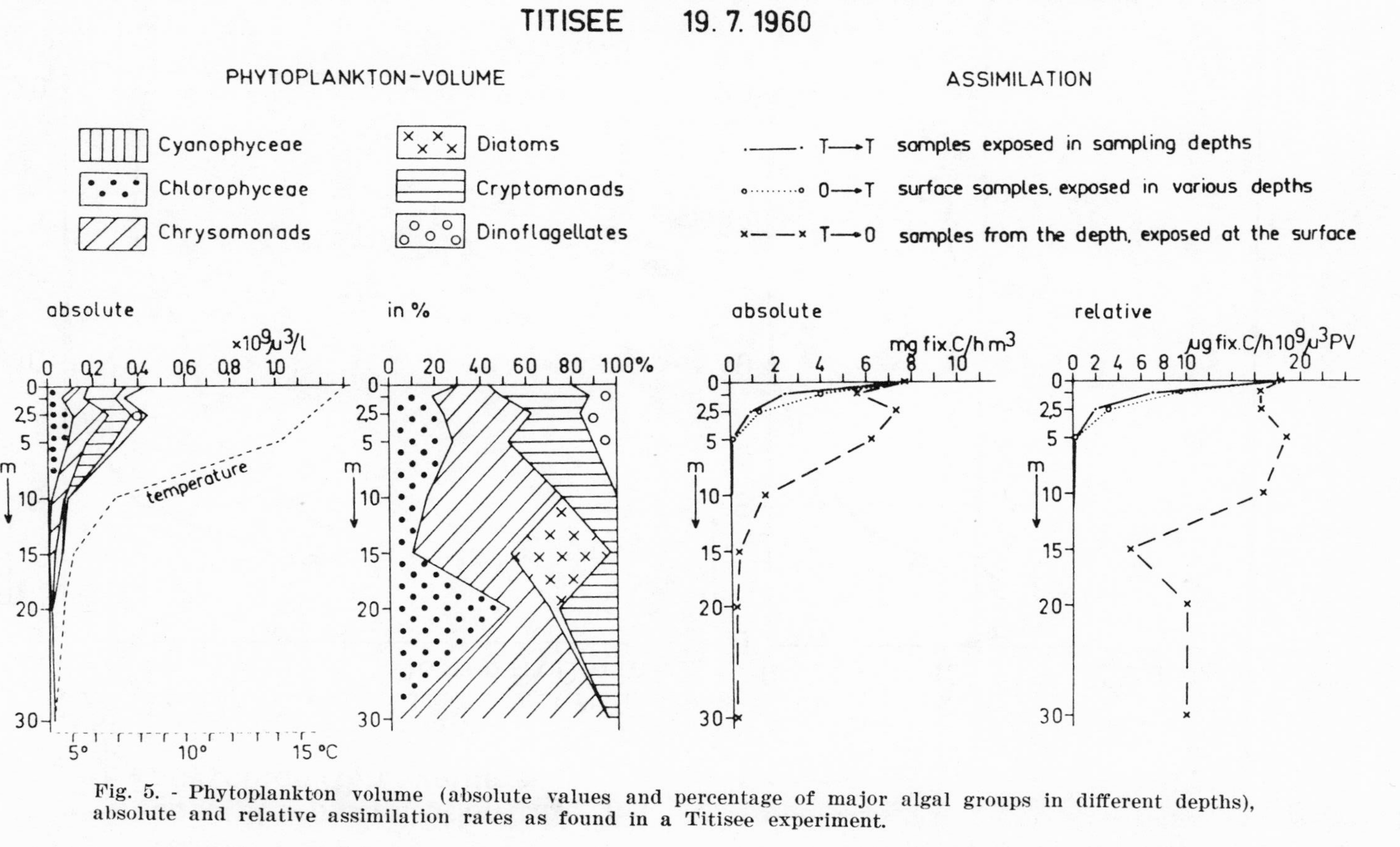

Fig. 5. - Phytoplankton volume (absolute values and percentage of major algal groups in different depths), absolute and relative assimilation rates as found in a Titisee experiment.

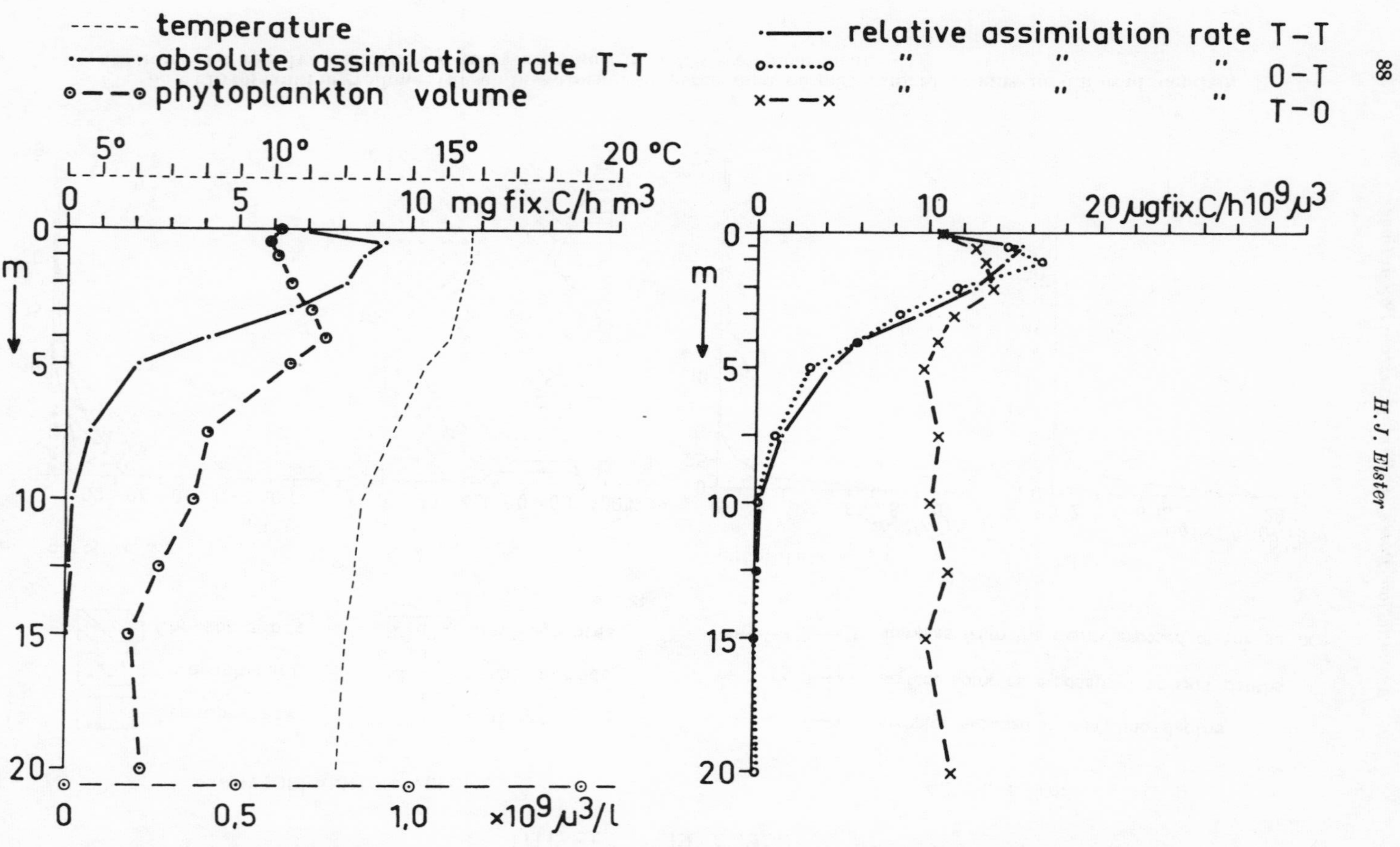

Fig. 6. - Average values of temperature, phytoplankton volume, absolute and relative assimilation rates in the Schluchsee (from 6 experiments).

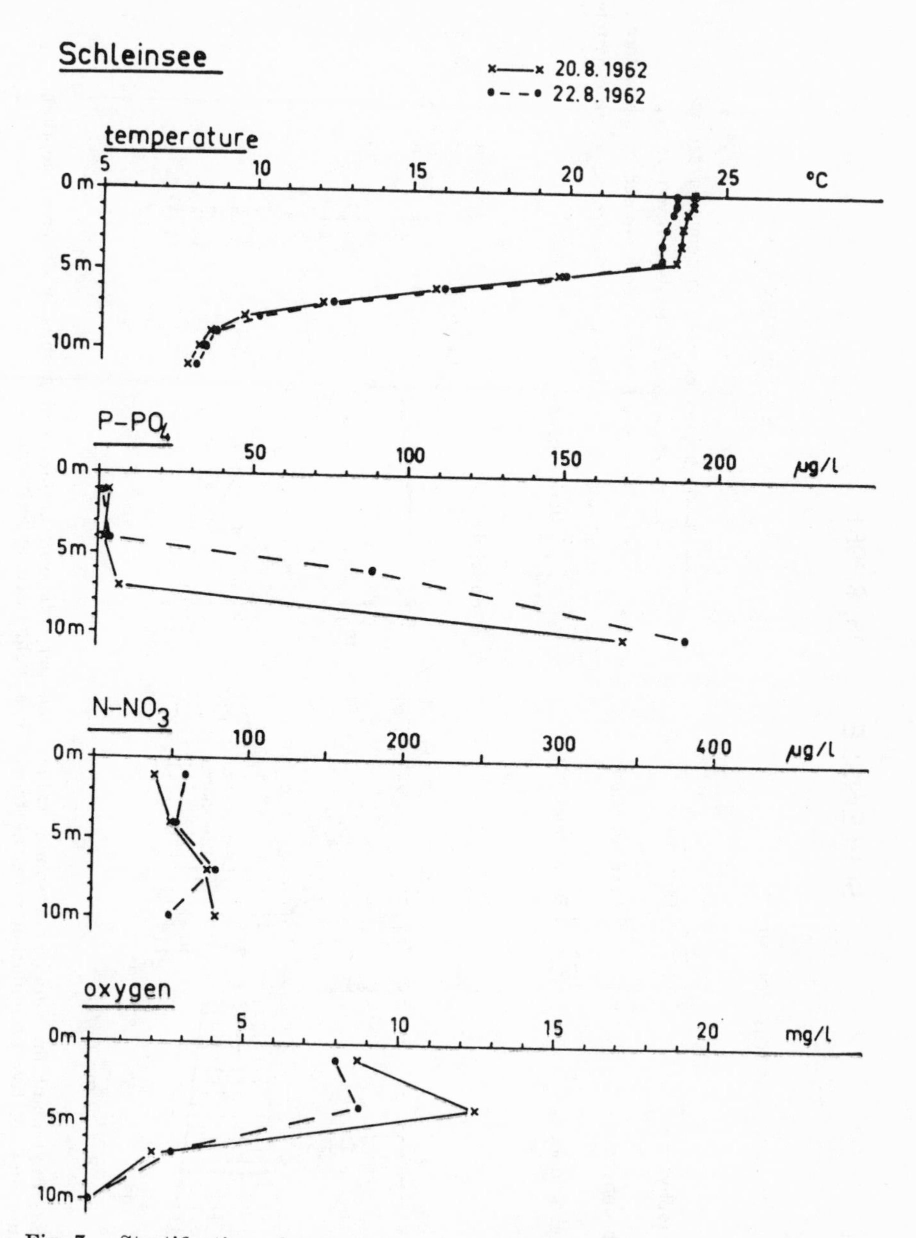

Fig. 7. - Stratification of temperature and chemical factors in the Schleinsee ($P\text{-}PO_4$ means phosphorus as PO_4^{---}; $N\text{-}NO_3$ means nitrogen as NO_3^-).

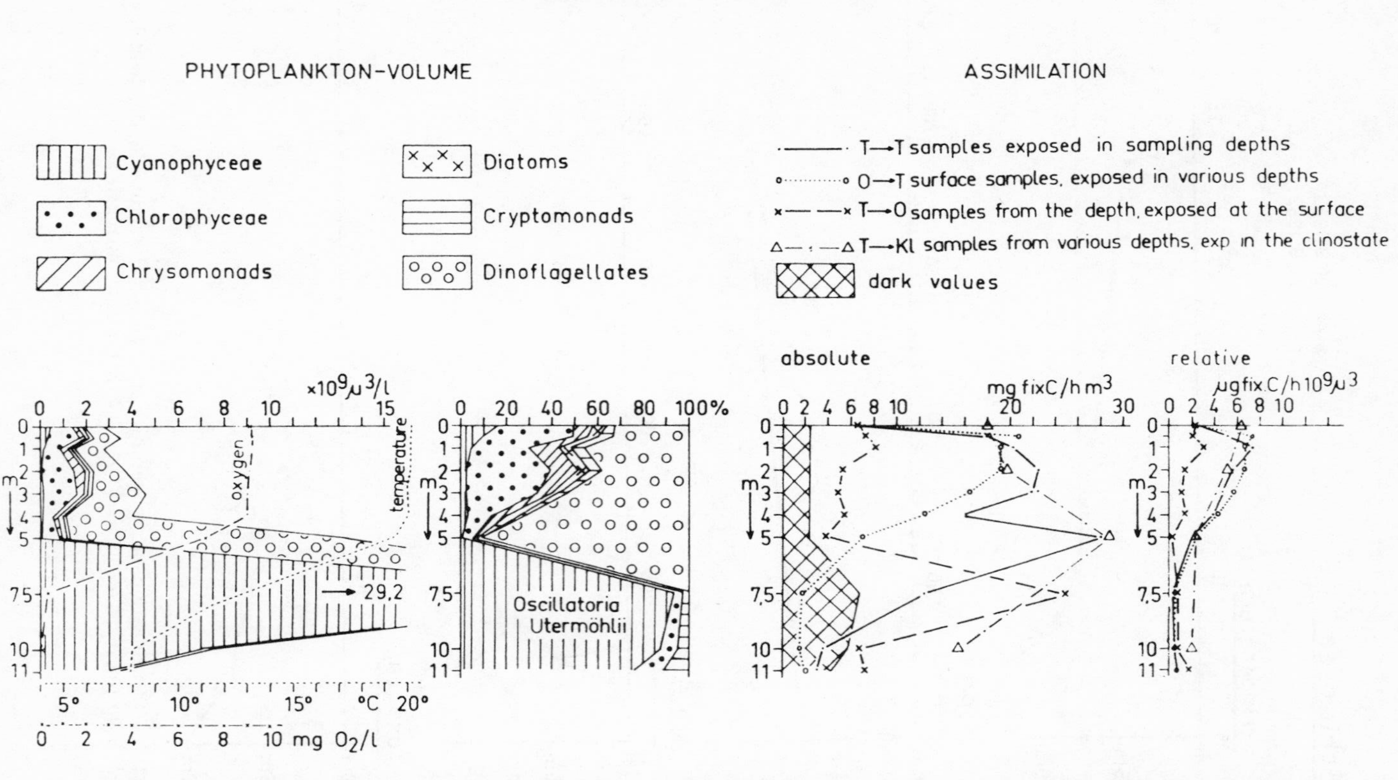

Fig. 8. - Phytoplankton volume (absolute values and percentage of major algal groups in different depths), absolute and relative assimilation rates as found in a Schleinsee experiment.

Because of nutrient (especially nitrogen) deficiency, it retreats in summer from the epilimnion and forms a new maximum in the upper metalimnion, which offers more nutrients. In the hypolimnic zone, deficient in oxygen and rich in organic and inorganic nutrients, we find an entirely different population. It consists of blue-green algae, sulfur bacteria, and other microorganisms, which could partially not be included in our counts. In part these hypolimnic organisms are heterotrophic or able to facultatively utilize organic substrates. The latter is indicated by high « dark fixation » of C^{14} [1].

The absolute assimilation rate ascends to a second peak at 5 m depth. Surface samples show the normal decline of activity with increasing exposure depth. The high absolute assimilation rate obtained from samples taken at 7.5 m and exposed at the surface is due to the conspicuous algal mass, but not to a stimulation of relative assimilation rate. The slopes of the relative assimilation rates follow the usual pattern.

Most Schleinsee series led to rather irregular curves (i.e., assimilation rates changed considerably from layer to layer). This might point to zones in which the fallout of damaged or dead phytoplankters accumulates, as has been stated in other lakes by Einsele and Grim (1938) and Grim (1951).

---

(1) *Control of dark values.* – During the years 1959-1961 the background value of the Friesecke and Hoepfner Geiger-Müller-counter kept a constant average of 31-32 ipm (extremes: 23 and 40 ipm).

The values for phytoplankton samples which had been kept in the dark bottles were independent of the filtrated volume:

filtrated volume	average values	number of samples
15 ml	47 ipm	14
25 ml	44 ipm	14
50 ml	45 ipm	14
75 ml	54 ipm	10

Hence, there is no dependance on the phytoplankton volume on the filter. In agreement with this finding no correlation could be detected between the magnitude of the dark values and the amount of phytoplankton (expressed as algal volume) in the respective plankton samples.

Dark values up to 50 ipm — i. e. after subtraction of the background count up to 20 ipm — can, therefore, not be due to the dark assimilation of $C^{14}O_2$ by the algae. The effect has apparently to be explained by adsorption of the isotope on the filter. This conclusion was verified in experiments with a solution containing C^{14} but no algae. A few samples gave, however, higher dark values. They probably contained bacteria which were able to assimilate radiocarbon in the dark.

In lakes like the Schleinsee the dark fixation of C^{14} can be influenced by chemical reactions which will be analyzed in subsequent investigations.

The Obersee of Lake Constance is the classical example of an oligotrophic lake in the older limnological literature. Today its assimilation curves (Figs. 9 and 10) display a typical « eutrophic » character in the terms of Findenegg. In March, a series of measurements in the Überlingersee revealed a maximum of absolute assimilation rate in the upper layers which can be attributed to a mass population of cryptomonads. In August, the maximum of algal mass has already shifted to a depth of 7.5 m and no longer contributes much to the productivity of the lake. Nevertheless, the algae of this zone photosynthesize sufficiently if brought back into the trophogenic zone. The stimulative effect of added nutrients shows that phosphate is a limiting factor in midsummer in the Obersee, despite the artificial eutrophication that has taken place.

The mean values of the Gnadensee, the relatively isolated northern part of the Untersee, indicate an even higher trophic level (Fig. 11). The high maximum of relative assimilation rate (22.5 μg C fixed/ $h \times 10^9$ μ^3) deserves further checking.

The Rhine flows through the southern part of the Untersee, the so called Rheinsee. In this basin there is a more vigorous and deeper extending turbulence, resulting in a rather dispersed distribution of phytoplankton (Fig. 12).

The composition and photosynthetic capacity of the Rheinsee phytoplankton result from a mixture of the Obersee and Gnadensee populations with the autochthonic organisms of the Untersee.

The major difference between the Obersee and the Untersee does not show up before winter or early spring (Fig. 13), and depends mainly on the contrast in depth of the two basins. At that time, when a peak of algal volume (mostly diatoms) is found in the 20 m deep Gnadensee, the population of the ten times deeper Obersee has not yet recovered from the winter turnover. The low absolute and relative assimilation rates apparently do not result from the temperature of 4°C but from the physiological status of the algae. For the winter series of February 14, 1963, samples were taken below and exposed under the ice cover. The experiments demonstrate that a conspicuous population of algae can build up under the ice (Fig. 13). Moreover, absolute and relative assimilation rates can approach the same order of magnitude as the values obtained during the summer.

These examples, and the comparison of the changes of temperature, algal volume, and assimilation rates, confront us with the following problem: which are the causes for the often accentuated stratifications of immobile algae in our lakes? These organisms which are incapable of active locomotion are exposed to the hydrographical conditions of the lake and subject to the mere caprice of fortune. They live in the continous danger of sedimentation, which leads to

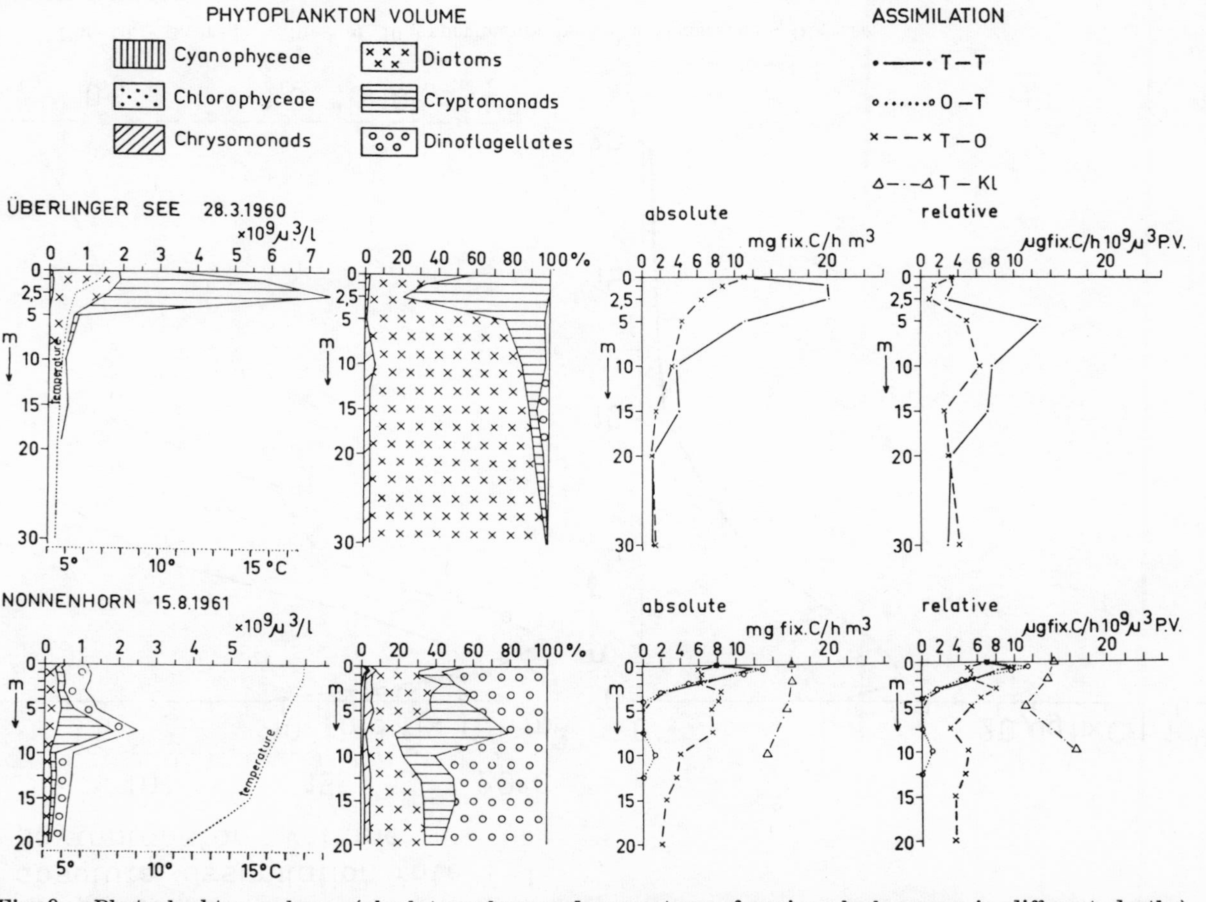

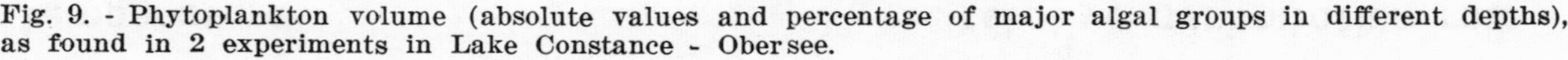

Fig. 9. - Phytoplankton volume (absolute values and percentage of major algal groups in different depths), as found in 2 experiments in Lake Constance - Obersee.

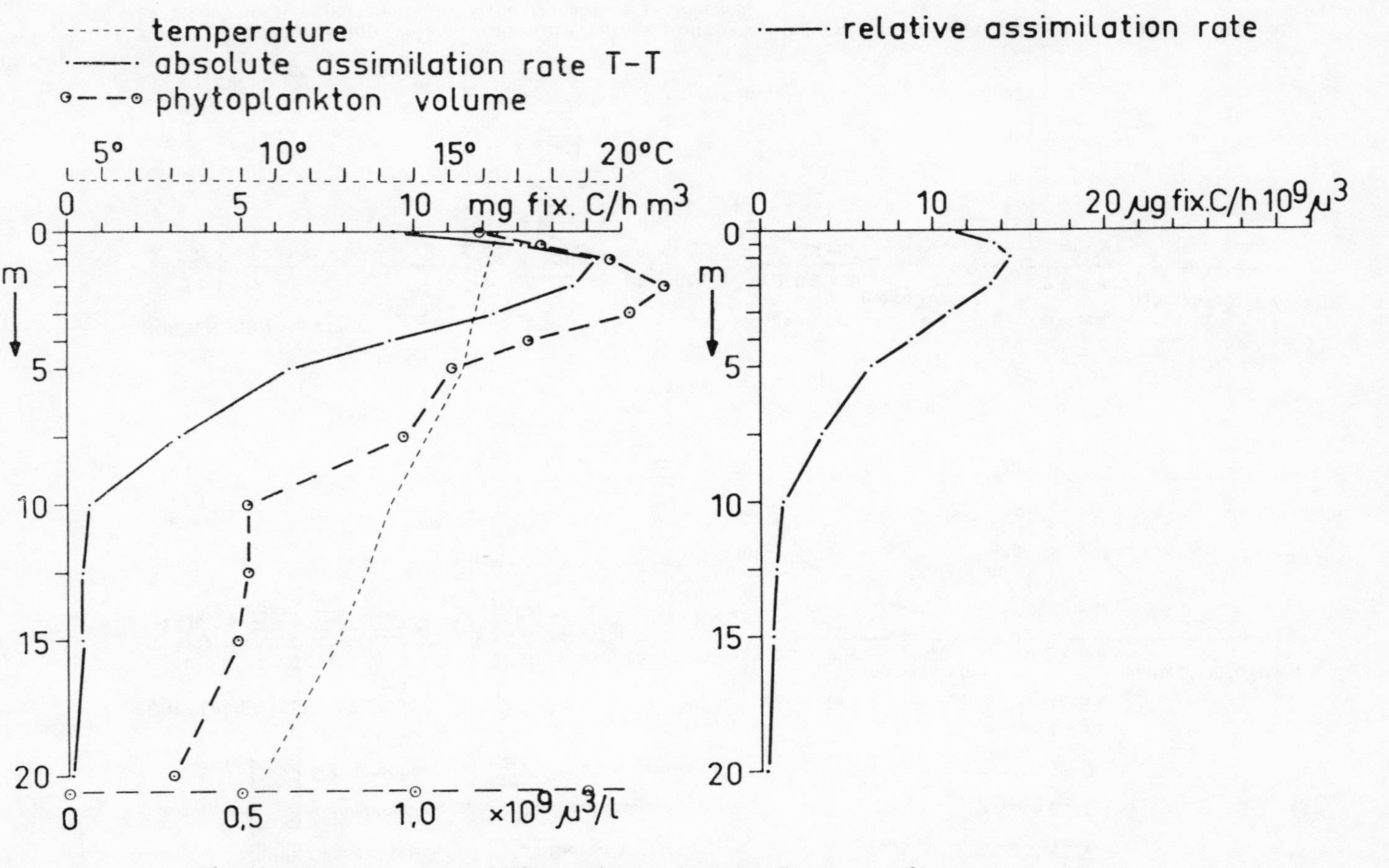

Fig. 10. - Average values of 10 experiments in Lake Constance - Obersee.

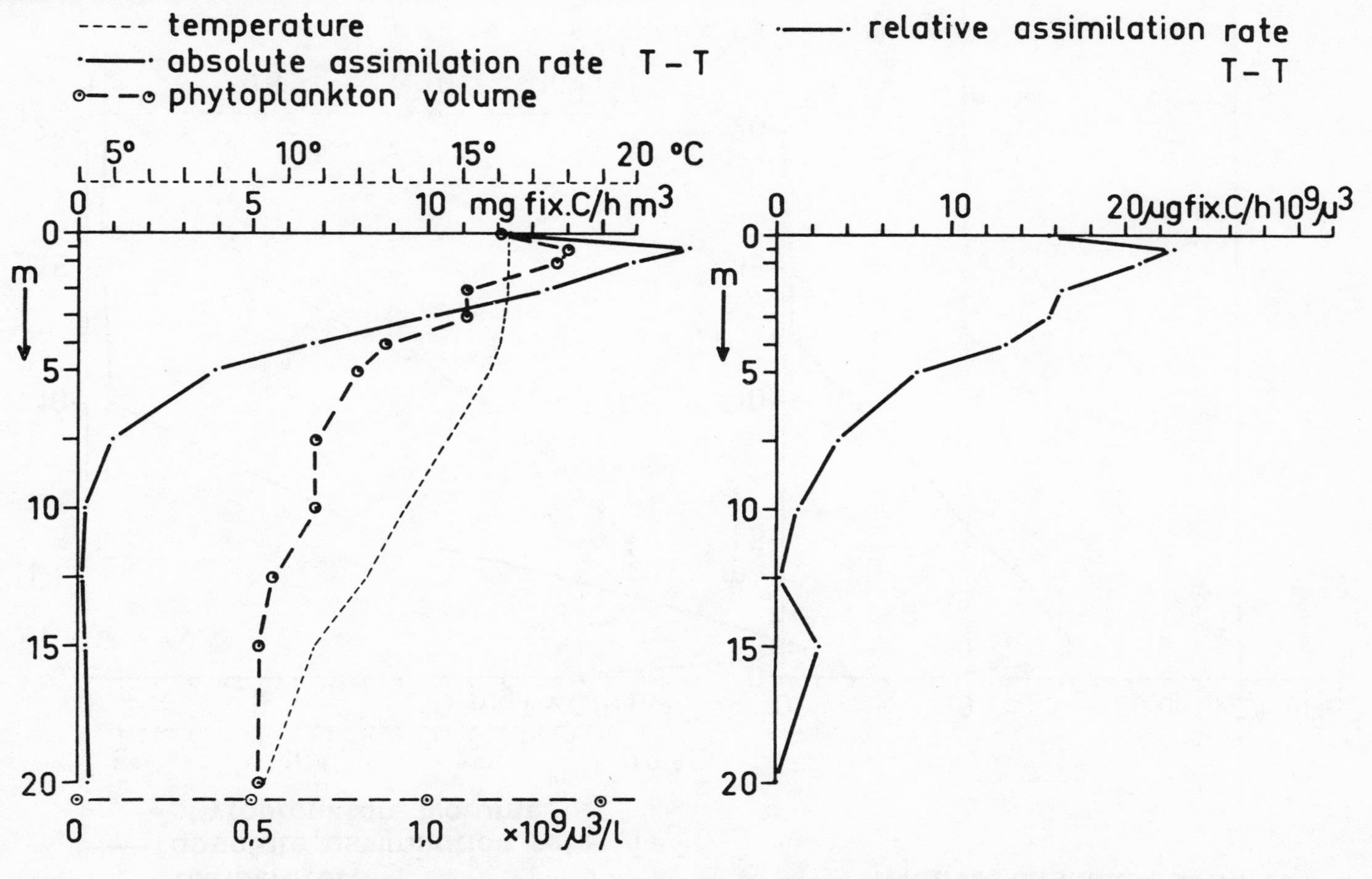

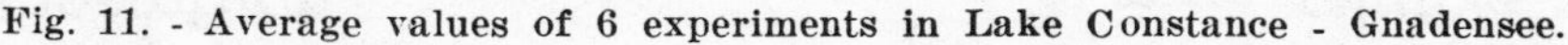
Fig. 11. - Average values of 6 experiments in Lake Constance - Gnadensee.

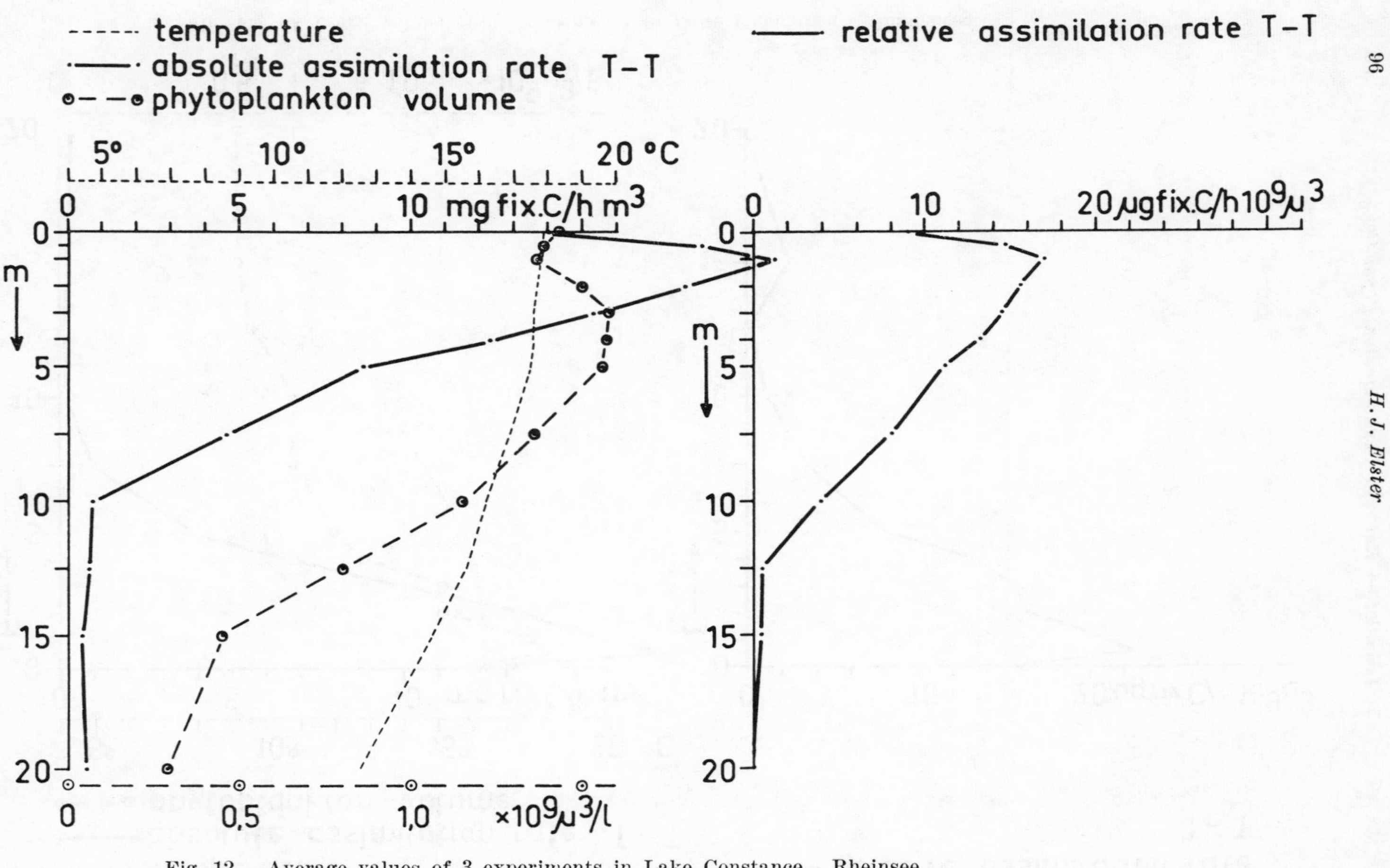

Fig. 12. - Average values of 3 experiments in Lake Constance - Rheinsee.

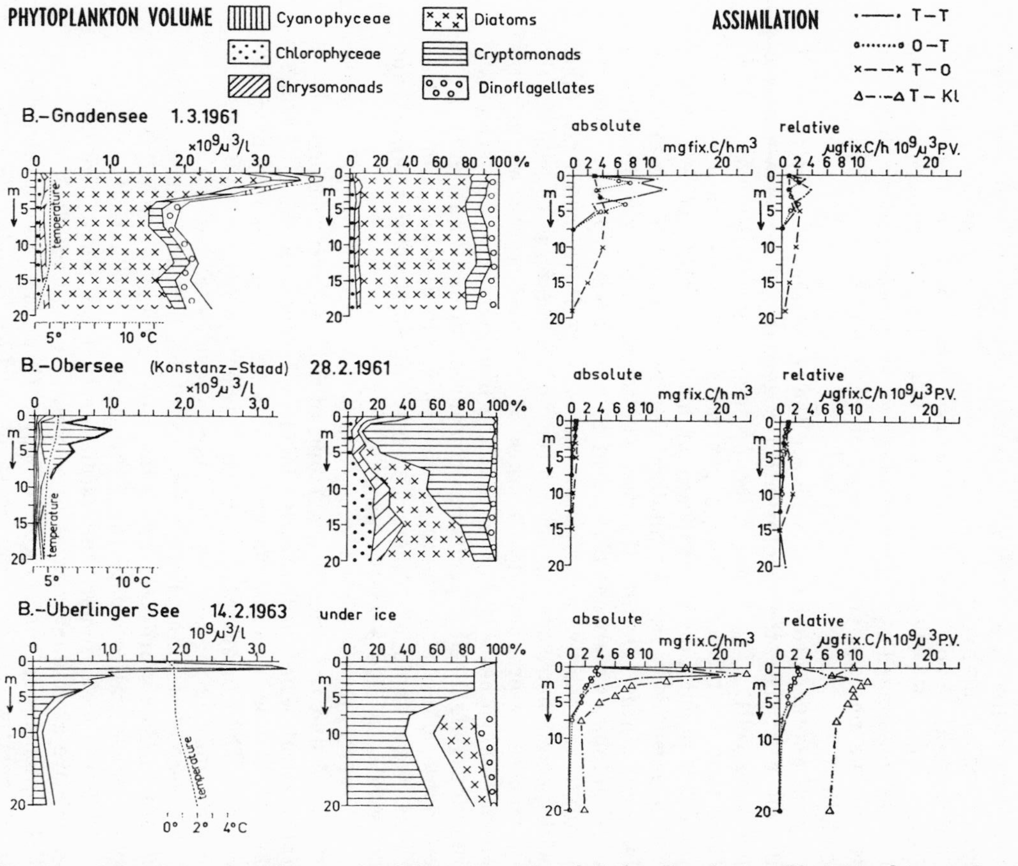

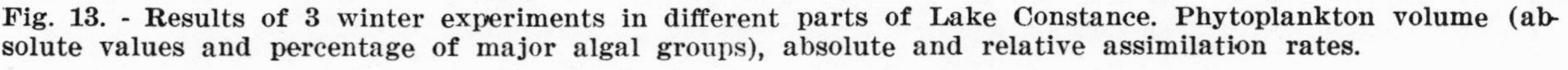
Fig. 13. - Results of 3 winter experiments in different parts of Lake Constance. Phytoplankton volume (absolute values and percentage of major algal groups), absolute and relative assimilation rates.

their frequent accumulation below the zone of optimal photosynthetic conditions or even below the compensation point. Nevertheless, they are able to build up dense populations. It is unknown which factors do grant the permanence of these phytoplankton maxima for longer periods of time.

Another open question is to which degree do the phytoplankton losses by sedimentation depend on the type of lake and the actual meterological or hydrographical conditions? These problems are of fundamental importance for the causal analysis of productivity in inland lakes. Furthermore, we have the following problem: which are the combinations of endogenous and exogenous factors that have led to the phylogenetical success of phytoplankters without active mobility?

Still another problem: which are the regulators of relative assimilation rate (i.e., the photosynthetic activity of a certain phytoplankton cell)? The formulation of such a question certainly indicates that we cannot expect a general answer: any natural phytoplankton population is composed of many different species, each of which presumably lives at different depths in a different mean physiological status. Despite the vast differences in the stratification of algal volume, we always observe almost the same type of curves for absolute and relative assimilation rates, if we forget about the lakes with pelagic heterotrophs among the phytoplankters. To explain the differences in productivity of lakes of different trophic nature, the simplest hypothesis would be that the photosynthetic activity of the individual algal cell depends on the concentration of nutrients. In an eutrophic lake, being richer in nutrients, the division rate of individual cells would be higher, the resultant algal volume bigger, and we would observe a higher productivity.

However, things are not that simple. In contrast to this hypothesis, we found the average maximum of relative assimilation rate of the vertical series to be always about the same. In the oligotrophic Feldsee it even exceeds the values obtained in the « eutrophic » Bodensee! The concentrations of the nutrients which we considered to be the limiting factors can thus not be exclusively decisive.

A second working hypothesis would be that the concentrations of the nutrients are saturating for algal growth in all of the lakes under investigation. Only the light regulates productivity.

Against a semilogarithmic scale we plot the percentage decrease of relative assimilation rates in comparison with the decline of total light intensity in the respective depths (Fig. 14). Below their maximum, relative assimilation rates of surface samples decrease approximately in parallel with light energy. Samples exposed in sampling depths produce a somewhat divergent pattern; we perceive the effect of an adaptation to the conditions in the depth and to its higher nutrient resources. Below 10 m our values are within the range of error and lack full significance.

Absolute assimilation rates increase more or less proportionally to the algal volume, while the relative assimilation rates decline with higher densities of phytoplankton (Fig. 15). The latter might be due to mutual shading of cells. On the other hand, the widely scattered values seem to indicate the general relationship between algal volume and photosynthetic rate accomplished in any particular case.

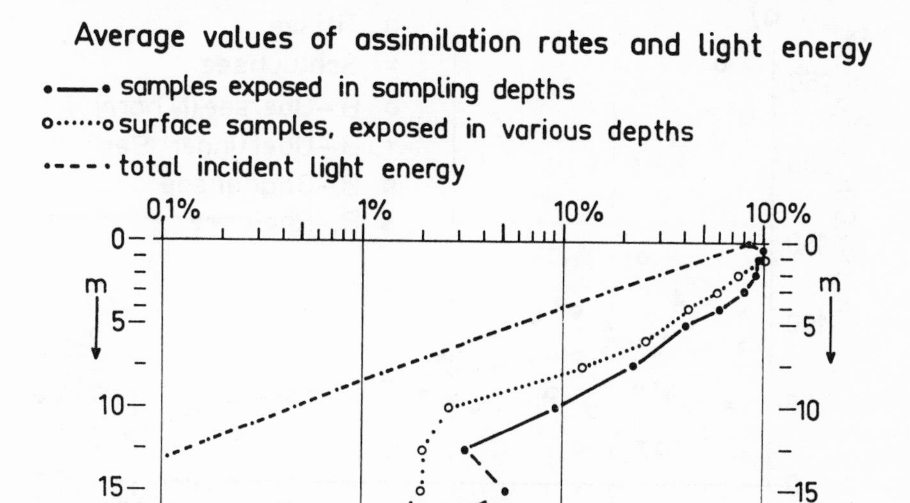

Fig. 14. - Average dependance of assimilation rates (in % of the maximum) on light intensity (in % of the surface values). The figure includes the data from all experiments with the exception of the Schleinsee values.

If we only take into account the mean values of several experiments, we obtain smooth curves of rather similar shape for the vertical profile of photosynthetic activity. If we consider each experimental series separately, big differences appear. Fig. 16 gives some examples for Feldsee samples taken from the surface and exposed in different depths. From the other lakes we obtained similar results. In the same light intensities or temperatures, but at different times, relative assimilation rates can vary over more than one order of magnitude.

« Light inhibition » and maximal relative assimilation rate are found at very different radiation energies (mcal values). Light intensities that were definitely lower than required in other experiments for maximal photosynthetic activity could cause « light inhibition ».

Maximal relative assimilation rates were frequently observed at 70 - 100 mcal/cm²min = 6000 - 10,000 Lux. The lowest comparable value was smaller by one order of magnitude, but even there we observed a decrease of assimilation rate towards the surface (« sur-

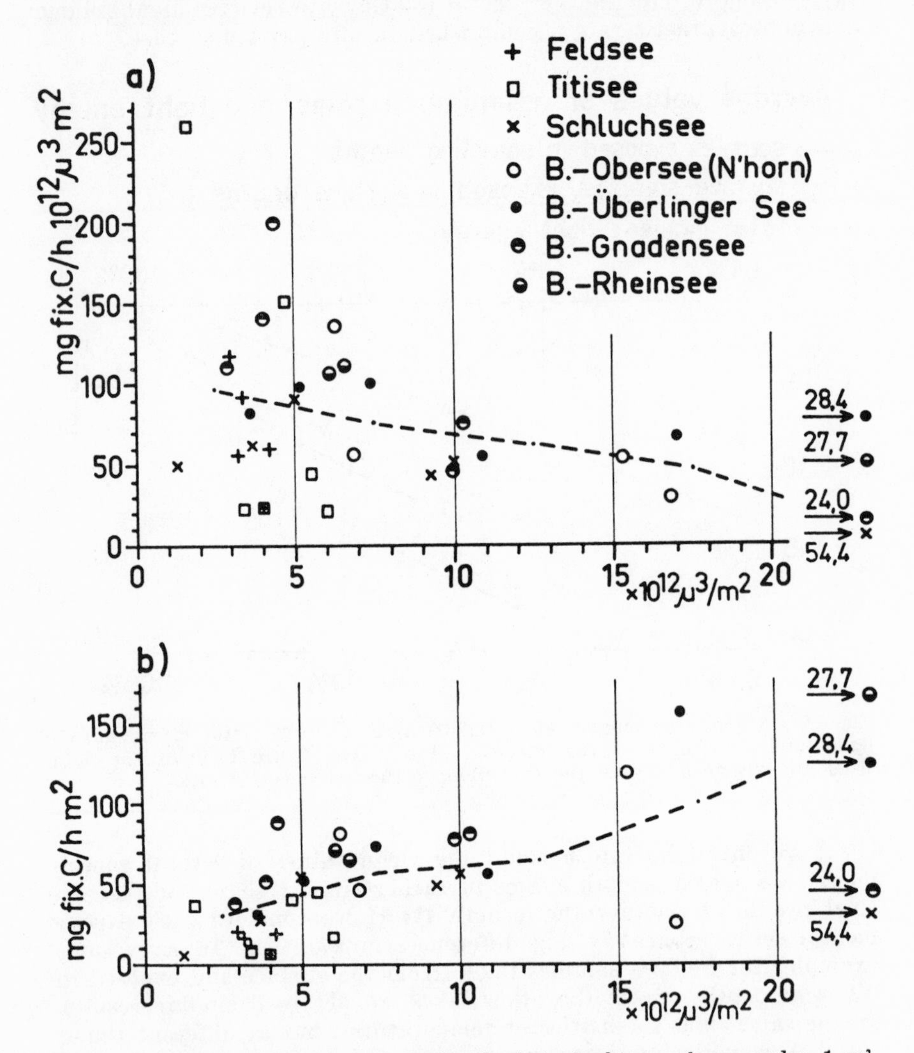

Fig. 15. - Assimilation rates per unit of phytoplankton volume under 1 m² lake surface between 0 and 10 m depth in the different lakes. a) relative assimilation rates; b) absolute assimilation rates. Figures and arrows at the right hand side of the diagrams refer to the respective assimilation rates at phytoplankton volumes greater than 20 × $10^{12}\mu^3/m^2$. The broken lines refer to the arithmetic mean of all values (from 34 experiments) and indicate the general trends of the results.

face inhibition »)! The highest light intensity at which a maximum of relative assimilation rate could be detected was at 400 mcal = = 35,000 Lux.

« Surface inhibition » is, therefore, widely independent of absolute light intensity. The quality of the radiation at the surface of a lake might be more essential than its absolute quantity. The short wavelength radiation is probably of special importance, as may be indicated by the maximum of assimilation rate close to the surface of the Titisee. In addition, the physiological status of the assimilants is critical.

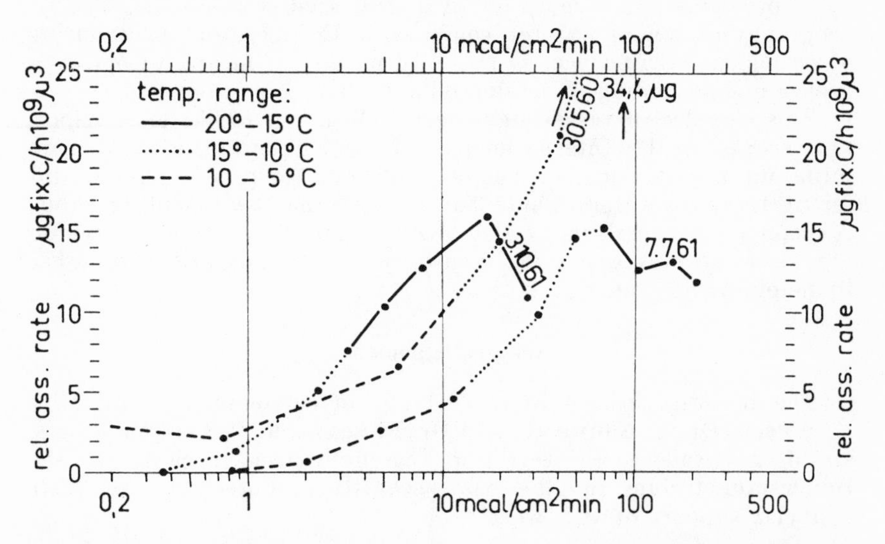

Fig. 16 - Surface samples of the Feldsee exposed at various depths showing the relation of relative assimilation rates to total light energy (3 experiments). The respective temperature ranges are indicated by solid, broken, or dotted lines.

The descending slopes of the assimilation curves do not converge into the same zero point but decline divergently. In most of the experiments the limit of traceability for a net assimilatory balance was at 0.1 to 1 mcal. There is consequently no absolute light intensity that determines the lower limit of the trophogenic zone, even in one and the same lake. The compensation level, i.e., the brink of the trophogenic zone in a lake, depends not only on physical and hydrographical conditions, but also on the physiological status of the respective phytoplankton community. The comparison of our numerous series of experiments have furthermore proven that neither the temperature nor the composition of the algal populations (separated into systematic groups) explain the differences in relative assimilation rates.

The reason for the conspicuous variability of relative photosynthetic activities at a certain light intensity and at similar temperatures in natural phytoplankton populations are, therefore, essentially unknown. I admit that we know from the literature many details of the action of isolated factors on algae grown under constant and more or less optimal laboratory conditions. Since we are not sure how the interactions of all these many factors act on the vast variety of physiological stages of different algae, we have to be extremely cautious in applying — if this is possible at all — these experimental data to the natural conditions, where we are confronted with our basic problems. Much more detailed field studies (including the hydrography of inland waters), combined with laboratory experiments of ecological significance, is the only promising means to obtain a deeper understanding of productivity problems in waters.

This complex of tasks might very well be one of the most important scopes of the International Biological Program, since a mere piling up of static descriptions of conditions for productivity in different types of waters would be of rather narrow scientific value. As long as the analysis of the effective causes is left aside, we do not know how to apply the results of our investigations to other limnological ecosystems.

Acknowledgements

The investigations were carried out in collaboration with Miss B. Motsch, Dr. A. Nauwerck, and Dr. R. Schröder. The author expresses his gratitude to the Deutsche Forschungs-Gemeinschaft and the Bundesministerium für die wissenschaftliche Forschung for their financial support of this work.

REFERENCES

Eckstein, H. 1963/64. Untersuchungen über den Einfluss des Rheinwassers auf die Limnologie des Schluchsees. - Arch. Hydrobiol., Falkau-Schriften *V*: 47-182.

Einsele, W. 1941. Die Umsetzung von zugeführtem anorganischen Phosphat im eutrophen See und ihre Rückwirkung auf seinen Gesamthaushalt - Zeitschr. Fischerei *39*: 407-488.

— and J. Grim. 1938. Über den Kieselsäuregehalt planktischer Diatomeen und dessen Bedeutung für einige Fragen ihrer Ökologie. - Zeitschr. Bot. *32*.

— and H. Vetter. 1938. Untersuchungen über die Entwichlung der physikalischen und chemischen Verhältnisse im Jahrescyclus in einem mässig eutrophen See. - Int. Rev. Hydrogr. Hydrobiol. *36*: 285-324.

Elster, H. J. 1962. Untersuchungen über die Rolle des Rheinwassers im Schluchsee und in den Zwischenstaubecken der Schluchseewerk-AG 1951-1954. Arch. Hydrobiol. Falkau-Schriften *IV*: 430-455.

— and B. Motsch. 1966. Untersuchungen über das Phytoplankton und seine Assimilation in einigen Seen des Hochschwarzwaldes, im Bodensee und Schleinsee. - Arch. Hydrobiol. Falkau-Schriften *V*: Heft 4 (im Druck).

-- and F. Schmolinsky. 1952/53. Morphometrie, Klimatologie und Hydrographie der Seen des südlichen Schwarzwaldes - Arch. Hydrobiol., Falkau-Schriften *I*: 157-211 und 375-441.

Findenegg, J. 1964. Types of planctic primary production in the lakes of the Eastern Alps as found by the radioactive carbon method. - Verh. Int. Ver. Limnol. *15*: 352-359.

Grim, J. 1939. Beobachtungen am Phytoplankton des Bodensees (Obersee) sowie deren rechnerische Auswertung - Intern. Rev. Hydrogr. Hydrobiol. *39*: 193-315.

— 1951. Ein Vergleich der Produktionsleistung des Bodensee-Untersees, des Obersees und des Schleinsees. - Abhandl. a. d. Fischerei, Lieferung *4*: 787-841.

Nauwerck, A. 1963. Die Beziehungen zwischen Zooplankton und Phytoplankton im See Erken. - Symb. Bot. Upsaliens. *XVII*, *5*: 1-163.

Schmolinsky, F. 1954. Einige Ergebnisse vergleichender Lichtmessungen an Seen des Hochschwarzwaldes und der Schweiz. - Arch. Hydrobiol., Falkau-Schriften *I*: 615-632.

Sorokin, J. I. 1956. The Use of Radioactive C^{14} for Measuring Production in Water Basins. - Trans. Hydrobiol. Soc. USSR 7: 271.

FACTORS CONTROLLING PRIMARY PRODUCTIVITY, ESPECIALLY WITH REGARD TO WATER REPLENISHMENT, STRATIFICATION, AND MIXING

INGO FINDENEGG

Biologische Station der österreich.
Akademie der Wissenschaften, Lunz am See, N.ö. Austria

For bibliographic citation of this paper, see page 10.

Abstract

In contrast to the sea, in lakes primary production generally takes place in non-homogeneous layers, and considerable gradients of temperature and concentration occur. Therefore production conditions are rather unstable and the « normal » assimilation curves of depth distribution are seldom found. In general, a rapid renewal of water, such as floods or vertical displacement by a tilted thermocline, depresses production; but slowly upwelling water rather favors it by supplying nutrients. Pronounced stratification may lead to an epilimnic nutrient depletion during summer; and oligo- or meromixis lowers production.

It is a well-known fact that primary production in lakes is controlled by the interaction of many factors which usually are divided into three groups: (1) physical factors originating directly or indirectly from solar radiation, such as light conditions, temperature, mixing and turbulence by the action of wind; (2) the content of nutrients in the euphotic zone of the lakes, and (3) the interaction of the organisms present in the plankton community which may promote or hamper the production of certain species. All these factors are mutually entangled, and interact to produce the distribution of organisms in space and time.

In the sea the euphotic zone comprises water masses of almost homogeneous character that extend largely in the horizontal but also in the vertical direction. Investigations on productivity are much facilitated by this fact. Production may be computed with some accuracy from a relatively small number of measurements. In lakes, however, the trophogenic zone often comprises layers of very different character, with remarkable gradients of temperature and chemical properties. Sometimes there are also layers of turbidity and different plankton communities because the thermocline may still be a part of the euphotic zone. In well stratified lakes with good light transmission two plankton communities as a rule are present: one consisting of polythermal species in the epilimnion, and another composed of oligothermal shade-forms in the metalimnion.

In addition freshwater quality in lakes changes within the yearly cycle, and sometimes also within shorter periods in response to weather and to floods. Vertical water displacement by tilting of the discontinuity layer or by circulation also causes unstable conditions.

If we are going to answer the question of the effect of water-replenishment on the productivity of lakes, we must remember that the ecosystem is altered in two different respects when water masses flow through a lake basin: on the one hand the « old » lake water is renewed by that of a river or brook which generally shows different physical and chemical qualities. On the other hand, a certain part of the standing crop of plankton will be pushed towards the outflow and carried away.

The effect of water replenishment depends primarily on the proportion of the volume of the lake to the volume of water leaving the basin within a year. The seasonal variations of water renewal are also of great importance. In some alpine lakes with considerable water replenishment, floods play a decisive role. In Traunsee, for example, it sometimes happens that the whole complex of the photosynthetic layer is replaced by the water of the Traun-river within two or three days. The same occurs in Lunzer See almost every year, but to a smaller extent. In one case this lake lost more than two thirds of its plankton biomass, and the total production decreased from 82 to 45 mg C assimilated per square meter and day, within about 4 days.

When a lake is flushed in that way, it is not only depopulated to a certain degree but as a matter of course total production is diminished for some time. If floods are a frequent event and the photosynthetic zone is deprived of essential parts of its phytoplankton repeatedly within one year, the lake seems to be very unproductive even though the ratio of assimilation rate to biomass of phytoplankton indicates a large relative production. In Lunzer See, for example, an average daily activity coefficient of 0.1 was found in the optimal cubic meter, and about 0.05 for the total water volume below the surface unit, by C^{14}-measurements. This is more than the average values of other Austrian lakes. Yet Lunzer See is extremely poor in phytoplankton over almost the whole year, and total assimilation rates seldom are higher than 80 mg $C/m^2/day$. In cases such as this the discrepancy between the different conceptions of production biology, namely high relative production rate and large standing crop, is most striking.

This discrepancy can be explained in part by the fact that « new water » that has just entered the lake my be relatively rich in phosphate and nitrate but contains only small amounts of organic compounds, whereas the « old » lake water has more or less large funds of phosphorus and nitrogen in the form of suspended or dissolved organic matter. When the nutrients of this new water are taken up by the algae no reserve is available and production is stopped until the algal population begins to decompose. On the contrary, in the « old water » an uninterrupted replenishment of nutrients goes on by means of the mineralisation of both dissolved and particulate organic matter, which is the more effective the higher the concentration of organic material. It is a well known fact that lake water generally contains much more dissolved organic matter than seston does (Birge and Juday 1934).

The difference in nutrient-contents between old and new lake water is valid only for pure and not for polluted inflowing water. Many other lakes do not answer this description, because they receive influents from upstream lakes. Examples are most of the Finnish lakes, where a chain of lakes often is linked by only short stretches of running water. A good example for both cases is given in Fig. 1, representing the assimilation rates in Lake Constance. This lake consists of two basins, Ober- and Untersee, connected by a short water course, the See-Rhein. In spring, when the snow is melting in the Swiss Alps, Obersee receives large masses of « new water » via the Alpenrhein. Primary production is relatively small therefore in June (lower part of Fig. 1). Untersee on the other hand gets « old water » from the Obersee and production is almost double. Towards fall (upper part of the graph) water replenishment in Obersee decreases and the total production rate in both basins shows only a small difference. The question of whether or not large water replenishment diminishes the production of a certain lake cannot be answered in a way that suits all cases. It is a problem of regional circumstances; in other words, it is a question of hydrographic and chemical differences between the old and the new water. In particular it is difficult to value the time of residence of « old water » in a lake (Piontelli and Tonolli 1964). In Lago Maggiore about a quarter of the lake volume is renewed every year; nevertheless 22% of the old water is still present after that period in the epilimnion.

Water replenishment may occur also in another way. The trophogenic layers may be displaced not only by water coming from outside, but also by upwelling water of deeper strata. This will be the case when the thermocline is tilted by the action of wind (Mortimer 1950, 1955). From recent investigations of Lehn (1963) in Lake Constance one can learn that surface water was forced down to a depth of 50 meters, while on the opposite side of the lake hypolimnic water flooded wide areas of the surface. When this displacement persists for some days the algae carried down into the dark zone will die. On the other hand, the areas covered with hypolimnic water also produce little phytoplankton, since propagation must start from cysts and spores.

On the other hand, if the welling up of meta- or hypolimnic water goes on slowly and continuously, production will be promoted by the supply of new nutrients which compensate the increasing depletion in the photosynthetic layers during summer. The importance of this supply is clearer, if we consider lakes in which eddy diffusion is hindered by a well developed thermal stratification. The influence on primary productivity of the separation of the epilimnion cannot be studied better than in Carinthian lakes. Because of their middle

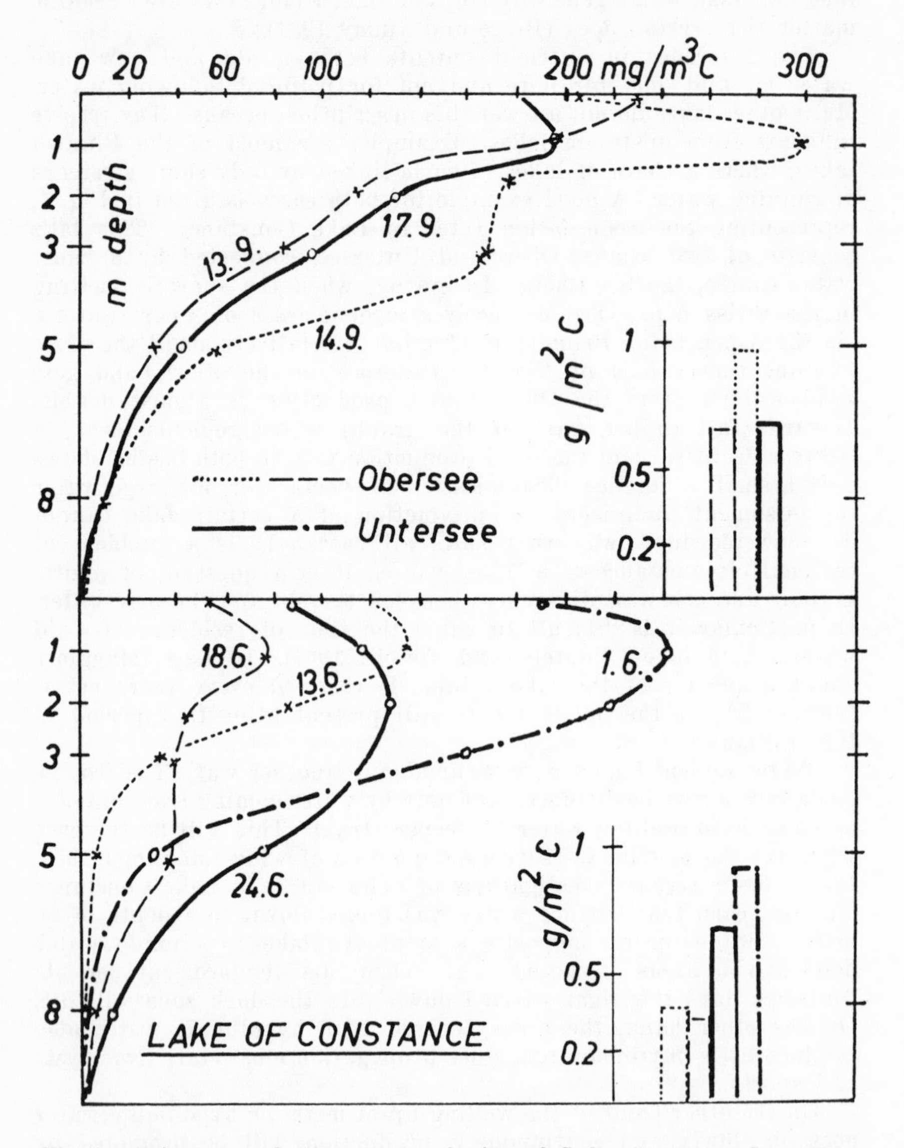

Fig. 1. - Primary production in two basins of Lake Constance. In autumn production is equal in both basins. In spring it is lower in the Obersee because of the entrance of Rhine water.

or small size, their wind sheltered position in an interalpine depression, and their climate rich in sunshine, they are typical examples of well stratified waters.

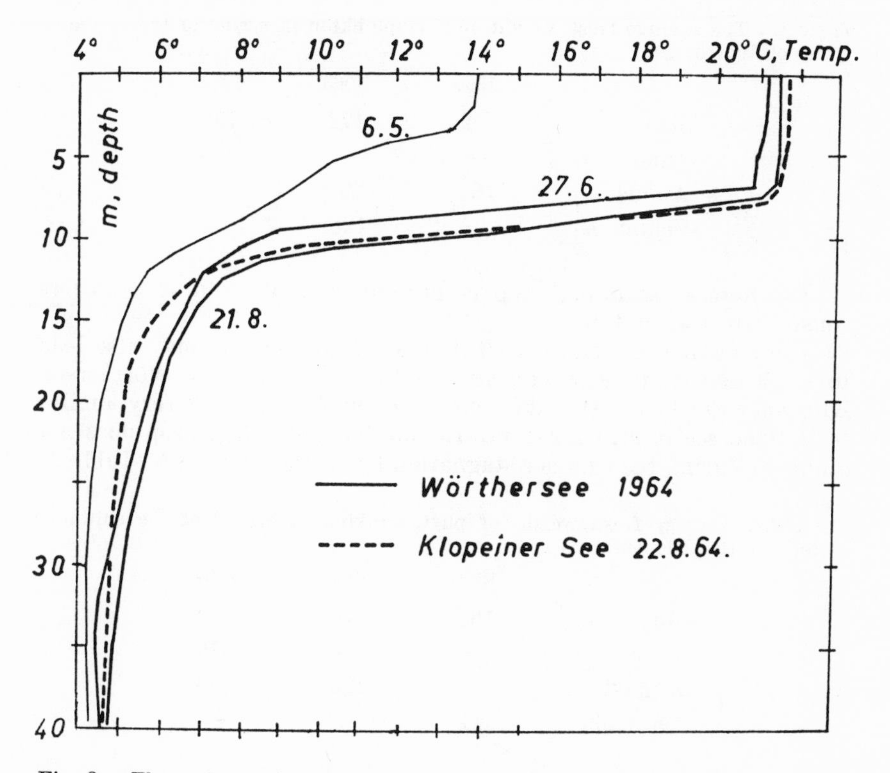

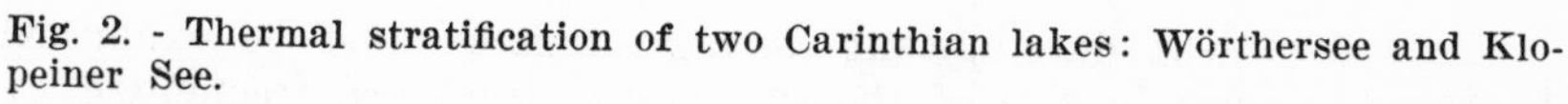
Fig. 2. - Thermal stratification of two Carinthian lakes: Wörthersee and Klopeiner See.

In Wörthersee the epilimnion is warmed to 22 or even 24°, and is sharply separated from the hypolimnion by a thermocline with a considerable gradient (Fig. 2). This thermocline is very stable and cuts off all exchange between epi- and hypolimnion. Therefore no supply of plant nutrients from deeper layers is possible, and the epilimnic primary production depends on the quantity of nutrients brought up by the spring circulation. As a large part of the dead spring plankters are mineralized in the epilimnion, subsequent generations of algae can reincarnate these substances. Although supplies are imported to some degree from the shore, they are not enough to compensate for the losses caused by the dead plankton sinking down into the hypolimnion. Thus a depletion of nutrients as well as of organic substance arises. This phenomenon has also been observed in lakes quite different from that in question (Ohle 1964). In Wörthersee the decrease is most conspicuous if the standing crops of the epilimnion in spring and summer are compared (Table 1).

Table 1. - The average fresh weight of phytoplankton in mg/m³ of the epilimnic water of Wörthersee.

	1960	1963	1964
May	752	472	750
June	—	—	310
August	164	290	72
September	98	128	—

The highest standing crop is present in May, and it decreases constantly toward fall.

The blocking effect of a well developed thermocline may also lead to a phenomenon quite opposite to that described in Wörthersee. Klopeiner See is a little lake roughly 1 km² in area, but very similar to Wörthersee in stratification. In this lake, standing crop does not diminish during the summer stagnation but rather increases (Table 2).

Table 2. - Average fresh weight of phytoplankton in mg/m³ of the epilimnic waters of the Klopeiner See.

	1961	1963	1964
May	168	370	280
June	—	—	310
August	—	420	450
September	244	—	—

This divergent development during summer must be explained by the difference in size of the two lakes. Both are frequented by tourists and polluted by the sewage of hotels. In the case of Wörthersee the imported organic matter is spread over an epilimnic water mass of an area of 19 km². Therefore it is diluted to such a degree that it does not compensate for the losses of phosphorus and nitrogen caused by the sinking plankton. In the small Klopeiner See, on the contrary, the epilimnic water is so fertilized that eutrophy increases during the summer. Thus isolation of the epilimnic layer from the hypolimnion by a well developed thermocline does not necessarily lead to a decrease of epilimnic production during summer. Small lakes may show an opposite development when they are affected by pollution. But in general it holds true that a sharply developed interface lowers production in the epilimnion for some months during summer stagnation. In this condition lakes are a small copy of the tropical ocean.

Nevertheless, small epilimnic production is not identical with deficiency in production of the whole lake. If the uppermost layers produce only small quantities of phytoplankton they allow radiation to penetrate into deeper layers. If this is the case, the whole metalimnion and even the uppermost part of the hypolimnion become part of the euphotic zone. As to the metalimnic production, it is true that it is handicaped with regard to the feeble illumination, but it is favored by the relatively high content of dissolved nutrients. This is due to several causes. The sinking movement of the seston is checked at a higher level by the increased specific weight and viscosity of the water, and decomposition goes on rapidly because of the intermediate temperature. Metabolic dynamics are most intensive in the epilimnic-metalimnic interface. Moreover, lake water enriched in nutrients from the sublittoral mud-water-interface is spread along this layer by horizontal water movements, and the same is the case with sewage that enters the lake. In both cases the water rich in nutrients is somewhat cooler than the epilimnic one and extends into the uppermost part of the metalimnion.

Therefore it is a quite common phenomenon in alpine lakes rather poor in nutrients for two maxima of production to develop: an epilimnic one mostly in 1-3 m, and a second in the metalimnion, as a rule between 12-20 m (Findenegg 1963, 1964 a). In Fig. 3 the vertical zonation of the phytoplankton biomass (g freshweight per m^3) as well as the assimilation quotient (mg C assim. per m^3 and day) in Klopeiner See are given. In the last years this initially small metalimnic maximum has increased considerably due to the invasion of the well known *Oscillatoria rubescens*, which appeared in this lake in 1961. From the curve for 1961 one can see also the vast quantities of this alga concentrated in the metalimnic strata, in spring mostly in 15-20 m depth, later on moving somewhat towards the surface. Previous to the invasion of *Oscillatoria*, the metalimnic population consisted of *Peridinium* and *Ceratium*, *Rhodomonas* and *Cryptomonas*, which also had a maximum between 8 and 12 m depth.

In contrast to the imposing biomass of *Oscillatoria*, the photosynthetic activity of the metalimnic population is rather low. The activity coefficient (ratio: mg C assimilated per m^3 and day/freshweight in mg/m^3) in the cubic meter of highest *Oscillatoria* density was in May at 20 m depth 0.03, in August at 18 m depth 0.03, and in October at 15 m depth 0.06.

There is no doubt that these low values are conditioned by the small percentage of radiation reaching deeper layers. In May and August it was 1-2%, in October 3% of surface intensity in the depth mentioned above. In late autumn, when *Oscillatoria* is distributed homogeneously in the euphotic zone by the autumnal circulation, activity coefficients up to 0.3 are reached in the optimal depth of 5 meters.

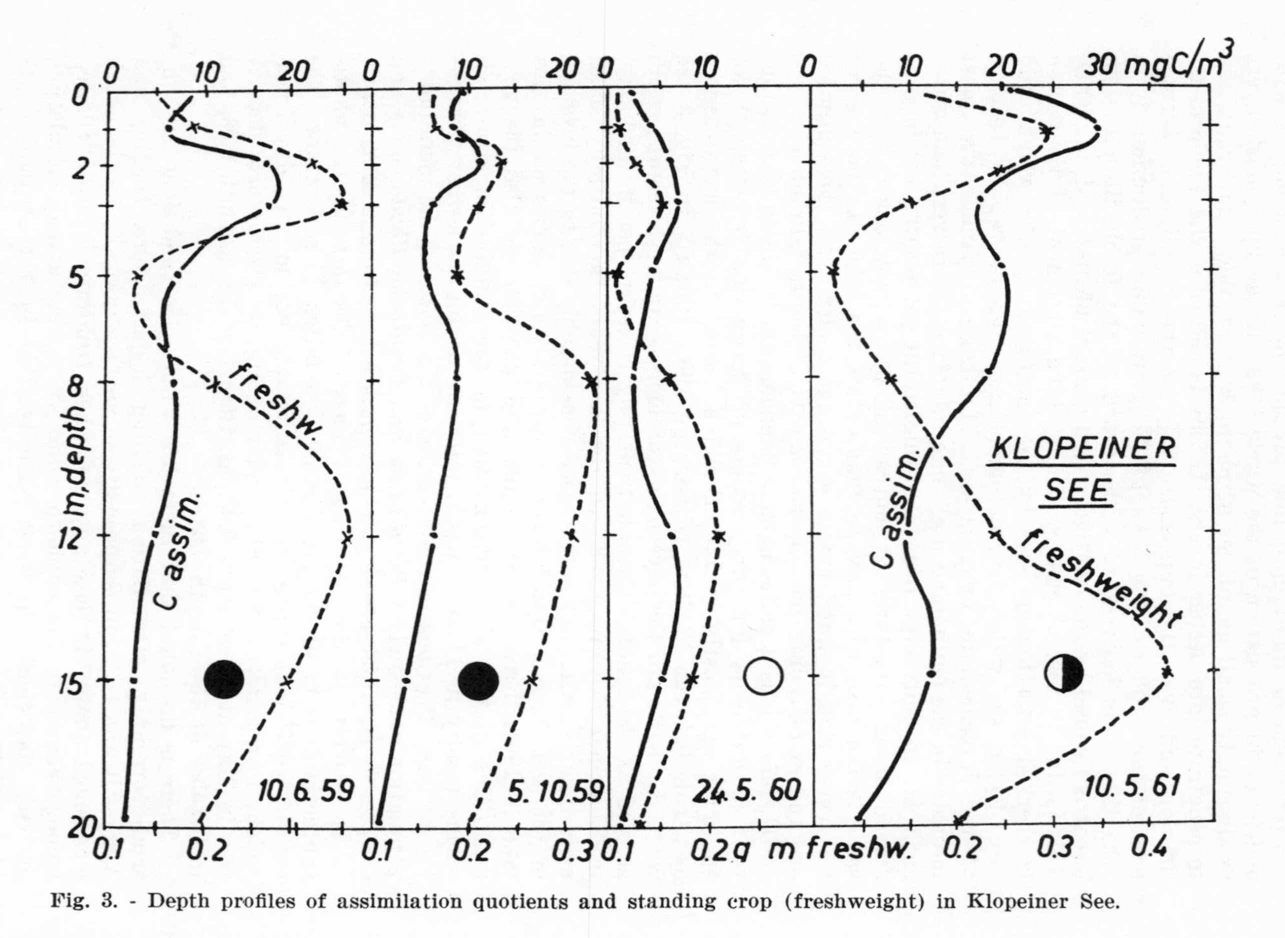

Fig. 3. - Depth profiles of assimilation quotients and standing crop (freshweight) in Klopeiner See.

The type of lake showing two centers of assimilation is relatively common, not only in Austria but also in Switzerland, because of the wind sheltered position of alpine valleys and inneralpine basins. It develops from the oligotrophic state which is characterized by the absence of a distinct optimum of assimilation. If pollution increases and light transmission remains good, a moderate epilimnic assimilation center appears, followed by a feeble metalimnic peak. This becomes larger when *Oscillatoria* invades the lake (Findenegg 1964 b). But not all alpine lakes develop in this way. The metalimnic phytoplankton community is missing in lakes with poor light transmission and a deep lying thermocline so that the metalimnion is below the euphotic limit. This is the case in Lake Constance, and in other large lakes because of the considerable motions of water. At the sametime in small ones with a higher content of humic substances or high productivity in the surface layers, conditions for production are similar.

The wind-protected situation of some alpine lakes may have still another consequence. If wind is lacking also in winter, the autumnal partial circulation does not lead to a complete turnover: they are meromictic. It is obvious that meromixis does not favor production. Part of the planktic biomass sinks down after death and is not mineralised before reaching the bottom. These nutrients, or at least part of them, do not come back into trophogenic layers. For this reason the upper strata are not as well supplied with phosporus and nitrogen as would be the case after a complete turnover. The quantities of nutrients stored in the monimolimnion of meromictic lakes may be remarkable in some cases. Fig. 4 shows the accumulation of total phoshorus and nitrogen in the deeper parts of the Wörthersee in 1964. From the curve of conductivity and that of oxygen one can see that the boundary between mixo- and monimolimnion is at a depth of about 45 m. The chemocline is marked also by a conspicuous increase of P and N. This consideration does not concern the meromictic lake type only but also the oligomictic one to some degree.

Of course there is also an exchange between mixo- and monimolimnion in meromictic lakes. Its efficiency depends on the energy of the water movements in the lake that cause turbulence and consequently eddy diffusion. The trophic level of the euphotic stratum will be lower as eddy diffusion becomes negligible. Therefore small and relatively deep waters of this kind are very poor in production. Krottensee in the Austrian Salzkammergut for example is 45 m deep and has an area of only 0.1 km^2. Moreover it lies in a wind-protected hollow and is therefore a standard meromictic lake, the mixolimnion being only 15 m thick. Production in the optimal cubic meter of this lake amounted to 20 $mgC/m^3/day$ at best, and the total assimilation never surpassed 80 mgC/m^2 assimilated.

In larger meromictic lakes the chemocline does not divide the water body into two strictly separated parts. The nutrients accumulated in the deepest layers can return into the productive strata to some degree by eddy diffusion, although it is very difficult to measure the percentage directly. This exchange of course depends on climatic factors and will not be the same every year. This may be seen from measurements of the oxygen stratification in Millstätter

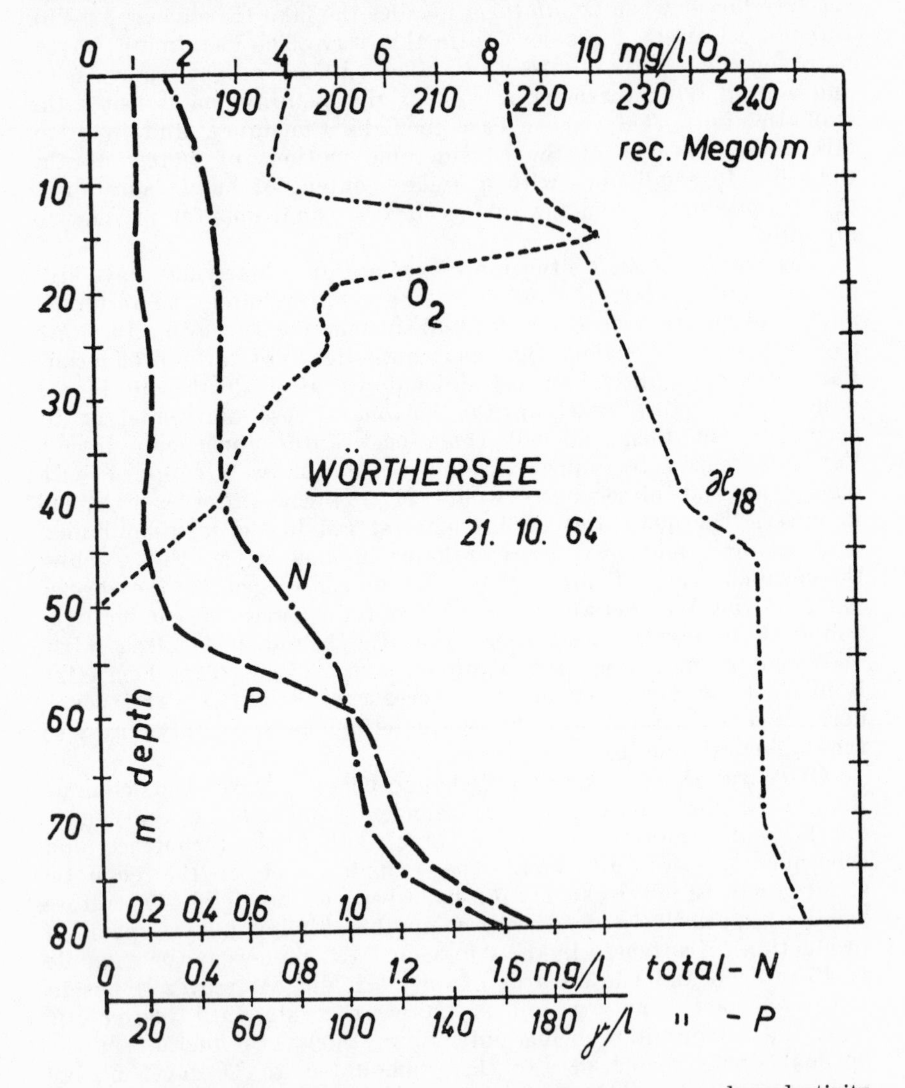

Fig. 4. - Stratification of total phosphorus, nitrogen, oxygen, and conductivity in the meromictic Wörthersee.

See since 1930 (Fig. 5). In this meromictic lake, relatively small climatic changes have caused a considerable increase in the exchange of water between the upper and deep layers, as can be seen from the rising O_2 values in the monimolimnion. During 1931-1935 hard winters occured, and as a rule the lake was frozen. The prolonged winter was followed by a quick warming up so that the phase of instability was short, and therefore the vertical water exchange was

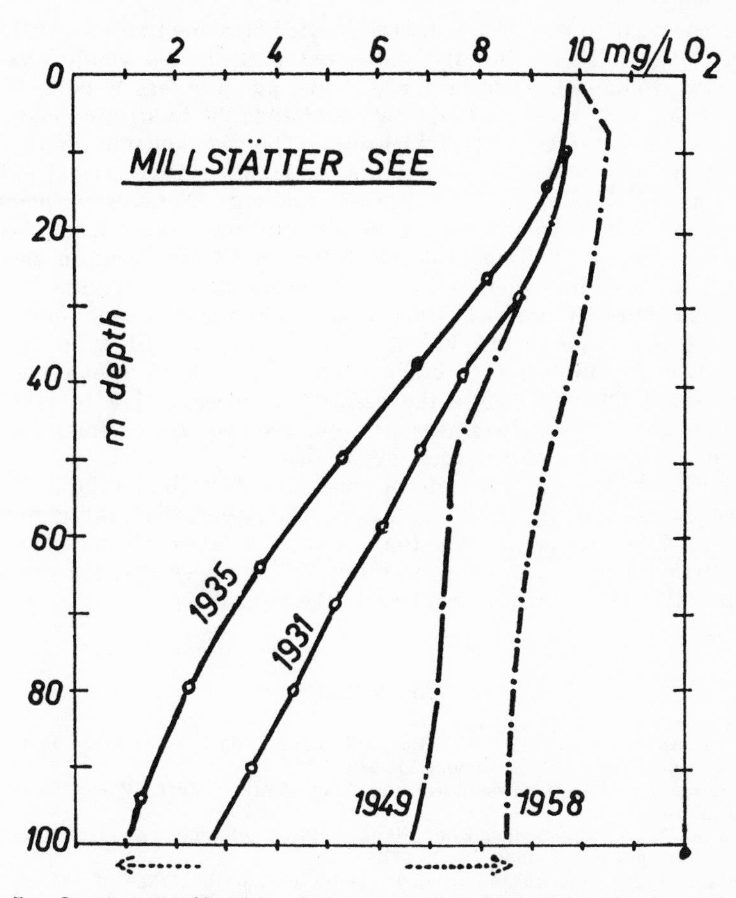

Fig. 5. - Oxygen stratification of the meromictic Millstätter See in periods of different climatic character.

small. In these years the oxygen supply from the surface layers to the monimolimnion was not sufficient to compensate for the O_2 consumption by decay of the dead plankton. Thus the oxygen content of the monimolimnion decreased from year to year.

In a later period, the climate became more oceanic (1949-1958) and the lake was more exposed to wind during winter because of the absence of an ice cover. Exchange increased and the oxygen content slowly grew, up to 8 mg/liter in the monimolimnion. We can assume that in that period of more intensive exchange also a relatively high percentage of the nutrients accumulated in the deep layers during the earlier years returned into the trophogenic zone.

In contrast to the circumstances in well stratified lakes mentioned above, large basins generally show less complicated conditions for primary production. The larger the lake and the less it is sheltered from wind, the more mixing and exchange is facilitated and the less the importance of stratification. The trophogenic layer will never be totally deprived of nutrients. In many cases, a deep epilimnic stratum will develop early in spring, and all processes of primary production will take place in a rather uniform water mass during « growing time » (Talling 1965). Similar conditions occur in smaller eutrophic lakes in which light transmission is so decreased by the crowded algae of water-blooms that the euphotic zone does not extend deeper than a few meters. Thus, with a shallow epilimnion production is confined to a homogeneous water body (Jonasson and Mathiesen 1959). Of course the results of investigation in lakes of this kind show a certain conformity, and lead easily to generalization and to conceptions of « normal depth profiles ».

It would be a satisfaction to the present writer if he has succeeded in focussing attention to the statement that these normal depth profiles seldom are to be found in alpine lakes. On the contrary, quite irregular stratification and large differences occur from lake to lake and are more the rule than the exception.

REFERENCES

Birge, E. A. and C. Juday. 1934. Particulate and dissolved organic matter in Wisconsin lakes. - Ecol. Monogr. *4* : 440-474.

Findenegg, I. 1963. Gewässergüte und Seenschutz. - Österr. Wasserwirtschaft *15* : 65-70.

— 1964 a. Produktionsbiologische Planktonuntersuchungen an Ostalpenseen. - Int. Rev. ges. Hydrobiol. *49* : 381-416.

— 1964 b. Types of planktic primary production in the lakes of the Eastern Alps as found by the radioactive carbon method. - Verh. int. Ver. Limnol. *15* : 352-359.

Jonasson, P. M. and H. Mathiesen. 1959. Measurement of primary production in two Danish eutrophic lakes, Esrom Sö and Furesö. - Oikos. *10* : 137-167.

Lehn, V. 1963. Einige Frühjahrsbefunde über die Isothermen-Phytoplankton-Relation im Bodensee. - Arch. Hydrobiol. *59* : 1-25.

Mortimer, C. H. 1950. The use of models in the study of water movement in stratified lakes. - Verh. int. Ver. Limnol. *11* : 254-260.

— 1955. Some effects of the earth's rotation on water movements in stratified lakes. - Verh. int. Ver. Limnol. *12* : 66-77.

Ohle, W. 1964. Interstitiallösungen der Sedimente, Nährstoffgehalt des Wassers und Primärproduktion des Phytoplanktons in Seen. - Helgol. Wiss. Meeresunters. *10* : 411-429.
Piontelli, R. and V. Tonolli. 1964. Il tempo di residenza delle acque lacustri in relazione ai fenomeni di arricchimento in sostanze immesse, con particolare riguardo al Lago Maggiore. - Mem. Ist. Ital. Idrobiol. *17* : 247-266.
Talling, J. F. 1965. The photosynthetic activity of phytoplankton in East African lakes. - Int. Rev. ges. Hydrobiol. *50* : 1-32.

MICRONUTRIENT LIMITING FACTORS AND THEIR DETECTION IN NATURAL PHYTOPLANKTON POPULATIONS

CHARLES R. GOLDMAN

University of California
Davis, California, U.S.A.

For bibliographic citation of this paper, see page 10.

Abstract

The requirements of natural phytoplankton populations for molybdenum, copper, vanadium, cobalt, manganese, zinc, boron, iron, sodium, and calcium are reviewed. Although emphasis is placed on positive response to micronutrient additions, the importance of inhibitory effects is also considered. Data is presented from cultures of Castle Lake water where significant photosynthetic stimulation, as measured by C^{14} assimilation, was observed with molybdenum addition. Cobalt, which is already at a high level in this lake, was found to be inhibiting. The sedimentary record of molybdenum, vanadium, cobalt, and manganese concentration is examined in relation to present conditions in the lake. The use of response surfaces is suggested as a useful *in situ* experimental approach for solving problems of interaction of various macro-and micronutrient factors in natural phytoplankton populations.

INTRODUCTION

The search for micronutrient limiting factors in natural populations of aquatic plants has really just begun. It has been long delayed by the concentration of investigators on the major nutrient requirements of planktonic algae, with little attention to micronutrients except in laboratory cultures. Interest in hydroponic culture of higher plants was probably a significant factor in directing laboratory studies to the micronutrient requirements of algae (Arnon 1958, Wiessner 1962). Nicholas (1963) has carefully reviewed both essential and nonessential elements, together with pure-culture methods. He lists sixteen elements essential for the growth of one or more organisms: N, P, Ca, Mg, Na, K, S, Fe, Mn, Cu, Zn, Mo, B, Cl, Co, and V; and notes that although Ni, Ti, Se, Pb, Ag, Au, Br, I, and others are often found in the ash of microorganisms, their requirement for growth is yet to be proved.

It is admittedly very difficult to distinguish between a specific action of a particular element and an absolute requirement. Classification of nutrients is usually made on the basis of quantitative requirements, which may vary greatly among the algae. Those elements which are required in relatively large quantities are referred to as major or macronutrients, while those required in small quantities are probably best termed micronutrients. The latter have also been called minor, trace, or oligoelements. The ambiguity of these classifications was recognized by Arnon (1958), who points out that calcium, required in large quantities by higher plants, should be considered a micronutrient for the green alga *Scenedesmus obliquus*, but a macronutrient for the blue-green alga *Anabaena cylindrica*. By the same reasoning used in the case of calcium, sulfur could be included among the micronutrients for most microorganisms. Nicholas (1963) feels that, because of the wide range of requirements within the micronutrient group, there is some justification for an « ultramicronutrient » group of metals, which would include both molybdenum and vanadium. Cobalt also should probably be placed in this group.

Investigation of the micronutrient requirements of algae by the isolation of growth factors in laboratory cultures has been the most frequent experimental approach; *e.g.*, Chu (1942), Rodhe (1948), Vollenweider (1950), Provasoli and Pintner (1953), Provasoli, McLaughlin, and Droop (1957), and Arnon (1958). This work has been reviewed by Lund and Taling (1957) and most recently by Lund (1965). Natural water from ponds and lakes has also been brought into the laboratory and cultured under various nutrient regimes and with various algal inoculations (Ström 1933, Fish 1955, Potash 1956, Eyster 1958, MacPhee 1961, and Lund 1964). Paralleling the development of laboratory culturing of natural phytoplankton populations, and stimulated by the widespread interest in primary productivity measurements, *in situ* studies were undertaken in a variety of environments. A great number of these studies have been designed to approximate the conditions under which algal photosynthesis occurs in nature. They have usually involved natural populations which are isolated by hard glass or plastic. In general, the shorter the experiment, the less time there is for these « microcosms » to change significantly from their surroundings. Much of this work has been reviewed by Talling (1961).

From the beginning of experimentation, one of the most serious problems in both laboratory and field methods has been contamination through culture containers and reagents. With improvements in glassware, plastics, reagent chemicals, re-crystallization procedures, ion absorption resins, and chelating agents, as well as the development of activation analysis and the availability of isotopes for checking contamination, reproducible studies on the requirements of algae in pure culture began to appear. It seems likely that other essential micronutrients may still be discovered as purification techniques are further advanced.

Various workers have made rate measures of photosynthesis in cultures. These experiments include the use of the oxygen method by Edmondson and Edmondson (1947) with fertilized salt water; and the use of the C^{14} method, by Goldman (1960 a) as a check on *in situ* assays of macronutrient limiting factors in Alaskan lakes, by

Schelske (1962) in four Michigan lakes, and by Wetzel (1965) in Indiana lakes. The application of the C^{14} method to cultures has been made in a number of marine studies, such as those of Ryther and Guillard (1959), Menzel and Ryther (1961), and Tanter and Newell (1963), and has now found rather general acceptance. The isotope approach represents a major advance in the rapid measurement of growth rates. It has the distinct advantage of providing an extremely sensitive measure of carbon assimilation which does not require the time necessary for cell division, can be accomplished before shifts in population diversity occur, and can easily be made in the field. Because of the extreme sensitivity of both the method and the physiological condition of the organisms concerned, great caution should be exercised in experimentation and in interpretation of results. Details on *in situ* culture procedures utilizing radioactive carbon are to be found in Goldman (1963).

The more field work that is done with micronutrients, the more obvious it becomes that the problem is extremely complex. It is evident from pure culture studies, which have shown a rather broad variation in the quantitative requirements of different species, that at a given time only certain portions of the natural phytoplankton community will be deficient in a given element, and respond to its addition. The measured response may therefore reflect the importance to the community of a deficient component as much as the intensity of the deficiency in the particular organisms concerned. Working in the field with natural light, temperature, and populations, one can measure a change in the metabolism of the total carbon-fixing community composed of autotrophic, heterotrophic, and mixotrophic organisms, but one cannot easily determine which particular organism is responding at a given time. Fractionation filtering has been used by marine workers, and by Sorokin (1964) to remove algae for studies of chemosynthesis by the smaller organisms. Goldman and Wetzel (1963) used 0.45 and 10 micron filters to determine the relative importance of photosynthetic organisms greater than and less than 10 microns in size. Some further insight into the organisms concerned and the metabolic pathways involved may be gained through the use of a variety of labeled compouds, and through comparison of uptake under conditions of light and dark. A measure of the response of individual species within the community will probably depend upon analysis under still more simplified culture conditions. In this respect, progress has recently been made by Wright and Hobbie (1965) in separating algal and bacterial response to dissolved organic matter. Some estimation of the degree to which axenic laboratory experiments can be used to explain the dynamics of natural populations certainly deserves particular attention by aquatic ecologists.

The reader should also be aware that it is not uncommon to find photosynthetic *inhibition* with even small additions of certain trace elements, and that, although this paper stresses positive responses to trace element additions, the negative responses encountered may also prove of considerable importance.

In presenting the evidence for the importance of micronutrients in a variety of environments, it is more convenient to consider the problem element by element because most aquatic environments remain as yet unexplored for micronutrient deficiencies.

MOLYBDENUM

Since the discovery by Arnon and Stout (1939) that molybdenum is among the elements essential for plant growth, this micronutrient has assumed considerable importance in agriculture. Plant physiologists have demonstrated its role in the formation of the enzyme nitrate reductase, as well as its involvement in nitrogen fixation. This dual role has been well illustrated in experiments with *Anabaena cylindrica,* which requires molybdenum when grown with nitrate or nitrogen gas but not when ammonia is used as a nitrogen source (Fogg and Wolf 1954). A good correlation between molybdenum deficiency and depressed chlorophyll content in both *Scenedesmus* and *Chlorella* has, however, been reported by Arnon *et al.* (1955) and Arnon (1958). The reduced photosynthesis in these cultures could

Fig. 1. Castle Lake, California, looking northeast from the cirque face towards Mount Shasta. A large development of the nitrogen-fixing mountain alder (*Alnus tenuifolia* Nutt.) may be seen along the shore line to the right.

be accounted for solely by this lowered chlorophyll content. This would seem to imply that molybdenum also assumes other roles in algal metabolism. Rather inefficient replacement of molybdenum by vanadium or tungsten has been reported for nitrogen-fixing Nostocaceae by Bortels (1940).

Hutchinson (1957), well known for his remarkable foresight, observed that the generally low concentration of molybdenum in natural waters might well be of biological importance. Castle Lake (Fig. 1) in the Klamath Mountains of California is the first lake where a deficiency of this element has been discovered (Goldman 1960 b). Fig. 2 illustrates the response of *in situ* cultures of Castle Lake water to molybdenum addition in June 1959. In October 1959 the molybdenum response was reduced. The culture with molybdenum added more nearly followed the control after the initial stimulation, and growth of the *in situ* culture was measured with frequent sampling over a four day period (Fig. 3). The lower growth rate of both

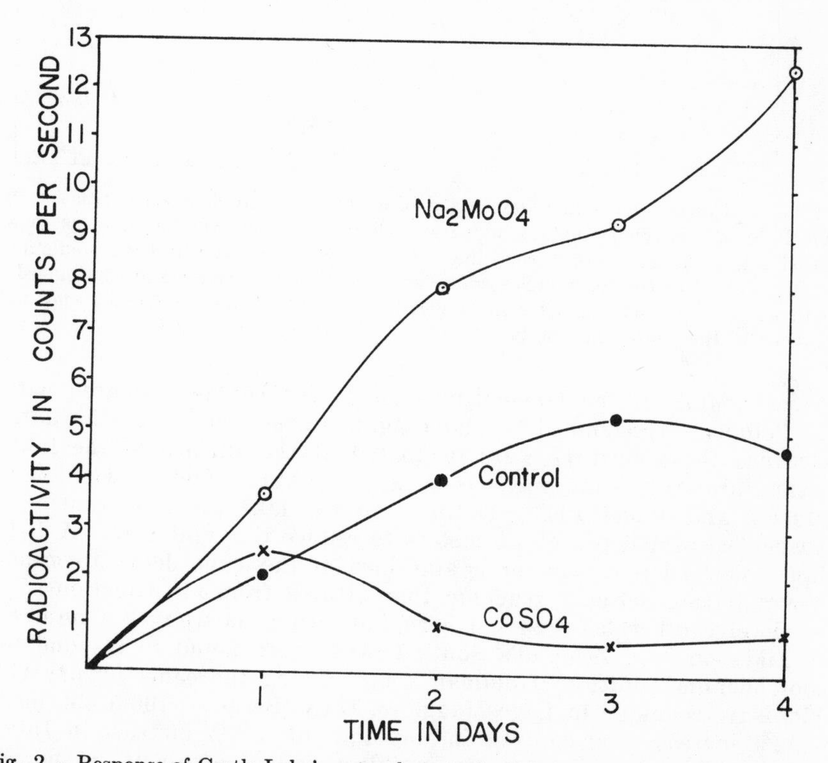

Fig. 2. - Response of Castle Lake's natural phytoplankton population to the addition of 100 ppb molybdenum and 5 ppb cobalt, as measured by C^{14} assimilation. The hard glass culture containers were maintained in the lake under natural conditions of light and temperature.

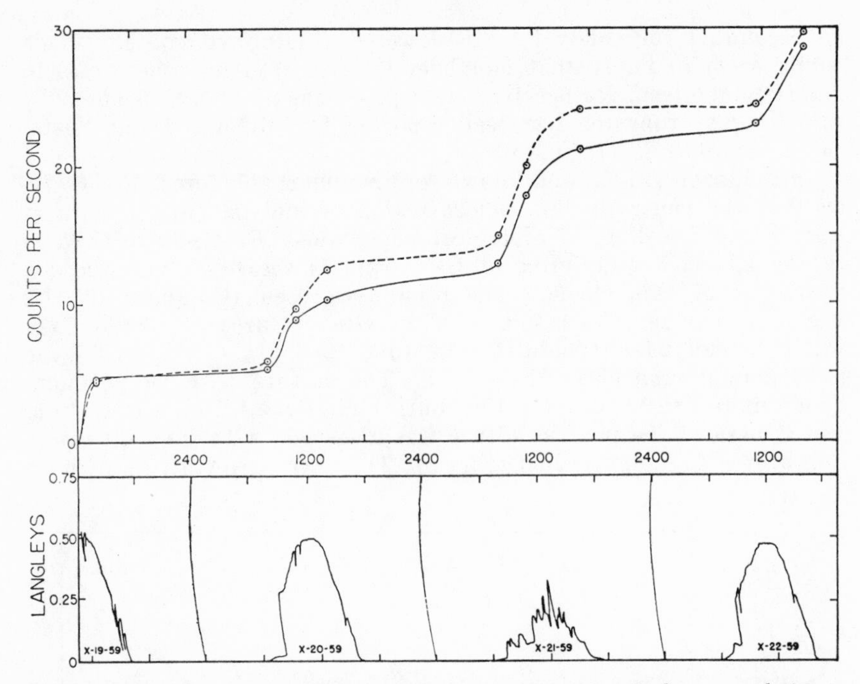

Fig. 3. - Growth over a four day period of a natural phytoplankton population to which 50 ppb molybdenum was added as sodium molybdate (broken line) as compared with a control culture (solid line). Growth was measured by C^{14} assimilation in the cultures at the times indicated. Light measurements were taken continuously with a recording pyrheliometer in langleys per minute (gm cal/cm²/min.) and are reduced to the photosynthetic portion of the spectrum (4000-7000 Å).

experimental and control cultures on 20 October as compared with 21 October corresponded to higher light intensity on the earlier date. Because these cultures were incubated at the surface of the lake, light inhibition may have been an important factor (Goldman, Mason, and Wood 1963). Cultures in the lake are now routinely suspended at a depth of 1.5 meters to reduce this inhibition. A Bellani spherical pyranometer is stationed at the same depth to measure radiation actually reaching the cultures from all directions.

Twelve other lakes in northern California, as well as a number of lakes on New Zealand's South Island, were found to respond to molybdenum addition (Goldman 1964). Using the same *in situ* C^{14} bioassay technique in Lago Maggiore, Italy, Becacos (1962) obtained a 12% increase over controls in May and only a 6% increase in July. In laboratory experiments, an inhibitory phase of about six hours was encountered, which agrees well with the author's experience in Castle Lake, California. To investigate total lake response to molybdenum fertilization, an application of sodium molybdate was made to Castle Lake in 1963. The distribution of molybdenum in the lake was followed through the use of a simple yet sensitive colorimetric technique which measured its concentration down to a level of 0.5 µg per liter of lake water (Bachmann and Goldman 1964). The conservative behavior of the element as observed in Castle Lake water made it ideal as an experimental micronutrient fertilizer and as a water mass marker. It has, to date, remained in solution for over a year, through two periods of lake turnover, with measurable depletion occurring only through outflow.

COPPER

As an essential constituent of many plant enzymes, copper is probably essential for most algae in minute quantities. To the author's knowledge, however, a definite requirement has been reported only for *Chlorella* (Walker 1953). Copper is particularly hard to remove as a contaminant of water distilled in a typical tin-lined copper still. Nicholas (1952) found 400 µg of copper per liter of water from such a still after thrice redistilling in pyrex glass. To date, copper has not been found to be a limiting factor in any studies of natural waters, but might be suspect as an inhibiting factor where environmental levels are high.

VANADIUM

Evidence for the requirement of vanadium for lower plant growth is presented by Arnon and Wessel (1953), and is reviewed by Arnon (1958). In their study, response in laboratory cultures was highest with the addition of 20 µg/liter. With *Scenedesmus obliquus*, the vanadium requirement was a thousand times as great as for molybdenum. Allen and Arnon (1955 a) were not able to substitute vanadium for the molybdenum requirement of *Anabaena*. In attempting to substitute vanadium for the molybdenum deficiency in Castle Lake cultures, the author has not been able to achieve any measurable stimulation. Vanadium requirements by different algae would appear to merit some further investigation in the laboratory.

COBALT

Cobalt has been definitely shown to be required by blue-green algae (Holm-Hansen, Gerloff, and Skoog 1954). Its essential nature in the formation of vitamin B_{12} and nitrate reductase make it one of the most interesting micronutrients. Benoit (1957) found cobalt levels in Linsley Pond, Connecticut, to be only one tenth of the optimum requirement reported for a typical chrysomonad. Only

2% to 13% of the element was tied up as vitamin B_{12} in the pond water. I have found cobalt to increase photosynthetic carbon fixation in eight out of the ten lakes tested on New Zealand's South Island (Goldman, in preparation), and in one Alaskan lake (Goldman 1964). In Castle Lake, where there is a high natural concentration, the addition of 5 ppb cobalt as $CoSO_4$ was inhibiting (Fig. 2). The slight initial stimulation in this experiment may reflect an immediate positive response to the small sulfur addition before the cobalt inhibition set in. This occurred well below the toxic level of 10 mg/liter reported by Arnon *et al.* (1955). Wetzel (1965) also reports cobalt toxicity in cultures of Crooked Lake, Indiana, but stimulation with B_{12}. Maximum response in the cobalt deficient New Zealand lakes was achieved at 20 μg/liter.

MANGANESE

Manganese and iron are frequently discussed together because of the similarity of their behavior in natural waters, especially with regard to the phosphorus cycle. Like iron, manganese is rapidly lost to lake sediments under aerobic conditions, and may be in short supply if natural chelation is inadequate. The first report of a manganese requirement was for *Chlorella* (Hopkins 1930). Although cell growth in cultures of *Scenedesmus* was arrested by manganese deficiency, chlorophyll content was not reduced (Arnon 1958). Cultures of four New Zealand lakes, two California lakes, and one Alaskan lake showed significant stimulation with manganese additions (Goldman 1964). An interesting suggestion of manganese and inorganic nitrogen toxicity by Gusseva, working on a Moscow-Volga canal reservoir, is reviewed by Wiessner (1962). In this study the blue-green *Aphanizomenon flos-aquae* appeared to be inhibited at natural environmental levels of manganese.

ZINC

Although zinc was the first essential micronutrient to be discovered (Raulins 1869), the only studies in natural waters the author knows of are his own. As might be expected where trace metals are often in short supply in the soil, cultures from 7 out of 10 lakes on New Zealand's South Island showed improved carbon assimilation with zinc addition. Photosynthesis in Lake Tahoe (California-Nevada) water was similarly stimulated by zinc addition, as was one Alaskan lake (Goldman 1964).

BORON

This element has been reported as necessary for the growth of *Nostoc muscorum* (Eyster 1952), and is sometimes present in fresh waters at very low levels. On the several occasions that the author has bioassayed with boron additions, only once, in Lake Nerka, Alaska, has he obtained slight and perhaps even then insignificant response (Goldman 1964). In Borax Lake, California, where environmental levels ranged from 440 to 850 mg/liter, some increase in carbon assimilation could be obtained with boron additions of 50 and 100 mg/liter (Wetzel 1964). A rather rapid adaption to increased boron concentration was evident in these experiments, which one would expect to be already in the toxic range. It is possible that the stimulation results from trace impurities in the reagents when large amounts are used, or that small pH changes affect the availability of other nutrients.

IRON

Iron may be considered as much a macro- as a micronutrient. Response of algal photosynthesis to iron addition has been reported for the Sargasso Sea (Menzel and Ryther 1961), for a Michigan lake (Schelske 1962), for Lake Tahoe, California-Nevada (Goldman 1964), and for two Indiana lakes (Wetzel 1965). Care must be given culture experiments with added iron which utilize the sensitive C^{14} method because of strong coprecipitation of the labeled carbon with ferric hydroxide (Goldman and Mason 1962). Prefiltered controls should be included as a measure of the radioactivity retained on the filter by this mechanism.

SODIUM

Allen and Arnon (1955 b) have shown sodium to be essential to *Anabaena cylindrica*, and *Microcystis aeruginosa* in culture benefited from its addition (Gerloff, Fitzgerald, and Skoog 1952). The availability of sodium in most natural environments, however, would preclude this as a trace element limiting factor. In land reclaimed from the sea, magnesium deficiency results from high sodium levels. In this case, and probably in most lake waters, the greatest importance of sodium presumbaly lies in its influence on the ratio of the monovalent to the divalent salts.

CALCIUM

As already noted, calcium can be considered a micronutrient for the green alga *Scenedesmus obliquus* but a macronutrient for the blue-green alga *Anabaena cylindrica*. Vollenweider (1950) demonstrated that a number of diatoms grow best with high calcium, and that there is interaction between calcium and potassium. As mentioned in relation to sodium, this probably reflects the great importance of monovalent to divalent ratios in the culture medium. An execellent laboratory study of this problem has been made with marine algae by Provasoli *et al.* (1957).

MICRONUTRIENTS IN THE SEDIMENT RECORD

From paleolimnological examinations of a four meter core of sediments from Castle Lake, California, it appears that lakes are reservoirs of history in the sense that they show remarkable variation in their trace element sediment composition (Fig. 4). It is of particular interest that molybdenum was once a reasonably abundant component of deposition but is now below the level of detection in the sediments. This may in part result from the buildup of the mountain alder (*Alnus tenuifolia*) on the watershed (see Fig. 1), which may be competing with the lake for available molybdenum (Goldman 1961). Vanadium is more constant but does not appear to be a satisfactory substitute for molybdenum in the ecosystem, while cobalt has been increasing together with manganese. As already noted (Fig. 2), the high cobalt levels today are reflected by inhibition with further additions to cultures of the lake water.

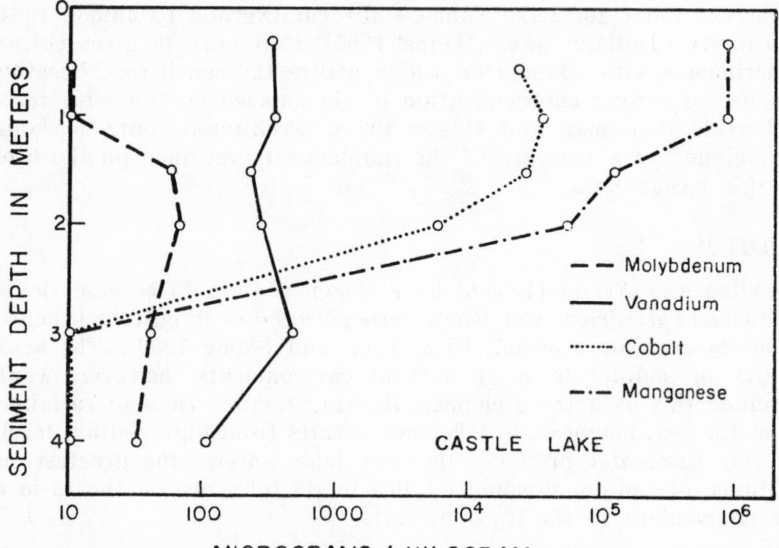

Fig. 4. - Micronutrient content in micrograms/kilogram of the sediment in the basin of Castle Lake, California. The four meters of sediments were collected with a plastic lined piston corer at a depth of 35 meters in the lake. Analysis was by emission spectrophotometry.

CONCLUSIONS

Although stressing positive responses in the introductory remarks, I have also touched on the problem of negative response to trace element addition in cultures of natural waters and the possibility of control of algal population in nature by the natural increase in one or more nutrients. In my experience, it is not at all uncommon to find inhibition by trace element additions at as low a level as 1 μg/liter. Although environmental levels of one or more trace elements may be at toxic or near toxic levels for most algae, we still find certain species which apparently thrive under these conditions. Adding more of an element already at an unfavorably high level is likely to produce an inhibition, and it is probable that particular species or races have developed which are especially adapted to these toxic levels, in the same manner that the low levels of certain elements may favor other species. Negative response in cultures should not be ignored but rather compared with trace element analyses of the waters involved, giving particular attention to possible interaction of the various metals.

The investigator should also concern himself with a positive response which may be indirect. As in the case of shifts in monovalent to divalent ratios, a population may well show positive response to a metal which simply competes for a binding site with a more toxic relative. Further, one element may substitute for another preferred element or otherwise indirectly affect carbon assimilation. In Castle Lake, where Mo, K, and S are all in low supply, a shift in the level of one element changes the response to the other two but not to the same degree. By using response surfaces (Box 1954), it is possible to more precisely describe the interaction of molybdenum with potassium and sulfur concentration (Goldman *in press*). The general approach can be extended to cover a spectrum of limiting factors, and is greatly facilitated by the use of a computer in the analysis of results. The importance of micronutrients in natural waters remains a fascinating area of ecological investigation which is still in the early stages of development.

Acknowledgements

The author wishes to acknowledge the support of the National Science foundation (Grant GB2436) and the John Simon Guggenheim Memorial Foundation, which awarded him a Fellowship during 1965. This paper was presented during his stay at the Istituto Italiano di Idrobiologia, and special thanks are extended to Professor Vittorio Tonolli and Dr. Livia Tonolli, not only for the use of the facilities there, but for their many kindnesses during his residence in Italy. The emission spectrophotometry was carried out by Mr. William Silvey. Mr. D.T. Mason provided valuable comment.

REFERENCES

Allen, M. B. and D. I. Arnon. 1955 a. Studies on nitrogen-fixing blue-green algae. I. Growth and nitrogen fixation by *Anabaena cylindrica* Lemm. - Plant Physiol. *30* : 366-372.

— — 1955 b. Studies on nitrogen-fixing blue-green algae. II. The sodium requirement of *Anabaena cylindrica*. - Physiol. Plant. *8* : 653-660.

Arnon, D. I. 1958. Some functional aspects of inorganic micronutrients in the metabolism of green plants. *In* A. A. Buzzati-Traverso [ed.], Perspectives in Marine Biology. - Univ. Calif. Press, Berkeley: 351-383.

— and P. R. Stout. 1939. Molybdenum as an essential element for higher plants. - Plant Physiol. *14*:599-602.

— and G. Wessel. 1953. Vanadium as an essential element for green plants. Nature *172*:1039-1041.

— P. S. Ichioka, G. Wessel, A. Fujiwara, and J. T. Woolley. 1955. Molybdenum in relation to nitrogen metabolism. I. Assimilation of nitrate nitrogen by *Scenedesmus*. - Physiol. Plant. *8*: 538-551.

Bachmann, R. W. and C. R. Goldman. 1964. The determination of microgram quantities of molybdenum in natural waters. - Limnol. & Oceanog. *9*:143-146.

Becacos, T. 1962. Azione di alcune sostanze inorganiche sull'attività fotosintetica in due laghi dell'alta Italia (Lago Maggiore e Lago di Mergozzo). - Mem. Ist. Ital. Idrobiol. *15*: 45-68.

Benoit, J. R. 1957. Preliminary observations on cobalt and vitamin B_{12} in freshwater. - Limnol. & Oceanog. *2*:233-240.

Bortels, H. 1940. Uber die Bedeutung des Molybdäns fur stickstoffbindende Nostocaceen. - Arch. Mikrobiol. *11*:155-186.

Box, G. E. P. 1954. Exploration and exploitation of response surfaces: Some general considerations and examples. - Biometrics *10*:16-60.

Chu, S. P. 1942. The influence of the mineral composition of the medium on the growth of planktonic algae. I. Methods and culture media. - J. Ecol. *30*:284-325.

Edmondson, W. T. and Y. H. Edmondson. 1947. Measurements of production in fertilized salt water. - J. Mar. Res. *6*: 228-246.

Eyster, C. 1952. Necessity of boron for *Nostoc muscorum*. - Nature *170*:755-756.

— 1958. Bioassay of water from a concretion-forming marl lake. - Limnol. & Oceanog. *3*: 455-458.

Fish, G. R. 1955. Chemical factors limiting growth of phytoplankton in Lake Victoria. - Jour. E. African Agric. *21*: 152-158.

Fogg, G. E. and M. Wolf. 1954. Nitrogen metabolism of blue-green algae. - Symposium Soc. Gen. Microbiol. *4*: 99-125.

Gerloff, C. G., G. P. Fitzgerald, and F. Skoog. 1952. The mineral nutrition of *Microcystis aeruginosa*. - Am. J. Bot. *39*: 26-32.

Goldman, C. R. 1960 a. Primary productivity and limiting factors in three lakes of the Alaska Peninsula. - Ecol. Monogr. *30*:207-230.

— 1960 b. Molybdenum as a factor limiting primary productivity in Castle Lake, California. - Science *132*: 1016-1017.

— 1961. The contribution of alder trees (*Alnus tenuifolia*) to the primary productivity of Castle Lake, California. - Ecology *42*: 282-288.

— 1963. The measurement of primary productivity and limiting factors in freshwater with C^{14}, pp. 103-113. *In* M. S. Doty [ed.], Proceedings of the conference on primary productivity measurement, marine and freshwater. U. S. Atomic Energy Commission, TID-7633.

— 1964. Primary productivity and micronutrient limiting factors in some North American and New Zealand lakes. - Verh. Int. Ver. Limnol. *15*: 365-374.

— Integration of field and laboratory experiments in productivity studies. - Proceedings Conference on Estuaries, Jekyll Island, Georgia, March 31-April 4, 1964 (*in press*).

— and D. T. Mason. 1962. Inorganic precipitation of carbon in productivity experiments utilizing C^{14}. - Science *136*:1049-1050.

— and R. G. Wetzel. 1963. A study of the primary productivity of Clear Lake, Lake County, California. - Ecology *44*: 283-294.

— D. T. Mason, and B. J. B. Wood. 1963. Light injury and inhibition in Antarctic freshwater phytoplankton. - Limnol. & Oceanog. *8*: 313-322.

Holm-Hansen, O., G. C. Gerloff, and F. Skoog. 1954. Cobalt as an essential element for blue-green algae. - Physiol. Plant. 7: 665-675.

Hopkins, E. F. 1930. The necessity and function of manganese in the growth of *Chlorella* sp. - Science *72*: 609-610.
Hutchinson, G. E. 1957. A treatise on Limnology. Vol. I. - John Wiley and Sons, New York. 1015 pp.
Lund, J. W. G. 1964. Primary production and periodicity of phytoplankton. - Verh. Int. Ver. Limnol. *15*:37-56.
— 1965. The ecology of the freshwater phytoplankton. - Biol. Rev. *40*:231-293.
— and J. F. Talling. 1957. Botanical Limnological methods with special reference to the algae. Bot. Rev. *23*: 489-583.
MacPhee, C. 1961. Bioassay of algal production in chemically altered waters. Limnol. & Oceanog. *6*:416-422.
Menzel, D. W. and J. H. Ryther. 1961. Nutrients limiting production of phytoplankton in the Sargasso Sea with special reference to iron. - Deep Sea Res. 7:276-281.
Nicholas, D. J. D. 1952. The use of fungi for determining trace metals in biological materials. - Analyst 77:629-642.
— 1963. Inorganic nutrient nutrition of micro-organisms. *In* F. C. Steward [ed.], Plant Physiology III. - Academic Press, New York and London: 363-447.
Potash, M. 1956. A biological test for determining the potential productivity of water. - Ecology *37*:631-639.
Provasoli, L. and I. S. Pintner. 1953. Ecological implications of *in vitro* nutritional requirements of algal flagellates. - Ann. N. Y. Acad. Sci. *56*:839-851.
— J. J. A. McLaughlin, and M. R. Droop. 1957. The development of artificial media for marine algae. - Arch. Mikrobiol. *25*:393-428.
Raulins, J. 1869. Etudes chimiques sur la vegetation. - Ann. Sci. Nat. V. Botan. *11*: 92-299.
Rodhe, W. 1948. Envinmental requirements of freshwater plankton algae. - Symb. Bot. Upsaliens. *10*: 149 pp.
Ryther, J. H. and R. R.L. Guillard. 1959. Enrichment experiments as a means of studying nutrients limiting to phytoplankton production. - Deep Sea Res. *6*: 65-69.
Schelske, C. L. 1962. Iron, organic matter, and other factors limiting primary productivity in a marl lake. - Science *136*: 45-46.
Sorokin, J. I. 1964. On the trophic role of chemosynthesis in lake bodies. - Int. Revue ges. Hydrobiol. *49*: 307-324.
Ström, K. M. 1933. Nutrition of algae. Experiments upon: the feasibility of the Schreiber method in fresh waters; the relative importance of iron and manganese in the nutritive medium; the nutritive substance given off by lake bottom muds. - Arch. Hydrobiol. *25*: 38-47.
Talling, J. F. 1961. Photosynthesis under natural conditions. - Ann. Rev. Plant Physiol. *12*: 133-154.
Tanter, D. J. and B. S. Newell. 1963. Enrichment experiments in the Indian Ocean. - Deep Sea Res. *10*: 1-9.
Vollenweider, R. A. 1950. Okologische Untersuchungen von planktischen Algen auf experimenteller Grundlage. - Schweiz. Z. Hydrol. *12*: 194-262.
Walker, J. B. 1953. Inorganic micronutrient requirements of *Chlorella*. I. Requirements for calcium (or strontium), copper and molybdenum. - Arch. Biochem. Biophys. *46*:1-11.
Wetzel, R. G. 1964. A comparative study of the primary productivity of higher aquatic plants, periphyton, and phytoplankton in a large, shallow lake. - Int. Rev. ges. Hydrobiol. *49*: 1-61.
— 1965. Nutritional aspects of algal productivity in marl lakes with particular reference to enrichment bioassays and their interpretation, - Mem. Ist. Ital. Idrobiol., 18 Suppl.: 137-157.
Wiessner, W. 1962. Inorganic micronutrients. *In* R. A. Lewin [ed.], Physiology and Biochemistry of Algae. - Academic Press, New York and London, pp. 267-286.
Wright, R. T. and J. E. Hobbie. 1965. The uptake of organic solutes in lake water. - Limnol. & Oceanog. *10*:22-28.

NUTRITIONAL ASPECTS OF ALGAL PRODUCTIVITY IN MARL LAKES WITH PARTICULAR REFERENCE TO ENRICHMENT BIOASSAYS AND THEIR INTERPRETATION [1]

ROBERT G. WETZEL [2]

Department of Zoology
Indiana University, Bloomington, Indiana

[1] Contribution No. 771, Department of Zoology, Indiana University, Bloomington, Indiana

[2] Present address: Department of Botany and Plant Pathology, W. K. Kellogg Biological Station, Michigan State University, Hickory Corners, Michigan

For bibliographic citation of this paper, see page 10.

Abstract

A brief summary is presented of some of the results of investigations of the causal relationships between nutrient factors in alkaline, highly calcareous marl lakes and low levels of growth of algal populations. By application of carbon-14 growth factor enrichment bioassays on simultaneous cultures of natural phytoplankton populations of several lakes, analyses were made of the role of organic and inorganic nutrients influencing seasonal fluctuations of rates of production. The low levels of algal productivity and associated low quantities of dissolved organic matter in extremely calcareous marl lakes are interrelated in a circular causal system that perpetuates low rates of growth. This situation is compounded by physiological unavailability of several inorganic nutrients present in insoluble forms for a large portion of the year.

The advantages and the problematic areas of the use of carbon-14 enrichment bioassays with natural populations of phytoplankton are emphasized. The sensitivity to growth responses, and the elimination of a lag in growth response, make the bioassay techniques ideal for the elucidation of the role of inorganic nutrients, as well as accessory organic growth factors, in regulation of population metabolism. The influence of minor elements is not always direct, and caution is necessary in interpretation of certain stimulatory responses.

Of the approximately 1,000 natural lakes in northern Indiana larger than two hectares, approximately 15 per cent are in the very generalized category of marl lakes. Such waters are characterized by high bicarbonate alkalinity and calcium concentrations, high pH, and marked deposits of calcareous materials. Productivity at all levels of organisms is moderate to extremely low. The marl lakes of Indiana represent waters nearest to what one might consider oligotrophic. For the past two years my interests have centered on the causal relationships between the low levels of growth of the algal populations and the nutrient factors of marl lakes.

In the brief presentation of some of the results of the investigations on algal productivity in marl lakes, some of the difficulties one encounters in the interpretation of C^{14} enrichment bioassays will be pointed out. Interpretation of bioassays is often difficult in waters of the chemical composition of marl lakes and where the growth responses of the phytoplankton are moderate in comparison to extremely infertile waters such as the Sargasso Sea.

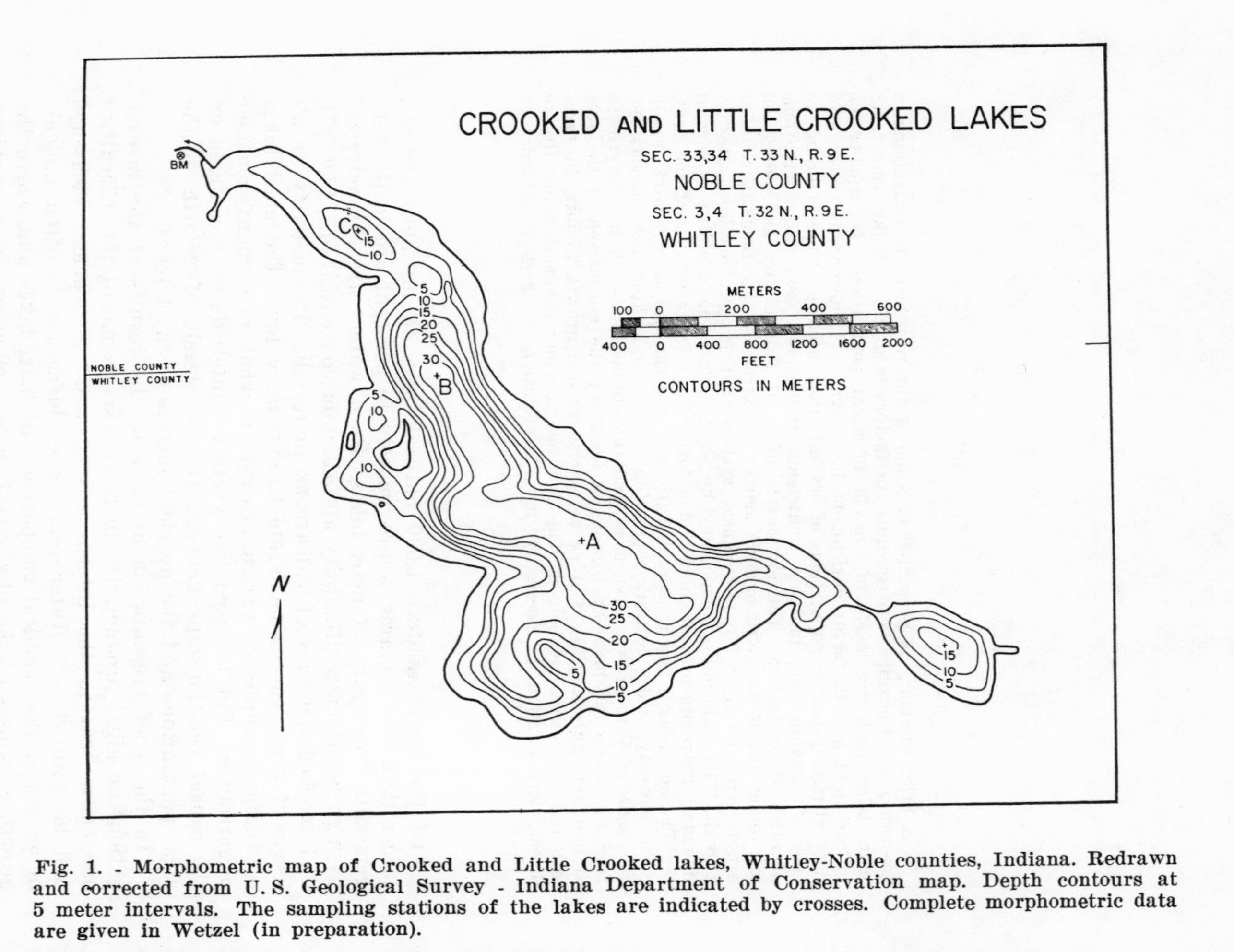

Fig. 1. - Morphometric map of Crooked and Little Crooked lakes, Whitley-Noble counties, Indiana. Redrawn and corrected from U. S. Geological Survey - Indiana Department of Conservation map. Depth contours at 5 meter intervals. The sampling stations of the lakes are indicated by crosses. Complete morphometric data are given in Wetzel (in preparation).

Of the spectrum of 13 lakes that were intensively investigated, mostly infertile marl lakes but some eutrophic in nature, two were found to be of particular interest. Crooked Lake is a large, relatively deep, extreme marl lake (Fig. 1), and exhibits low rates of productivity at all trophic levels. Connected to Crooked Lake by a few meters of channel is Little Crooked Lake, considerably smaller in size, of much greater productivity, but of quite similar water chemistry. As one would expect, there are morphometric and edaphic factors that influence the productivities of the two lakes to a considerable extent.

The primary productivity of Crooked Lake is moderately low, with relatively minor fluctuations on a seasonal basis (Fig. 2). The annual mean productivities were fairly similar during the two years, 469.2 and 358.7 $mgC/m^2/day$ in 1963 and 1964, respectively. The productivity of Little Crooked Lake per square meter of water exhibited much more violent, rapid fluctuations (Fig. 3) than did Crooked Lake. In general, although not completely, there was synchrony between the two lakes. The smaller lake exhibited very similar annual mean values of productivity in 1963 and 1964, 618.1 and 598.2 $mgC/m^2/day$ respectively. The trophogenic zone of Crooked Lake varied between the surface and 15-20 meters and in Little Crooked Lake between 5 and 7 meters. Moreover, in Little Crooked Lake a majority of the primary productivity occurred in three meters of the metalimnion, resulting in a considerable metalimnetic oxygen maximum. While there are numerous similarities as well as many dissimilarities, suffice it to say at this point that Crooked and Little Crooked lakes provided an excellent situation to compare nutritional responses experimentally by simultaneous culturing techniques. Complete details of the C^{14} methodology, auto- and heterotrophic algal productivity, and limnology of Crooked and Little Crooked Lakes are in preparation (Wetzel, MS).

One could attack the problem as has been done many times in the past, namely, by determining as many physical and chemical parameters as possible and seeking correlation between these fluctuating parameters and oscillations in growth of the composite photosynthetic populations. A few examples of some of these parameters are interjected here to demonstrate the limits to which such correlation may be taken.

The seasonal distribution of soluble iron is complex and varies rapidly in the epilimnetic waters of Crooked Lake (Fig. 4). Certain periods of minimal concentrations of iron, as in early June, correlate nicely with minimal periods of growth. Some minimal occurrences of iron, for example in early August, are difficult to relate to rapid fluctuations in growth. The difficulties encountered in this and other examples in attempts at correlation indicate that the relationship is not simply related to a single dominant influencing factor. Moreover, instantaneous concentrations may indicate actual influencing quan-

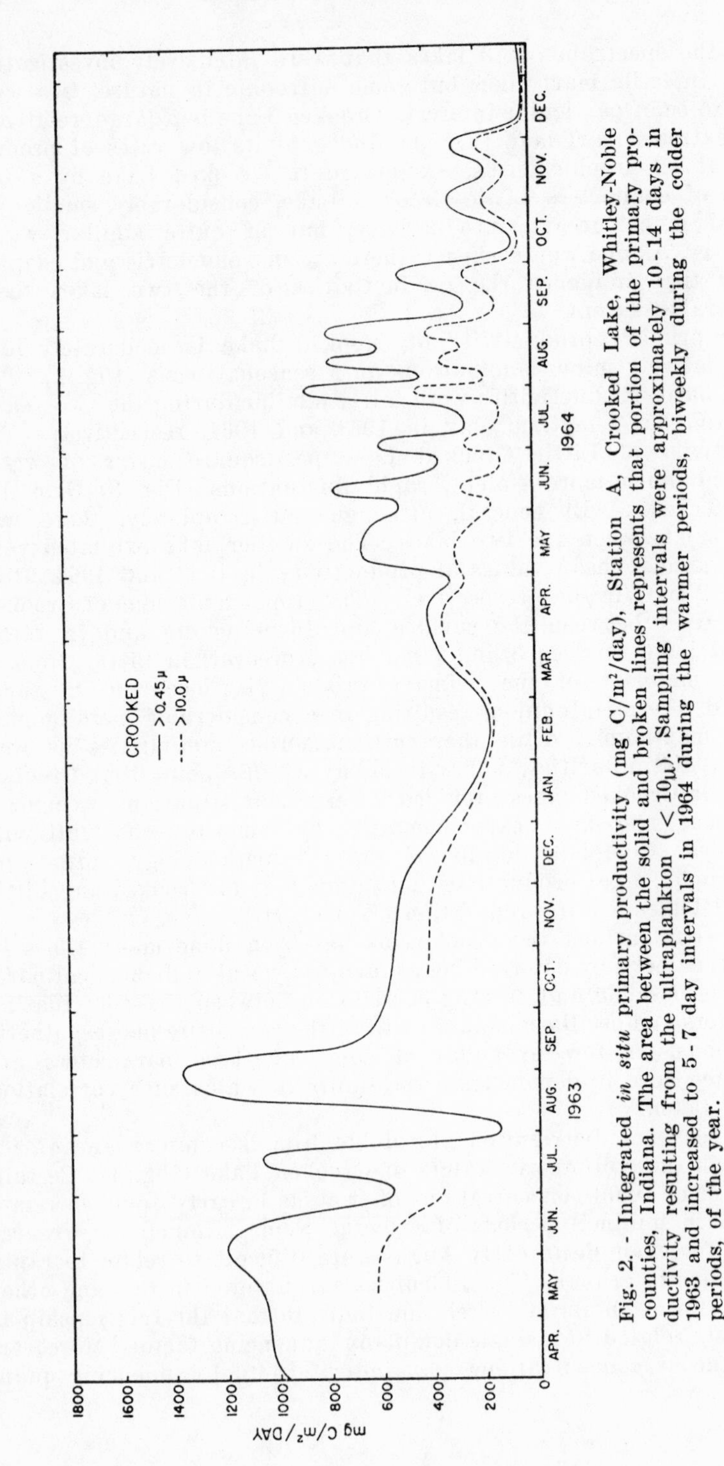

Fig. 2 - Integrated *in situ* primary productivity (mg C/m²/day), Station A, Crooked Lake, Whitley-Noble counties, Indiana. The area between the solid and broken lines represents that portion of the primary productivity resulting from the ultraplankton ($< 10\mu$). Sampling intervals were approximately 10-14 days in 1963 and increased to 5-7 day intervals in 1964 during the warmer periods, biweekly during the colder periods, of the year.

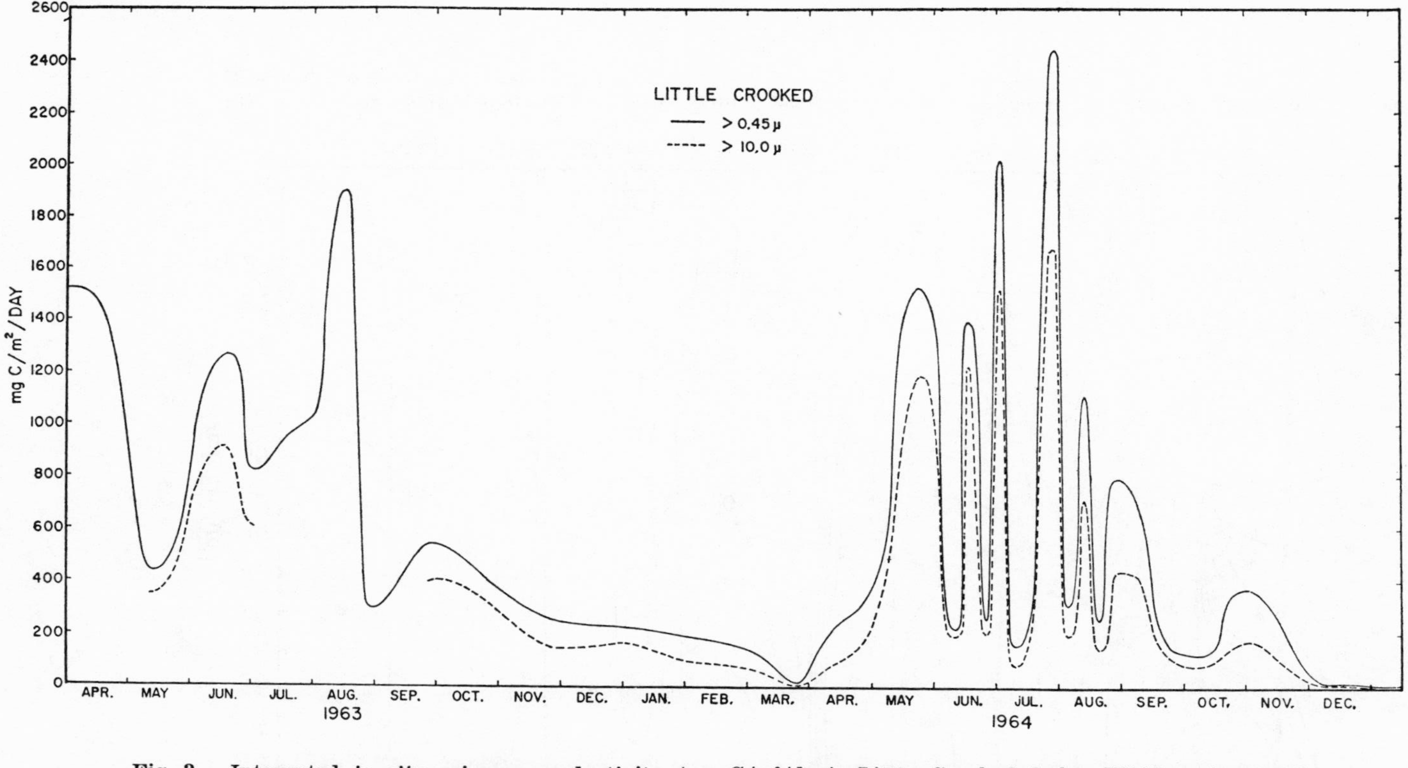

Fig. 3. - Integrated *in situ* primary productivity (mg C/m^2/day), Little Crooked Lake, Whitley County, Indiana. The area between the solid and broken lines represents that portion of the primary productivity resulting from the ultraplankton ($< 10\mu$). Sampling intervals were approximately 10 - 14 days in 1963 and increased to 5 - 7 day intervals in 1964 during the warmer periods, biweekly during the colder periods, of the year.

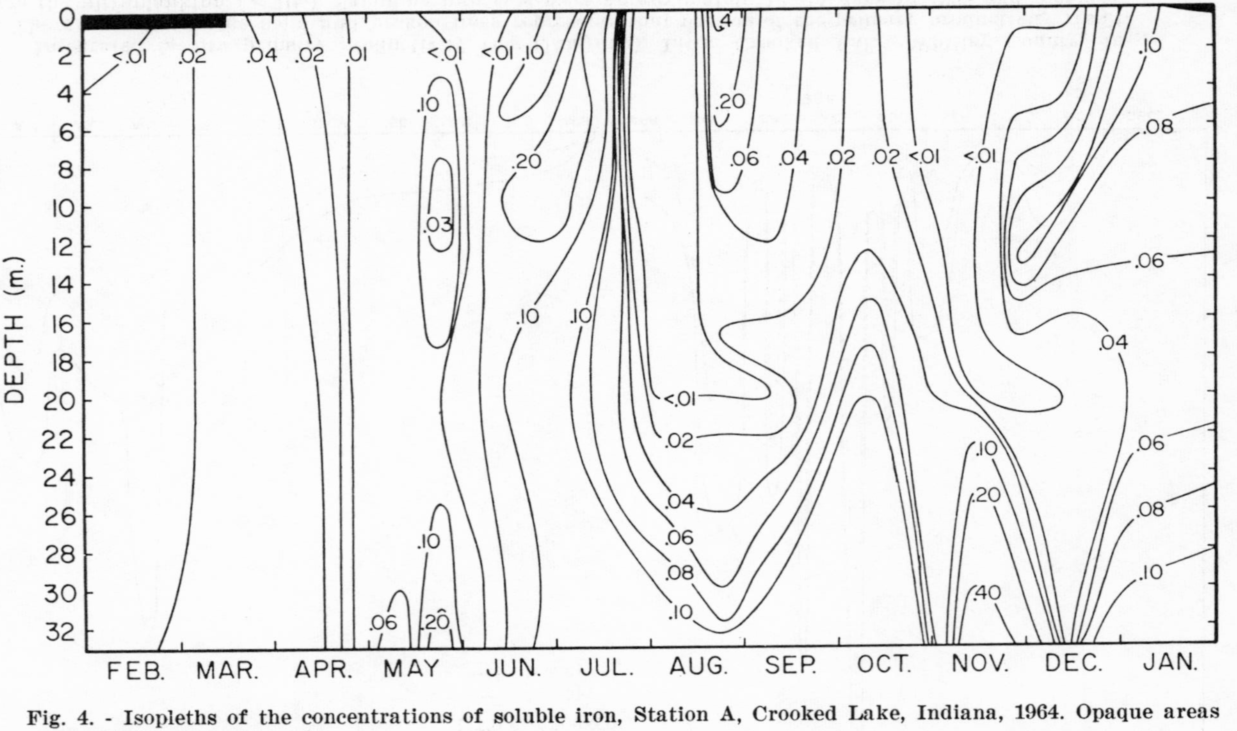

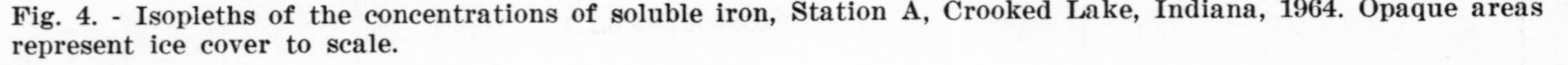
Fig. 4. - Isopleths of the concentrations of soluble iron, Station A, Crooked Lake, Indiana, 1964. Opaque areas represent ice cover to scale.

tities of a particular limiting factor or be the resultant product of the regeneration from a declining population peak.

Seasonal distribution of total phosphate (Fig. 5) can be related to seasonal pulses in productivity. Peaks in early July and low growth in early September fit the corresponding levels of phosphorus well. Again, the relationship is weak at other periods of the year. Problems of rapid turnover and re-utilization, and in some cases storage, of materials such as phosphorus negate the value of phosphorus determinations to a large extent in seeking such relationships. The distribution of manganese (Fig. 6) serves as an example of the situation where « normal » chemical techniques are not sufficiently sensitive for aerobic conditions and concentrations of the more productive zone of the lake. Using advanced spectrochemical techniques, concentrations in near-surface waters were found to be in the range of 10-20 μg/liter. It is unlikely that manganese concentrations seriously limit growth in this lake, as shown in enrichment experiments of the type discussed below. One could go on much further with this line of analysis, covering NO_3, other cations and anions, trace elements, numerous physical parameters (light, turbidity, temperature, stratification, etc.), and other chemical parameters. Even when one systematically analyses the relationships of some thirty parameters to detailed analyses of the vertical and seasonal growth and biomass distribution correlation exists only where many influential factors overlap. There is almost always extensive interaction, and many problems of interpretation. In brief, one still can only speculate, admittedly in an educated fashion, from such descriptive correlations. Seldom are distinct relationships between growth and a particular influencing factor as apparent, for example, as the diatom and silica periodicity of Lake Windermere (Lund 1964). Moreover, many of the potentially most critical factors, such as dissolved organic compounds, are largely ignored because of our present inability to determine their concentrations rapidly. In natural situations, limiting factors are not only relative but highly dynamic and changing in their effects upon growth.

It is of the utmost importance in turning to an experimental approach to work as closely as possible with natural plankton populations. This statement by no means implies that pure culture work is not important in these studies, for obviously it serves as a crucial basis for interpretation of results of the work with natural populations. But the distance is large between pure culture systems and the dynamic interactions of a heterogeneous plankton population. A great deal may be said for both approaches (Fogg 1965). Ideally, but rarely in fact, one combines both types of analyses.

The use of enrichment bioassays on phytoplankton populations has a long history but only recently have they achieved very extensive application. Experimental enrichment of pure cultures in filtered marine and lake water was perhaps first employed by Schreiber (1927)

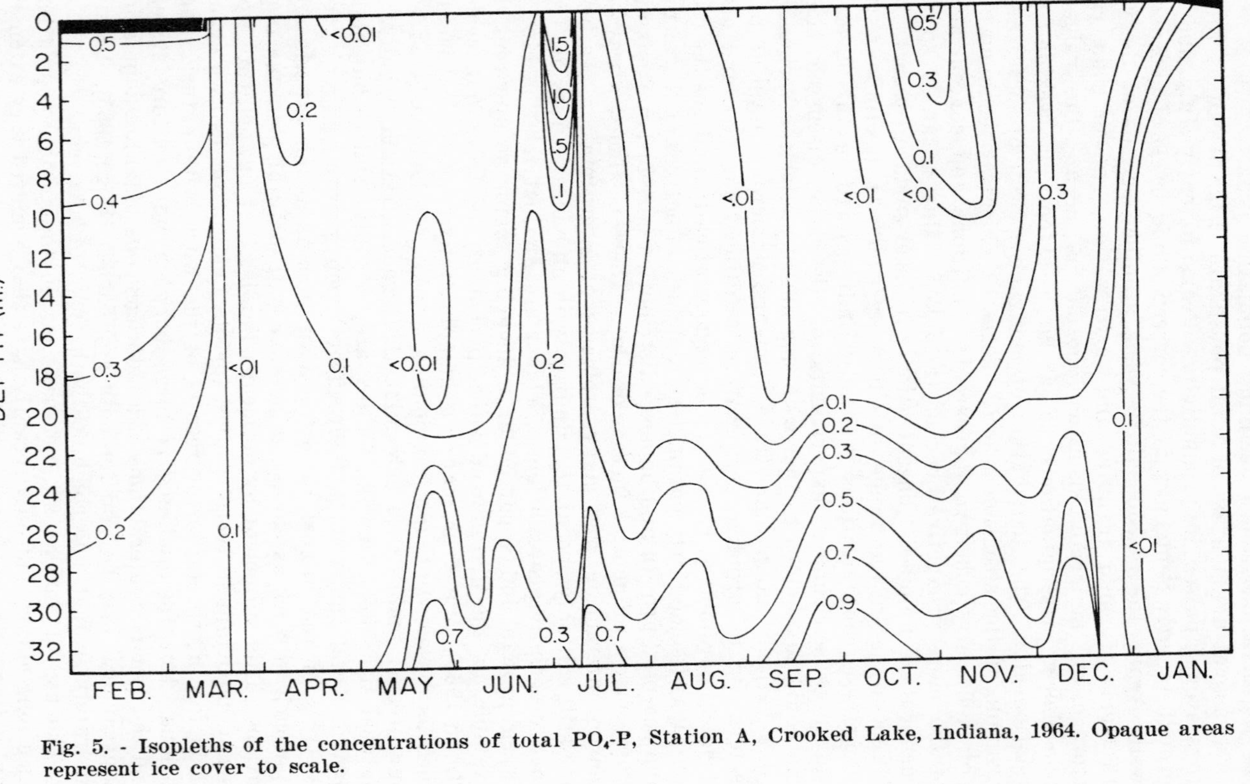

Fig. 5. - Isopleths of the concentrations of total PO_4-P, Station A, Crooked Lake, Indiana, 1964. Opaque areas represent ice cover to scale.

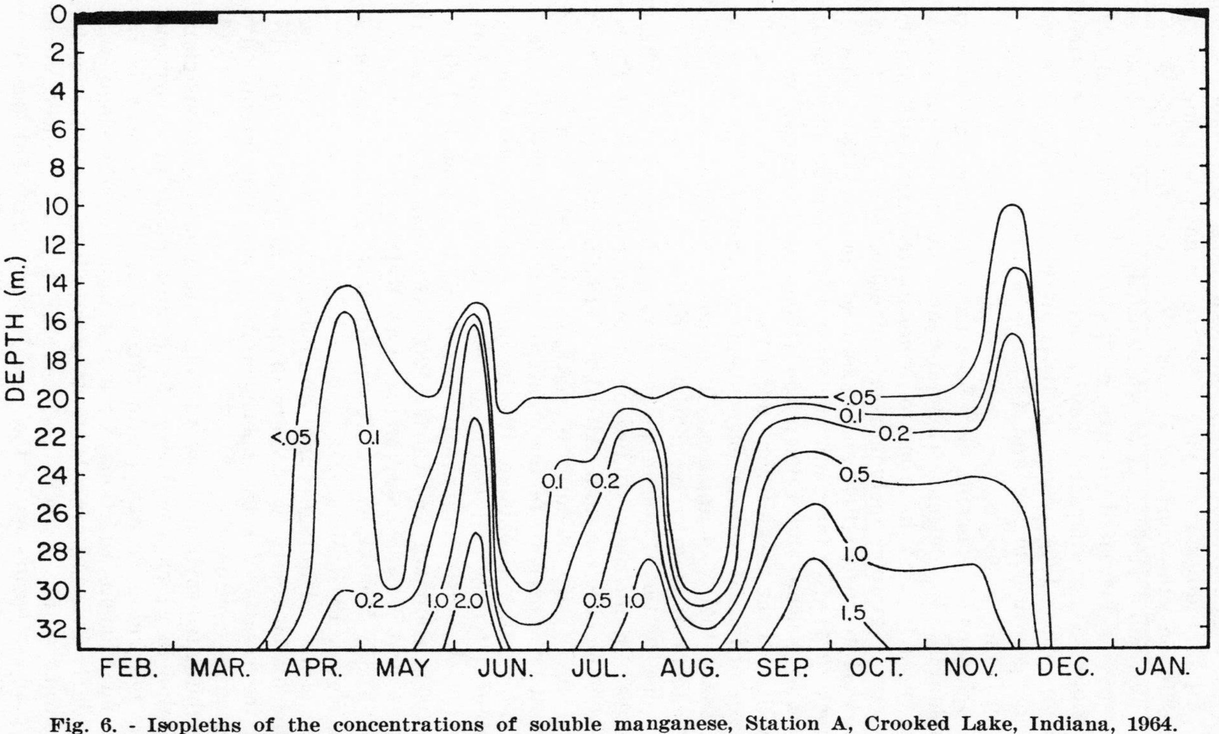

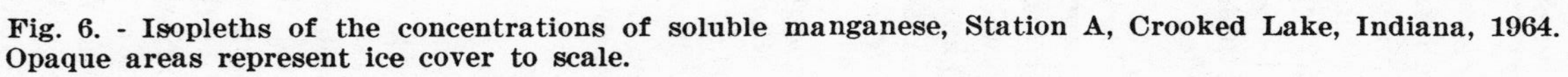
Fig. 6. - Isopleths of the concentrations of soluble manganese, Station A, Crooked Lake, Indiana, 1964. Opaque areas represent ice cover to scale.

and Strøm (1933), where increases in biomass were determined. In his classical studies of the nutritional requirements of certain planktonic algae, Rodhe (1948) extended these techniques considerably. Other studies employing lake water as a medium for pure cultures include the works of Eyster (1958), Potash (1956), MacPhee (1961), and Lund (1964), all of which used increases in biomass as a measure of increased growth after various lengthy periods of incubation. This approach has recently been used extensively (Johnston 1963, 1964; Smayda 1964) in marine waters in which both changes in biomass and C^{14} uptake have been employed to assay growth.

Enrichment experiments have been used for some time on natural phytoplankton populations. Generally long incubation periods were needed in early work in order to demonstrate significant differences in growth. Examples include assays of changes in pigment concentrations (Hutchinson 1941) and in numbers (Guseva 1938, Fish 1956, Štěpánek 1961) after inorganic enrichment. Edmonson and Edmonson (1947) and Nelson and Edmondson (1955) used changes in oxygen production as the assay technique in two brief studies on the effects of phosphorus and nitrate enrichment on growth. The need for a rapid assay tool was apparent in order to circumvent deleterious bottle effects of lengthy incubation periods.

The use of the C^{14} technique as a rapid bioassay tool was first introduced in nutrient studies of laboratory cultures of *Chlorella* by Steemann Nielsen and Al Kholy (1956). Application of this method to *in situ* cultures of natural phytoplankton populations for limiting nutrient factor analyses was developed in 1957 in freshwater lakes (Goldman 1960 a) and similarly used in marine situations (Ryther and Guillard 1959). Recently these techniques have been variously and widely used (Goldman 1960 b, 1961, 1963, 1964, 1965; Goldman and Carter 1965; Goldman and Wetzel 1963; Goldman and Mason 1962; Menzel and Ryther 1961; Tranter and Newell 1963; Menzel, *et al.* 1963; Schelske 1962; Johnston 1963, 1964; Wetzel 1964, MS).

Numerous advantages and disadvantages exist in the use of enrichment bioassays with natural plankton populations. In all cases, the results of such experimentation must be intepreted with considerable caution.

As indicated earlier, one is working with a heterogeneous plankton population in which a majority of the multiplicity of environmental factors governing growth are in effect. The C^{14} bioassay is rapid; if the enrichment additions fill a nutritional requirement that is particularly limiting growth of the populations, growth stimulation is apparent within a few hours (see Figs. 9 and 11, lower portions). No lag phase occurs as is the case in some pure cultures because of need for intra- and extracellular equilibrium of certain organic compounds (*e.g.* glycolate, Nalewajko, *et. al.*, 1963). In some instances the effect of an addition is not apparent immediately in that the element must first be modified by bacterial action into an organic compound required by the organisms. Cobalt is an example in its relationship to cobalamin. In Crooked Lake, where epilimnetic concentrations of cobalt are less than 2 μg/liter, small concentrations exert increasingly inhibitory effects (Fig. 7). In Little Crooked Lake, where natural concentrations are similar, inhibitory responses were not evident, possibly due to some complexing mechanism with greater concentrations of organic matter. Vitamin B-12, on the other hand, consistently resulted in rapid and marked growth responses in Crooked Lake and only slight or negligible stimulation in the simultaneous cultures in Little Crooked Lake (Fig. 8).

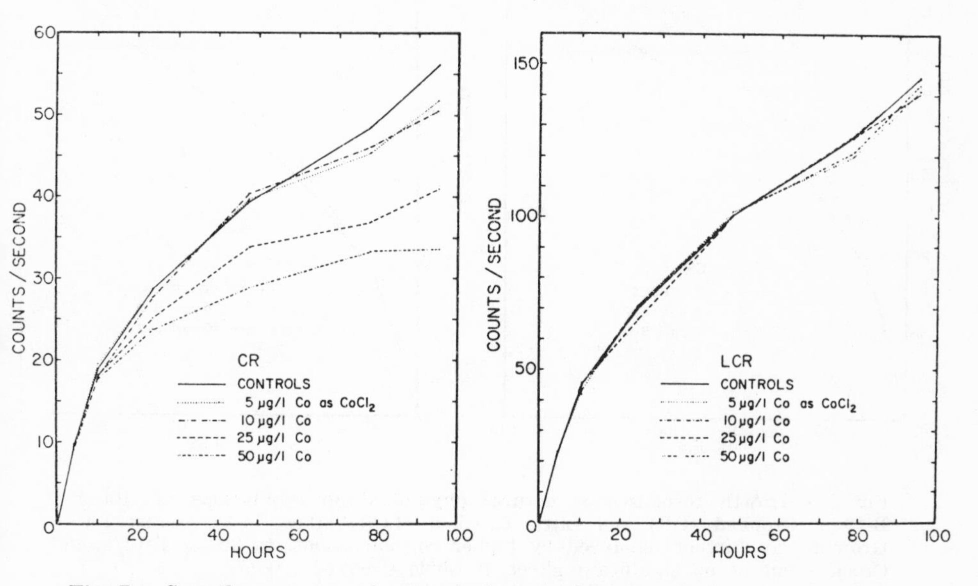

Fig. 7. - Growth responses of natural phytoplankton populations to micro-concentrations of cobalt as $CoCl_2$ in Crooked (CR) and Little Crooked (LCR) lakes, 3-6 August 1964.

Enrichment bioassays permit the experimental application of synthetic compounds of known physical and physiological function as well as natural compounds of known nutritive value but for which natural concentrations are unknown and determined with difficulty. Iron alone frequently results in large stimulation of growth in the extreme marl water Crooked Lake (Fig. 9). This response is doubled by the addition of a chelating agent which effectively prevents reactions of iron in the alkaline water which precipitate it in insoluble forms. Growth responses to iron in the more productive Little Crooked Lake were considerably less than in Crooked and often negligible. The addition of a chelating agent in combination with iron was of no additional value to growth in Little Crooked water. It is likely that the greater concentrations of natural organic matter were already functioning in a chelation capacity in the more productive lake.

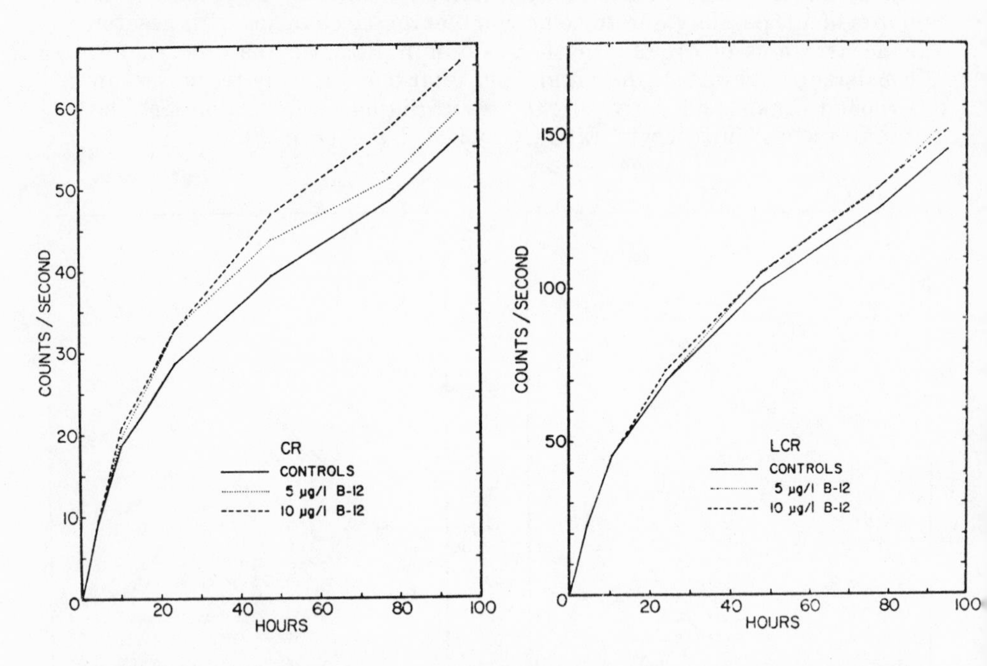

Fig. 8. - Growth responses of natural phytoplankton populations to vitamin B-12 in Crooked (CR) and Little Crooked (LCR) lakes, 14-18 August 1964. Growth was further enhanced by higher concentrations, to 20 µg B-12/l., in Crooked but of no significant effect in Little Crooked Lake.

Problems, and areas requiring caution, in the interpretation of the results of enrichment bioassays are many. Rapid exhaustion of critical nutrients is probably circumvented in most cases in which short incubation periods are used. Relative abundance of limiting factors (Verduin 1952), and shifting from the threshold level of one controlling factor to another, are other problems that are difficult to determine in heterogeneous cultures. Inorganic precipitation of C^{14} by some compounds, *e.g.*, iron, is a source of possible error and can be accounted for by prefiltered controls (Ryther and Guillard 1959, Goldman and Mason 1962), and by cell enumeration (Menzel, *et al.* 1963).

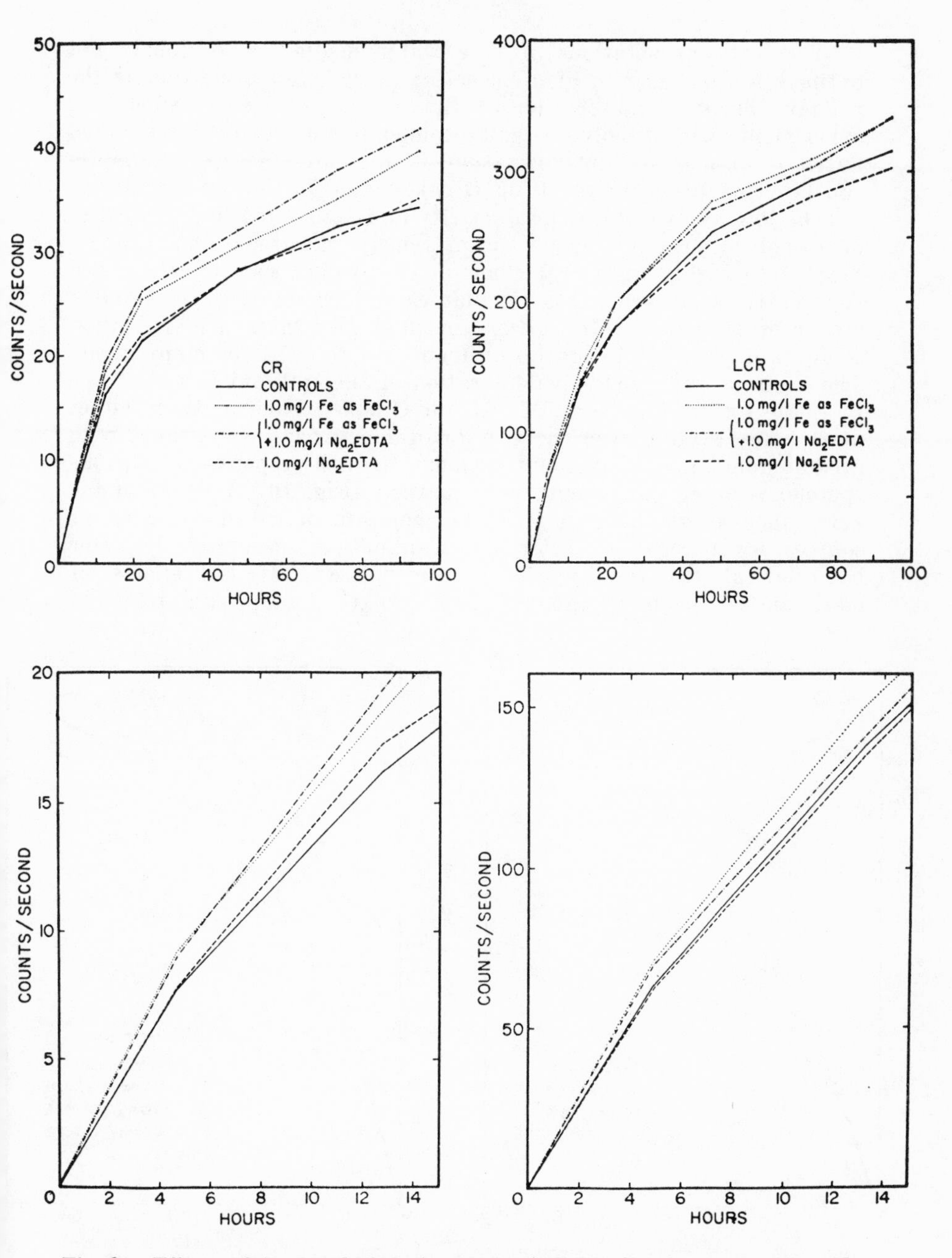

Fig. 9. - Effects of iron and chelating agent (ethylenediaminetetraacetic acid) on growth of natural phytoplankton of Crooked (CR) and Little Crooked (LCR) lakes, 30 July - 3 August 1964. Lower figures are expanded portions of the upper figures during the initial period of the incubation. Activity of prefiltered controls subtracted from those of the experimental.

The nutrient solutions of lake water media are obviously not optimal for the growth of all species; indeed, this condition is the primary basis of application of the bioassays. Hence, when one experiments with dissolved organic compounds of several interrelated nutritive properties, interpretation is difficult. Several examples can be cited from the work on Crooked and Little Crooked lakes.

The low levels of algal productivity and associated low quantities of dissolved organic matter in extremely calcareous marl lakes, coupled to highly unfavorable mono- to divalent cation ratios, are apparently interrelated in a circular causal system that perpetuates low rates of growth. In culture, optimal growth is achieved with mono- to divalent cation ratios of about 2 : 1 to 30 : 1 for many algae. The mean annual mono-divalent ratios of the epilimnion were 1 : 5.8 in Little Crooked Lake and 1 : 8.5 for Crooked Lake in 1964. Other marl lakes exhibit even more unfavorable growth conditions with greater dominance of divalent cations. Growth responses to various complex-forming compounds were marked (Fig. 10). Certain amino acids such as glycine can serve as chelators of cations as well as sources of nitrogen, directly by deamination or indirectly after bacterial alteration. By simultaneous assays of the effects of inorganic nitrogen it can be shown that at approximately the

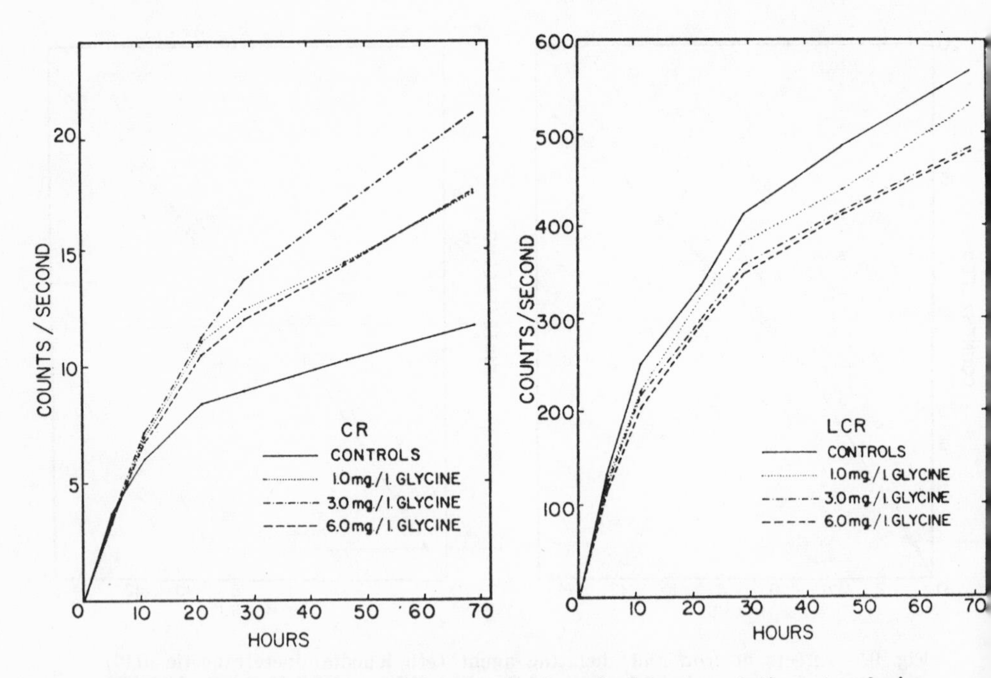

Fig. 10. - Growth responses of natural phytoplankton populations to glycine additions in Crooked Lake (CR) and Little Crooked Lake (LCR), 23-26 July 1964.

1 mg/liter level, glycine is serving primarily as a source of nitrogen and effects above this are likely related to complexing of divalent cations. At high concentrations, *e.g.*, 6 mg/liter, an inhibitory response begins to appear. Where additions of inorganic nitrogen have no effect, as in Little Crooked, inhibitory responses were consistently found with glycine and other amino acids, such as alanine, possibly the result of selective complexing of other organic compounds and especially trace elements.

Hydrolyzates of casein and other proteins, such as peptone and tryptone, contain mixtures of amino acids, some of which can be utilized by certain algae (Miller and Fogg 1958; Fogg and Miller 1958). Tryptone provides a relatively poor source of nitrogen and a good one of phosphorus, but it has been shown (Fogg and Miller 1958) that a primary effect is the complexing of cations (Fogg and Westlake 1955) and production of a more favorable monovalent: divalent cation ratio. Very large and rapid increases in growth were observed on addition of tryptone to cultures of natural plankton (Fig. 11). While some stimulatory response was evident on the additions of nitrogen, phosphorus, and other compounds, that of tryptone always greatly exceeded that of the others. Stimulatory growth responses also occurred to a lesser extent in simultaneous cultures from Little Crooked Lake where additions of nitrogen, phosphorus, and other compounds had little significant effect and natural concentrations of dissolved organic matter and extracellular compounds are considerably higher than in Crooked Lake (Wetzel MS). Responses to citrate additions, another good chelator of calcium, were significant but did not occur to the degree that was found with other organic compounds and indicate that much of the stimulatory effect of the latter were due to sources of nitrogen and phosphorus. The response to citrate was again greater in Crooked Lake than in Little Crooked.

The algal concentrations of pure culture conditions differ enormously from most natural population densities. Moreover, the complex and dynamic chemical composition of natural water solutions often interacts with enrichment additions to the point where effective concentrations may differ markedly from actual. The need of greater concentrations than are known to be required from pure culture conditions is frequently observed.

The initial growth response to a particular limiting factor can differ greatly from that of a few days later. The experiments of Menzel, *et al.* (1963), related both to shifts in nutrient concentrations and rapid changes in species composition, demonstrated dramatically the interaction between N, P, Fe, Si and species alterations in the progression of 9-day experiments. But of perhaps greater significance was the suggestion that some limiting factors, such as iron, can function less as a nutritional requirement than in a catalysing capacity in the utilization of P and N for growth.

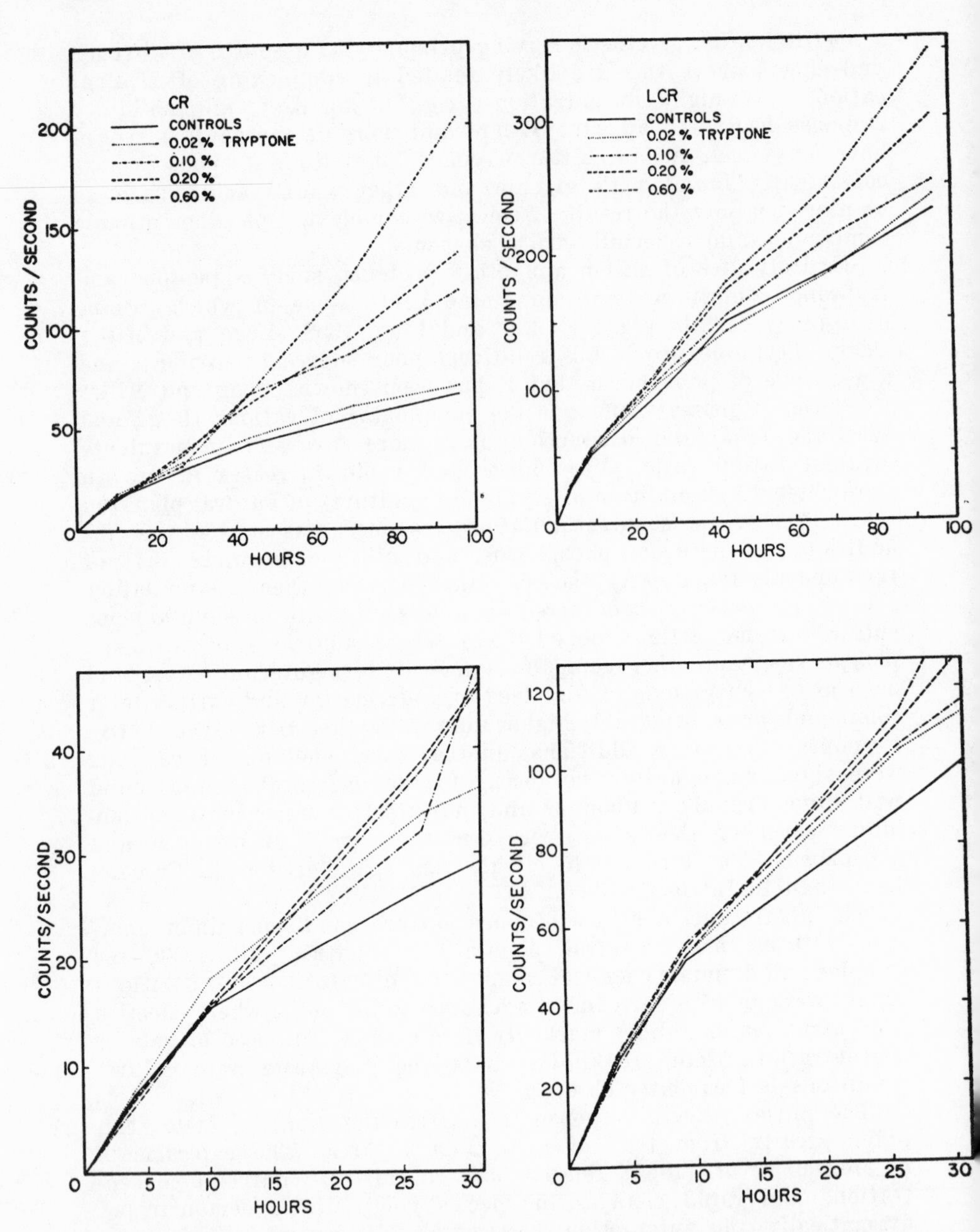

Fig. 11. - Effects of tryptone on growth of natural phytoplankton populations of Crooked Lake (CR) and Little Crooked Lake (LCR), 9-13 July 1964. Lower figures are expanded portions of the upper figures during the initial period of incubation.

Stimulatory responses such as have been observed with small additions of iron in combination with N and P were similarly observed with aluminum which has no known metabolic function. Also to be pointed out from their investigations in the Sargasso Sea is the slow response to additions for several days, which is related to shifts in species composition in cultures of lengthy duration.

In situ incubation of cultures has advantages of natural temperatures, etc. over artificial simulation of conditions. Difficulties are encountered when attempting to simulate natural light, not only quantitatively but especially qualitatively. Darkness periodicity is known to be essential to certain algal forms. These difficulties have led Fogg (1965) to recommend *in situ* incubation in many cases in spite of the inconvenience. I have found that growth response in enrichment experiments is more rapid and consistent if incubation is done at one to two meters depth rather than at the surface in floating cribs, presumably because of the rapid attenuation of the photoinhibitory effects of high surface light.

In conclusion, I would like to mention briefly that enrichment bioassays of natural plankton populations are ideally adapted to the sensitive and rapid assay of the effects of inorganic and exotic organic toxic substances, other pollutional aspects, studies of lake eutrophication (*e.g.*, Goldman and Carter 1965), etc. The C^{14} techniques of Wetzel (1964) are also adaptable to nutritional studies in submerged macrophytes. The potential of enrichment bioassays as an effective tool is great, but interpretation must be made in cognizance of the numerous factors influencing growth that may be in effect simultaneously.

Acknowledgments

The financial assistance of the National Science Foundation (Grant GB-1452) and Federal Aid Projects (F-7-R-2, -3, and -4) is gratefully acknowledged. Appreciation is also extended to Drs. D. G. Frey and S. D. Gerking, Indiana University, and Dr. C. R. Goldman, University of California, Davis, for numerous periods of fruitful discussion. Prof. G. E. Fogg, Westfield College, London, kindly read the manuscript, and offered helpful comments.

REFERENCES

Edmondson, W. T. and Y. H. Edmondson. 1947. Measurements of production in fertilized salt water. - J. Mar. Res. *6*: 228-246.

Eyster, C. 1958. Bioassay of water from a concretion-forming marl lake. - Limnol. Oceanog. *3*: 455-458.

Fish, G. R. 1956. Chemical factors limiting growth of phytoplankton in Lake Victoria. - East African Agric. J. *21*: 152-158.

Fogg, G. E. 1965. Algal cultures and phytoplankton ecology. - Madison, Univ. of Wisconsin Press. 126 pp.

— and J. D. A. Miller. 1958. The effect of organic substances on the growth of the freshwater alga *Monodus subterraneous*. - Verh. int. Ver. Limnol. *13* : 892-895.
— and D. F. Westlake. 1955. The importance of extracellular products of algae in freshwater. - Verh. int. Ver. Limnol. *12* : 219-232.
Goldman, C. R. 1960 a. Primary productivity and limiting factors in three lakes of the Alaska Peninsula. - Ecol. Monogr. *30* : 207-230.
— 1960 b. Molybdenum as a factor limiting primary productivity in Castle Lake, California. - Science. *132* : 1012-1013.
— 1961. The contribution of alder trees (*Alnus tenuifolia*) to the primary productivity of Castle Lake, California. - Ecology. *42* : 282-288.
— 1963. The measurement of primary productivity and limiting factors in fresh water with carbon-14. *In*: Proc. Conf. on Primary Productivity Measurement (1961). - U.S.A.E.C. TID-7633. pp. 103-113.
— 1964. Primary productivity and micro-nutrient limiting factors in some North American and New Zealand lakes. - Verh. int. Ver. Limnol. *15* : 365-374.
— 1965. Micronutrient limiting factors in natural phytoplankton populations. - Mem. Ist. Ital. Idrobiol. 18 Suppl. : 121-136.
— and R. C. Carter. 1965. An investigation by rapid carbon-14 bioassay of factors affecting the cultural eutrophication of Lake Tahoe. - J. Water Poll. Control Fed. *37* : 1044-1059.
— and D. T. Mason. 1962. Inorganic precipitation of carbon in productivity experiments utilizing carbon-14. - Science. *136* : 1049-1050.
— and R. G. Wetzel. 1963. A study of the primary productivity of Clear Lake, Lake County, California. - Ecology. *44* : 283-294.
Guseva, K. A. 1938. Gidrobiologicheskaia proizvoditelnost i prognoz tsveteniia vodoemov. (Hydrobiologic productivity and the prognoses of blooms of reservoirs). - Mikrobiologiia. *7* : 303-315. (Russian; English summary).
Hutchinson, G. E. 1941. Limnological studies in Connecticut. IV. The mechanisms of intermediary metabolism in stratified lakes. - Ecol. Monogr. *11* : 21-60.
Johnston, R. 1963. Sea water, the natural medium of phytoplankton. I. General features. - J. Mar. Biol. Ass. U. K. *43* : 427-456.
— 1964. Sea water, the natural medium of phytoplankton. II. Trace metals and chelation, and general discussion. - J. Mar. Biol. Ass. U. K. *44* : 87-109.
Lund, J. W. G. 1964. Primary production and periodicity of phytoplankton. - Verh. int. Ver. Limnol. *15* : 37-56.
MacPhee, C. 1961. Bioassay of algal production in chemically altered waters. - Limnol. Oceanog. *6* : 416-422.
Menzel, D. W. and J. H. Ryther. 1961. Nutrients limiting the production of phytoplankton in the Sargasso Sea, with special reference to iron. - Deep-Sea Res. *7* : 276-281.
— E. M. Hulburt, and J. H. Ryther. 1963. The effects of enriching Sargasso Sea water on the production and species composition of the phytoplankton. - Deep-Sea Res. *10* : 209-219.
Miller, J. D. A. and G. E. Fogg. 1958. Studies on the growth of Xanthophyceae in pure culture. II. The relation of *Monodus subterraneus* to organic substances. - Arch. Mikrobiol. *30* : 1-16.
Nalewajko, C., N. Chowdhuri, and G. E. Fogg. 1963. Excretion of glycollic acid and the growth of a planktonic *Chlorella*. *In*: Studies on Microalgae and Photosynthetic Bacteria (Tokyo). pp. 171-183.
Nelson, P. R. and W. T. Edmondson. 1955. Limnological effects of fertilizing Bare Lake, Alaska. - U.S.F.W.S., Fish. Bull. *56* (102) : 415-436.
Potash, M. 1956. A biological test for determining the potential productivity of water. - Ecology. *37* : 631-639.
Rodhe, W. 1948. Environmental requirements of fresh-water plankton algae. Experimental studies in the ecology of phytoplankton. - Symb. Bot. Upsal. *10*(1). 149 pp.

Ryther, J. H. and R. R. L. Guillard. 1959. Enrichment experiments as a means of studying nutrients limiting to phytoplankton production. - Deep-Sea Res. *6* : 65-69.

Schelske, C. L. 1962. Iron, organic matter, and other factors limiting primary productivity in a marl lake. - Science. *136* : 45-46.

Schreiber, W. 1927. Die Reinkultur von marinem Phytoplankton und deren Bedeutung für die Erforschung der Produktionsfähigkeit des Meerwassers. - Wissensch. Meeresunters. N. F. Abt. Helgol. *16* (10) : 1-34.

Smayda, T. J. 1964. Enrichment experiments using the marine centric diatom *Cyclotella nana* (Clone 13-1) as an assay organism. - Occas. Publ. Narragansett Mar. Lab. *2* : 25-32.

Steemann Nielsen, E. and A. A. Al Kholy. 1956. Use of C^{14}-technique in measuring photosynthesis of phosphorus or nitrogen deficient algae. - Physiol. Plant. *9* : 114-153.

Štěpánek, M. 1961. Limnologická studie o Nádrži Sedlice u Želiva. XV. Praktická modifikace Potashova biologického testu na stanovení potenciální produktivity vody. (Limnological study of the Reservoir Sedlice near Zeliv. XV. Practical modification of the Potash's biological test for the estimation of the potential productivity of water). - Sci. Pap. Inst. Chem. Technol., Prague, Technol. Water. 5 : 227-239. (Czech.; English and Russian summaries).

Strøm, K. M. 1933. Nutrition of algae. Experiments upon: the feasibility of the Schreiber method in fresh waters; the relative importance of iron and manganese in the nutritive medium; the nutritive substance given off by lake bottom muds. - Arch. Hydrobiol. *25* : 38-47.

Tranter, D. J. and B. S. Newell. 1963. Enrichment experiments in the Indian Ocean. - Deep-Sea Res. *10* : 1-9.

Verduin, J. 1952. Limiting factors. - Science. *115* : 23.

Wetzel, R. G. 1964. A comparative study of the primary productivity of higher aquatic plants, periphyton, and phytoplankton in a large, shallow lake. Int. Rev. ges. Hydrobiol. *49* : 1-61.

— (MS.) Auto- and heterotrophic planktonic productivity and controlling nutrient factors in Crooked and Little Crooked marl lakes of northern Indiana. (In preparation).

III

PRODUCTION AND UTILIZATION OF ORGANIC SOLUTES BY BACTERIA AND PHYTOPLANKTON

DISSOLVED ORGANIC MATTER IN SURFACE WATER AS A PARAMETER FOR PRIMARY PRODUCTION

EGBERT KLAAS DUURSMA

Present address: I.A.E.A., Musée Océanographique, Monaco

For bibliographic citation of this paper, see page 10.

Abstract

Dissolved organic matter in the sea originates from living organisms, mainly the phytoplankton of the surface layers. The formation of a great bulk of new dissolved organic substances occurs some weeks after a phytoplankton bloom, from which it may be concluded that the substances were derived from decomposing detritus by bacteria or from digested detritus (plus living plankton) by zooplankton.

In relation to the primary production of a closed system the production of dissolved organic compounds can only be a fraction of this amount. The conversion factors from primary production, which is the first trophic level to the formation of dissolved organic matter, which is a later trophic level, are not known. Should they be of the order of 5 to 10%, this would mean that the primary production should be at least 10 to 20 times higher than the formation of dissolved organic constituents.

For one North Sea station Duursma (1963) came to a lower limit of 52 g $C/m^2/year$ for this formation, which should mean that the primary production should be at least 520-1040 g $C/m^2/year$ (conversion factor 10-5%). This last value may agree with the fact that the primary production at this station is higher than the known mean value of the North Sea, which is 58-135 g $C/m^2/year$ (Steemann Nielsen, personal communication).

However, in open ocean water throughout a year, about the same variation in dissolved organic matter content may occur as for the North Sea station and possibly in some thicker surface water layer. This may indicate the lower level of primary production if more was known about the value of the conversion factor.

REFERENCES

Duursma, E. K. 1963. The production of dissolved organic matter in the sea, as related to the primary gross production of organic matter. - Neth. J. Sea Res. *2*: 84-94.

THE KINETICS OF RELEASE OF EXTRACELLULAR PRODUCTS OF PHOTOSYNTHESIS BY PHYTOPLANKTON

G. E. FOGG and W. D. WATT
Department of Botany
Westfield College, London, N. W. 3.

For bibliographic citation of this paper, see page 10.

Abstract

The effects of light intensity, carbon dioxide concentration and other factors on the release of extracellular glycollate in cultures of *Chlorella pyrenoidosa* are described. Extracellular products may amount to a considerable proportion of the total carbon fixed in photosynthesis by phytoplankton in lakes, the proportion tending to increase as the population density decreases and becoming high when photosynthesis is inhibited by high light intensity. The kinetics of this excretion resemble those of the release of glycollate by *Chlorella* and there is evidence that glycollate is sometimes, at least, the major extracellular product of natural phytoplankton.

The release of products of photosynthesis from phytoplankton cells is of obvious significance for the interpretation of data from radiocarbon determinations of primary productivity, and it is important to understand the effects of environmental conditions and physiological state of the cells on the extent to which it occurs. In attempts to learn something of the kinetics of these relations we have used two different approaches:

1) Laboratory experiments with pure cultures, concentrating on glycollic acid as an extracellular product. We have confined our attention to this particular substance because, whereas other extracellular products which have been studied seem to be liberated irreversibly in relatively small amounts, glycollic acid appears to pass freely between cell and medium, sometimes appearing extracellularly in quite large amounts; it is possible, therefore, that glycollic acid is of greater quantitative importance under natural conditions. In these experiments glycollic acid has been determined colorimetrically by the Calkins method using 2: 7-dihydroxynaphthalene with precautions to avoid interference by nitrate and ammonium ions (Watt and Fogg, in press).

2) *In situ* experiments with natural phytoplankton in which the total amount of extracellular products of photosynthesis has been estimated from the amount of radiocarbon appearing in organic matter in filtrates from samples allowed to photosynthesize in the presence of C^{14}-bicarbonate in the usual manner (Fogg, Nalewajko and Watt 1965).

Our findings regarding the release of glycollic acid in laboratory studies of *Chlorella pyrenoidosa,* which are reported in detail elsewhere (Watt and Fogg, in press), suggest that the situation is complex, the amount being affected by light intensity, carbon dioxide concentration and cell concentration, as well as by the physiological state of the cells. During growth in culture of limited volume the concentration of glycollate in the filtrate increases but the pattern is different according to whether carbon dioxide is limiting or saturating. In the former case the rate of production of glycollate per cell diminishes with time whereas in the latter case it increases. Short-term experiments, in which cells are transferred from one set of conditions to another, are perhaps most instructive. Glycollate release then occurs only when cells grown under low light intensity and high carbon dioxide concentration are transferred to conditions of high light intensity and relative low carbon dioxide concentration. The effects of such changes on the release of glycollate in a growing culture are illustrated in Fig. 1. No other combination of these factors results in any appreciable release of glycollate. Other characteristics of this short-term release are that the quantity of glycollate is greater the greater the relative growth rate of the cells used, and that the rate of glycollate release falls after 50 to 100 minutes, suggesting exhaustion of the glycollate precursor.

These findings are in agreement with the hypothesis, put forward on the basis of biochemical studies (Pritchard, Griffin and Whittingham 1962), that glycollate is derived from ribulose diphosphate, the carbon dioxide acceptor in the photosynthetic fixation cycle. Under low light intensity and high carbon dioxide concentration the concentration in the cells of ribulose diphosphate is evidently high; when the cells are transferred to high light intensity and low carbon dioxide concentration the excess is broken down to glycollate.

This situation seems to occur in other algae besides *Chlorella pyrenoidosa.* Hellebust (1965) detected glycollic acid by chromatography among other extracellular products in cultures of all 22 species of marine phytoplankton which he examined. Nalewajko (in press) has carried out experiments by the radiocarbon method on the short-term release of photosynthetically fixed carbon with cultures of 24 species of freshwater phytoplankton. The amount of extracellular product liberated was affected by light intensity and by carbon dioxide concentration in all cases in the way just described for *C. pyrenoidosa.*

From these laboratory results one may predict that glycollate release from natural phytoplankton should occur and be especially marked when low concentrations of actively growing cells are subjected to high light intensity and carbon dioxide deficiency.

Results obtained in *in situ* experiments in Windermere and Blelham Tarn, English Lake District, are being reported elsewhere (Fogg, Nalewajko and Watt, 1965). In these the average excretion

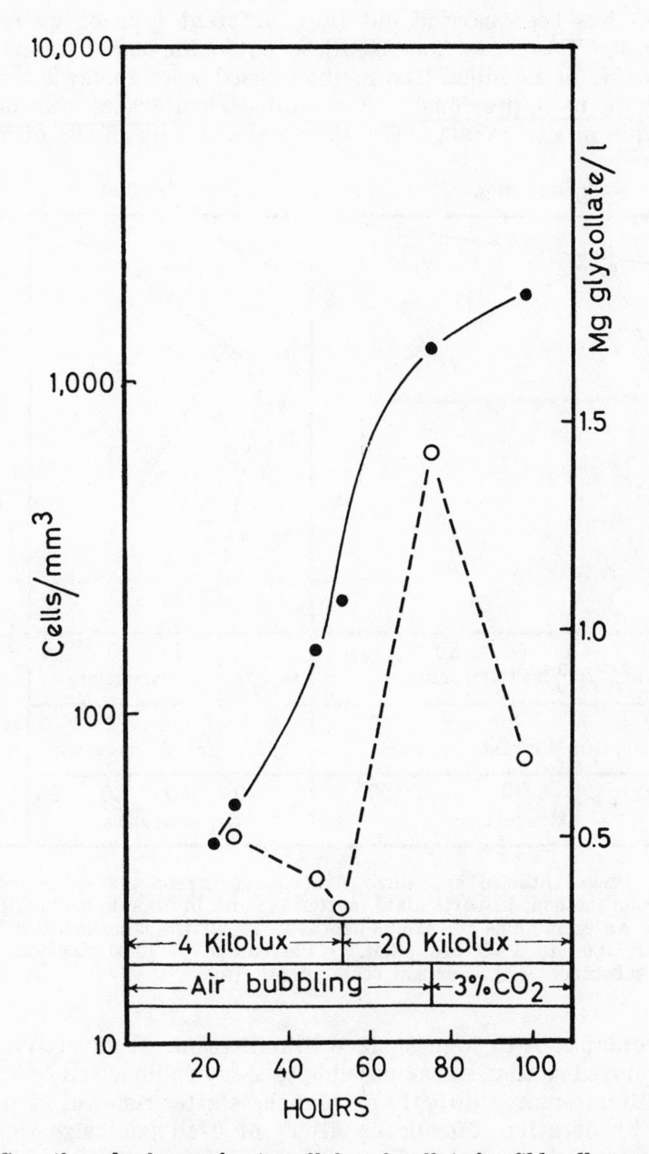

Fig. 1. - Growth and release of extracellular glycollate by *Chlorella pyrenoidosa* in a culture of limited volume under various conditions of light intensity and carbon dioxide concentration.

was found to be 35% of the total carbon fixed. Further work (Watt, in press) has been carried out in a different type of water, Tring Reservoirs. These are four shallow eutrophic canal reservoirs 35 miles N. W. of London. The methods used were somewhat different from those used previously. The radiocarbon tracer was added to the sample *in situ* (Watt 1965), thus avoiding any effects of exposure

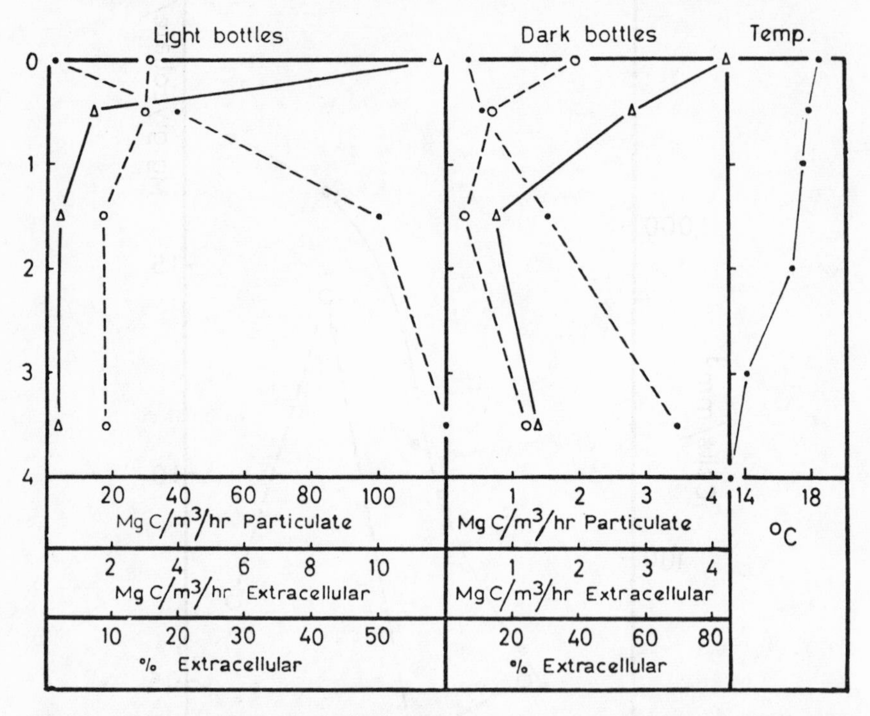

Fig. 2. - Photosynthetic and dark fixation of carbon, as estimated by the radiocarbon method, in particulate matter (●) and in organic matter in filtrates (O) from water samples at various depths in an *in situ* experiment in Tringford Reservoir, 1000 to 1400 hrs (G.M.T.) Percentage of total fixation in extracellular substances (Δ). Vertical scale: depth (m).

of phytoplankton to high surface illumination. High activity tracer was employed so that it was possible to assay radioactivity on aliquots of the filtrate plated directly on planchets after removal of inorganic carbon by aeration. Membrane filters of 0·4μ pore size were used.

In a typical example when surface illumination was intense the maximum in fixation in particulate matter was just below 3 metres. Radiocarbon in organic matter in the filtrate was less than 2% of the total fixation below 1·5 metres but rose to more than 50% at the surface. Extracellular radiocarbon in filtrates from dark bottles was relatively high but less absolutely than in the light (Fig. 2).

High rates of excretion have consistently been observed both in Windermere and in the Tring Reservoirs when light intensities are high enough to inhibit photosynthesis (Fig. 3). Increases in the relative amounts of extracellular products at low light intensities have also been encountered, in laboratory cultures as well as in *in situ* experiments. From the evidence available it appears that the dark

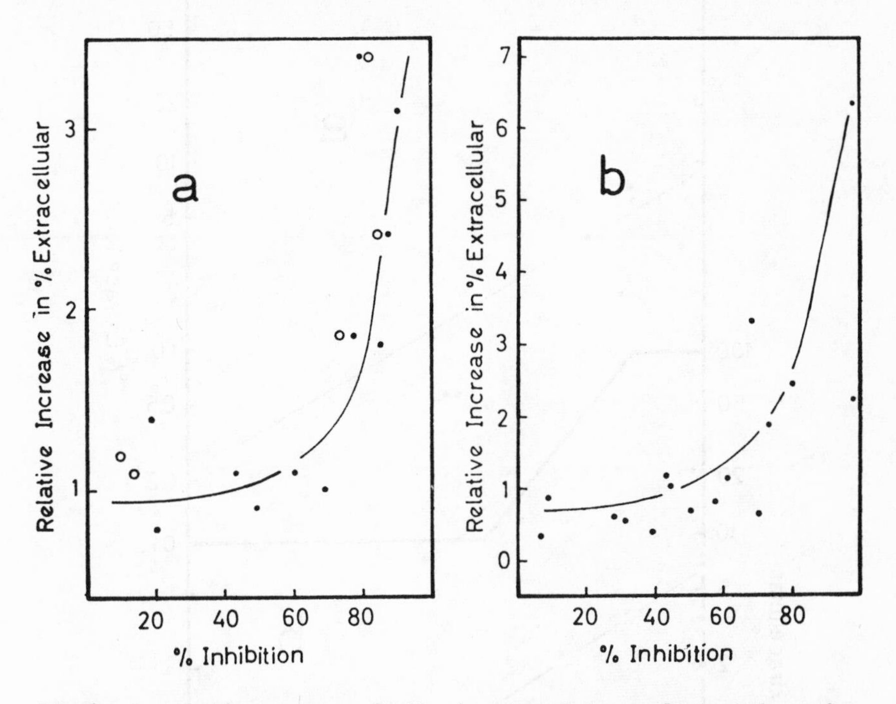

Fig. 3. - Increase in percentage fixation in extracellular products at the surface relative to that at the depth at which maximum photosynthesis occurred as a function of percentage inhibition of photosynthesis at the surface; a) Tring Reservoirs (•) *in situ*, (○) in tank (results of Watt); b) Windermere (results of Fogg, Nalewajko and Watt).

excretion process does not involve glycollate and that it is inhibited by light. A further important finding is that the relative amount of extracellular product increases as the plankton population density decreases. Fig. 4 shows the percentage of extracellular products for many sets of observations in the Tring Reservoirs plotted against the chlorophyll *a* and *b* concentrations in the water. The range is from 0·2% up to 20% at the lowest chlorophyll concentrations. In the relatively oligotrophic lake Windermere we found the average value to be higher, 35%. This is in agreement with the trend in the Tring results.

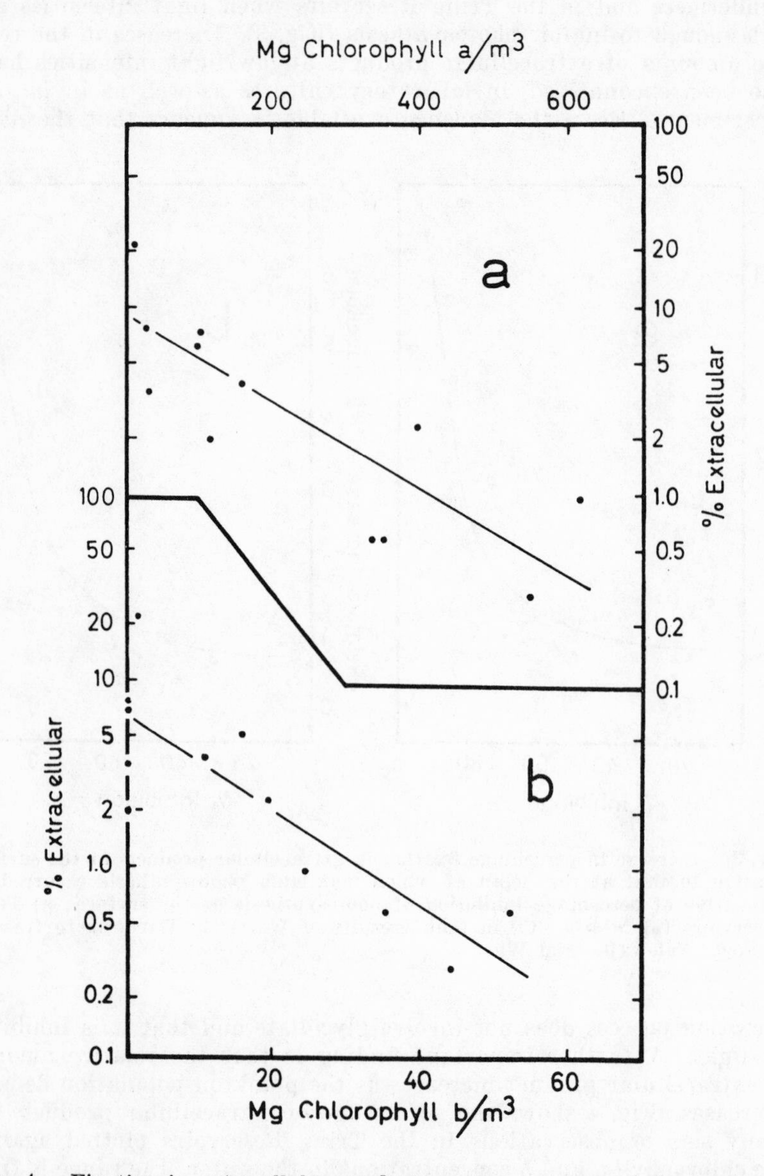

Fig. 4. - Fixation in extracellular products as a percentage of the total as a function of the concentrations of chlorophylls *a* and *b* in the Tring Reservoirs.

It now seems justifiable to say that liberation of extracellular products by phytoplankton is a normal phenomenon and that it becomes more important the more oligotrophic the water. The results of the *in situ* observations just described fit rather well with those of the laboratory experiments. For example, the maximum rates of excretion occur in the surface waters at high light intensities when carbon dioxide may be limiting. The generally higher rates of extracellular production observed in lakes as compared with cultures is to be expected from the population density effect.

To bring our two lines of enquiry together it is necessary to demonstrate that glycollate is the major extracellular product of phytoplankton growing under natural conditions. Some evidence that glycollate is released by natural phytoplankton and that it

Fig. 5. - Autoradiograph of two-dimensional paper chromatogram of the extracellular products formed during a 3-hour *in situ* experiment in a Tring Reservoir with a natural population consisting largely of *Stephanodiscus hantzschii*. The radioactivity at the origin was less than 10% of the total. The substance forming the larger spot ran with glycollic acid in four different solvent systems.

occurs in concentrations of the order of 0·1 mg 1^{-1} in fresh-water has already been put forward (Fogg and Nalewajko 1964). Methods for analysis for glycollate in natural waters are, however, far from satisfactory and, to add to other difficulties, it readily forms a dimer ester which cannot be estimated by the Calkins method. Hellebust (1965) found with most of the phytoplankton species which he studied that glycollate was present in only relatively small amounts among other extracellular products. His experiments, however, lasted for a period at least as long the generation time of the alga and were carried out under constant conditions. According to our findings this would not lead to extensive release of glycollate and any that was released might have been reabsorbed or transformed. Recently, Watt (in press) has carried out short-term *in situ* experiments, with high activity C^{14}-bicarbonate, in which extracellular products have been separated by chromatography. From these it is clear that glycollic acid is sometimes, at least, the major extracellular product of photosynthesis (Fig. 5). It remains for further work to establish whether this is generally so.

Acknowledgements

We are grateful to the Development Commission for a grant in support of this work and to Dr. J. Rzóska and to the Nature Conservancy for providing facilities for work on the Tring Reservoirs.

REFERENCES

Fogg, G. E. and C. Nalewajko. 1964. Glycollic acid as an extracellular product of phytoplankton. - Verh. int. Ver. Limnol. *15*: 806-810.

— C. Nalewajko, and W. D. Watt. 1965. Extracellular products of phytoplankton photosynthesis. - Proc. Roy. Soc. B. *162*: 517-534.

Hellebust, J. A. 1965. Excretion of organic compounds by marine phytoplankton. Limnol. Oceanog. *10*: 192-206.

Nalewajko, C. In press. Photosynthesis and excretion in various plankton algae. - Limnol. Oceanog.

Pritchard, G. G., W. J. Griffin, and C. P. Whittingham. 1962. The effect of carbon dioxide concentration, light intensity and *iso*nicotinyl hydrazide on the photosynthetic production of glycollic acid by *Chlorella*. - J. exp. Bot. *13*: 176-184.

Watt. W. D. and G. E. Fogg. 1965. A convenient apparatus for *in situ* primary production studies Limnol. Oceanog. *10*: 298-300.

— In press. Release of dissolved organic material from the cells of phytoplankton populations. - Proc. Roy. Soc. B.

— In press. The kinetics of extracellular glycollate production by *Chlorella pyrenoidosa*. - J. exp. Bot.

COMPETITION BETWEEN PLANKTONIC BACTERIA AND ALGAE FOR ORGANIC SOLUTES

JOHN E. HOBBIE [1] and RICHARD T. WRIGHT [2]

The Institute of Limnology
Uppsala, Sweden

[1] Present adress. Department of Zoology, North Carolina State University, Raleigh, N. C.

[2] Present address: Department of Biology, Gordon College, Wenham, Mass.

For bibliographic citation of this paper, see page 10.

Abstract

New methods are presented to measure the uptake of organic solutes by planktonic microorganisms. By measuring the velocity of uptake of radioisotopes over a range of substrate concentrations, two mechanisms of uptake can be differentiated. One of these, attributable to the bacteria, appears to be a transport system that is very effective at substrate concentrations below 100 µg/liter. The other, in the algae, follows the kinetics of simple diffusion and is effective only at higher substrate concentrations (above 500 µg/liter). In natural waters the bacteria are so effective that they keep the substrate concentration below 20 µg/liter and so drastically limit the heterotrophy of the algae.

INTRODUCTION

The heterotrophic bacteria in lakes are certainly living on the organic material dissolved in the water. The problem remains, however, of just what compounds they are using, the rate of use, and the concentration of these compounds. If there are heterotrophic algae also subsisting on these compounds, the same questions would apply to them.

We have developed methods of studying these problems that use C^{14} labeled organic compounds. These methods are a direct result of the findings of Parsons and Strickland (1962) that the uptake of organic compounds by plankton could be analysed by the formulae of enzyme kinetics. Some results are reported here of the competition between planktonic bacteria and algae for these dissolved organic compounds.

Another use of these methods is to describe the size and physiological activity of the bacterial population taking up the substrate. This approach to aquatic microbiology has been called for by several authors in recent years (Heukelekian and Dondero 1964), and could be a valuable addition to the traditional methods of direct counts and enrichment cultures.

METHODS

The methods have already been presented in detail by Wright and Hobbie (1965), and Hobbie and Wright (1965). Small quantities of labeled and unlabeled substrate are added to 50 ml samples of lake water and the samples incubated on a shaking table from ½ to 10 hours in a constant temperature room. After incubation the samples are filtered onto membrane filters ($< 0.5\ \mu$ pore size) and the activity of the filters is measured. The original methods have been slightly modified so that the uptake of organic solutes at low substrate concentrations is measured by adding fractions of a μc directly to 50 ml samples with a micropipette. Also, it has been found that some loss of labeled material from the cells may occur when Lugol's acetic acid solution is used for fixation. Tests should be made to see if this loss introduces any significant errors.

RESULTS AND DISCUSSION

The general formula for uptake of any solute is

$$v = (S_n + A)\ c/C\mu t \tag{1}$$

where v is the velocity of uptake (mg liter^{-1}hr^{-1}). $c/C\mu t$ is the per cent of the added radioactive isotope taken up by the organisms per hour (Wright and Hobbie 1965), S_n is the natural substrate concentration (mg/liter), and A is the substrate added in the experiment (mg/liter).

A planktonic *Chlamydomonas sp.* (undescribed, $9 \times 6\ \mu$) that grows very well on glucose in the dark was used for a laboratory study of uptake velocity kinetics in algae. A bacteria-free culture grown at 3C under fluorescent light was placed in the dark 2 days before an experiment and 50 mg glucose/liter added to adapt the algae to heterotrophic uptake. Just before the experiment, which was run at 3C, the algae were washed in glucose-free medium. The final concentration of algae was 10 million cells/liter (dry weight of 1600 μg, oxidizable carbon content of 550 μg).

When the velocity of uptake of glucose (v_d) was measured at different concentrations of added substrate (A), it was found (Fig. 1) that the velocity was directly proportional to the substrate concentration up to 1 mg glucose/liter. This uptake may be a form of simple diffusion where the substrate is used up as fast as it enters the cell (Wright and Hobbie 1965) and can be described by the general formula

$$v_d = k_d\ (S_n + A) \tag{2}$$

In this laboratory experiment the S_n was zero. The diffusion constant, k_d(hr^{-1}), is the slope of the line.

A culture of planktonic bacteria (motile rods) from Lake Erken was also used in this experiment. This culture was grown at 23C on a mineral medium plus 500 mg glucose/liter, harvested during the exponential phase of growth, and washed in glucose-free medium. Fifty million bacteria/liter were used (dry weight 6 μg, oxidizable carbon of 3 μg). The velocity of uptake of glucose (v_t) increased

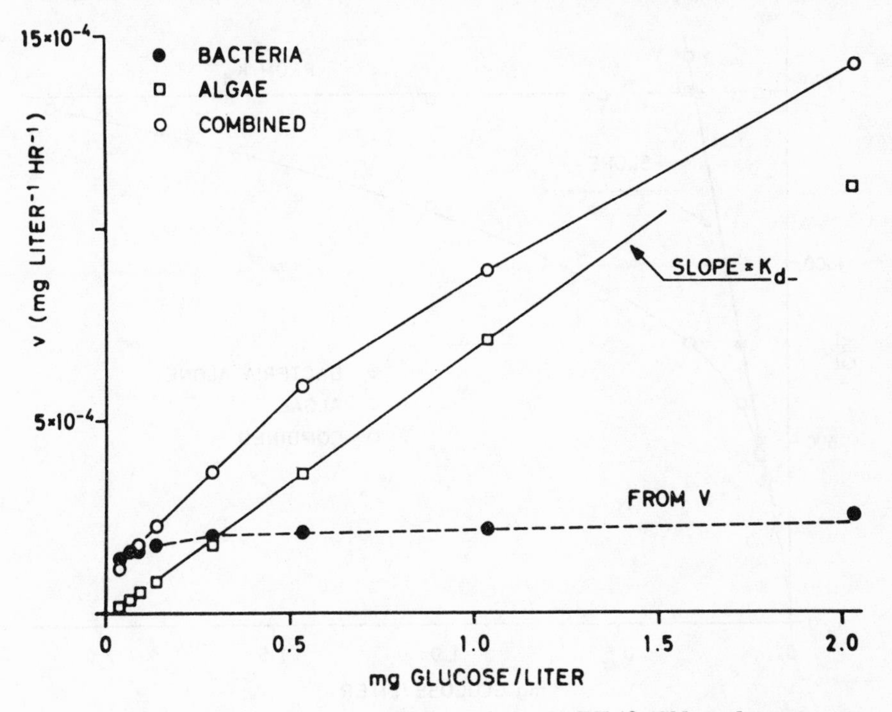

Fig. 1. - The velocity of glucose uptake for bacteria EY-19, *Chlamydomonas* sp., and the two combined.

rapidly at first as the substrate concentration increased, but soon reached a maximum velocity, V, above which point increasing amounts of substrate had no effect (Fig. 1). This type of uptake is probably due to a transport system in the bacterial cell wall (Kepes 1963), and follows the Michaelis-Menten equation:

$$v_t = V\,(S_n + A)/(K_t + S_n + A) \tag{3}$$

K_t (mg/liter) is a transport constant similar to the Michaelis constant. Again, S_n was zero in this experiment. Equation (3) is inverted and multiplied by $(S_n + A)$:

$$(S_n + A)/v_t = (K_t + S_n)/V + A/V \tag{4}$$

From equation (1) it is seen that $C\mu t/c$ may be substituted for $(S_n + A)/v_t$ in (4). Therefore, when $C\mu t/c$ is plotted against A the slope of the line is $1/V$ while the intercept on the ordinate is $(K_t + S_n)/V$ and on the abscissa $-(K_t + S_n)$. The same data used in Fig. 1 have been plotted in Fig. 2 according to the modification of equation (4). In this type of plot, the uptake by bacteria gives

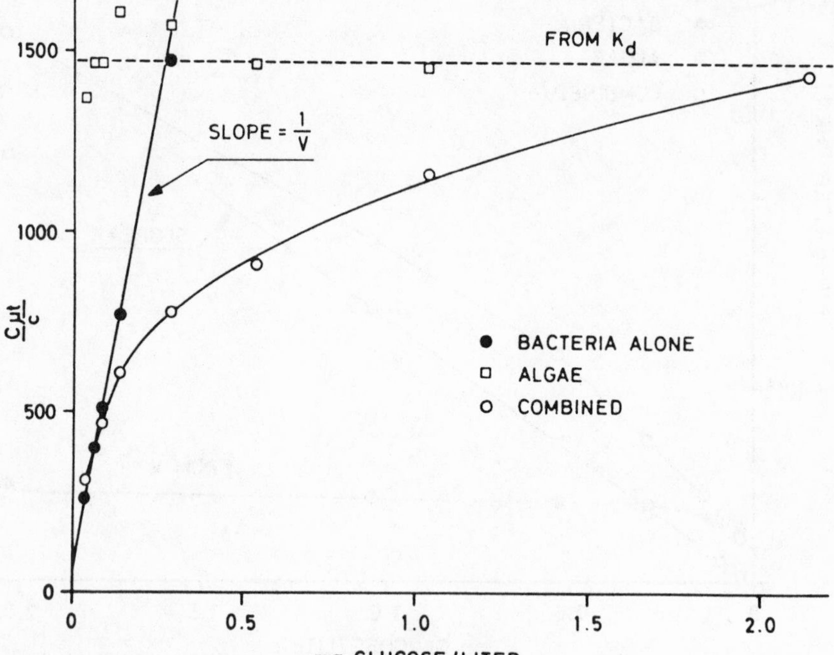

Fig. 2. - Uptake of glucose for bacteria EY-19, *Chlamydomonas* sp., and the two combined. Plotted according to equation (4).

a straight line, while the combined uptake gives a line that continually curves to the right. The V is calculated from the straight line.

From this laboratory experiment, it appears that different uptake systems are present in bacteria and algae. When the velocity of uptake of natural plankton is measured these two systems will be operating, and the resulting velocity versus substrate concentration curves should resemble the combined curve in Fig. 2. In Lake Erken, this type of curve has been found at all depths and at all seasons of the year (Fig. 3). This figure also shows the good agreement between the observed curve from Lake Erken and the curve calculated from measurements of the V and k_d.

However, it is not necessary to measure uptake velocities over such a wide range of substrate concentrations to differentiate between the bacterial and algal uptake. When a concentration range of 0.5 to 2.0 mg substrate/liter is used, the change in velocity of uptake will be due to the diffusion of the algae and the k_d can be determined. When very low concentrations of substrate are used, < 60 µg/liter, the uptake will be mostly due to the bacteria (Fig. 4). The $K_t + S_n$, 9 µg acetate/liter in Fig. 4a and 20 µg in Fig. 4b sets

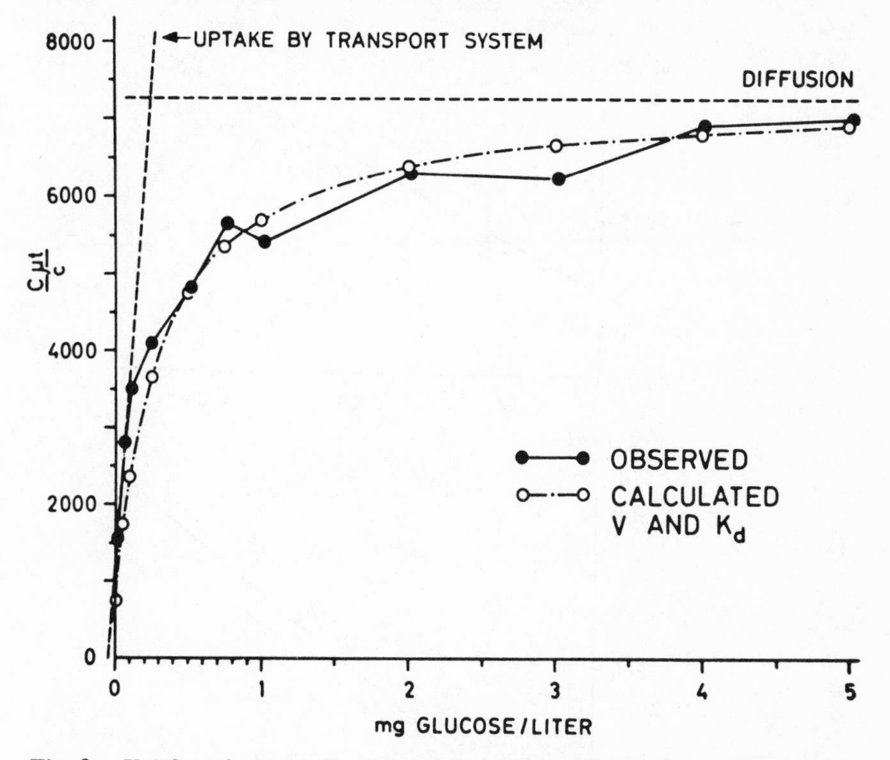

Fig. 3. - Uptake of glucose in Lake Erken (1 m, 14 February 1964) plotted according to equation (4). Theoretical curves calculated from V and k_d.

an upper limit for the S_n, while the V, 13×10^{-5} mg liter^{-1} hr^{-1} (Fig. 4 a) and 2×10^{-5} (Fig. 4 b) is the upper limit for uptake velocity.

This method of finding V and $K_t + S_n$ can also serve as a bioassay. This is done by adding a bacterial culture with a known K_t to a sample of filtered lake water and measuring the $K_t + S_n$ (Hobbie and Wright 1965). Values of up to 60 µg glucose/liter have been found in polluted waters, but in unpolluted waters the quantities are usually less than 10 µg. Acetate has been studied only in Lake Erken, where the concentrations are always less than 20 µg.

At these very low substrate levels found in nature, the transport systems of the bacteria are much more effective than the diffusion-like mechanisms of the algae. This may be illustrated by comparing the time required for the two populations to take up all of the substrate. This turnover time (T) assumes a constant supply and is given by the general formula:

$$T = S_n/v \tag{5}$$

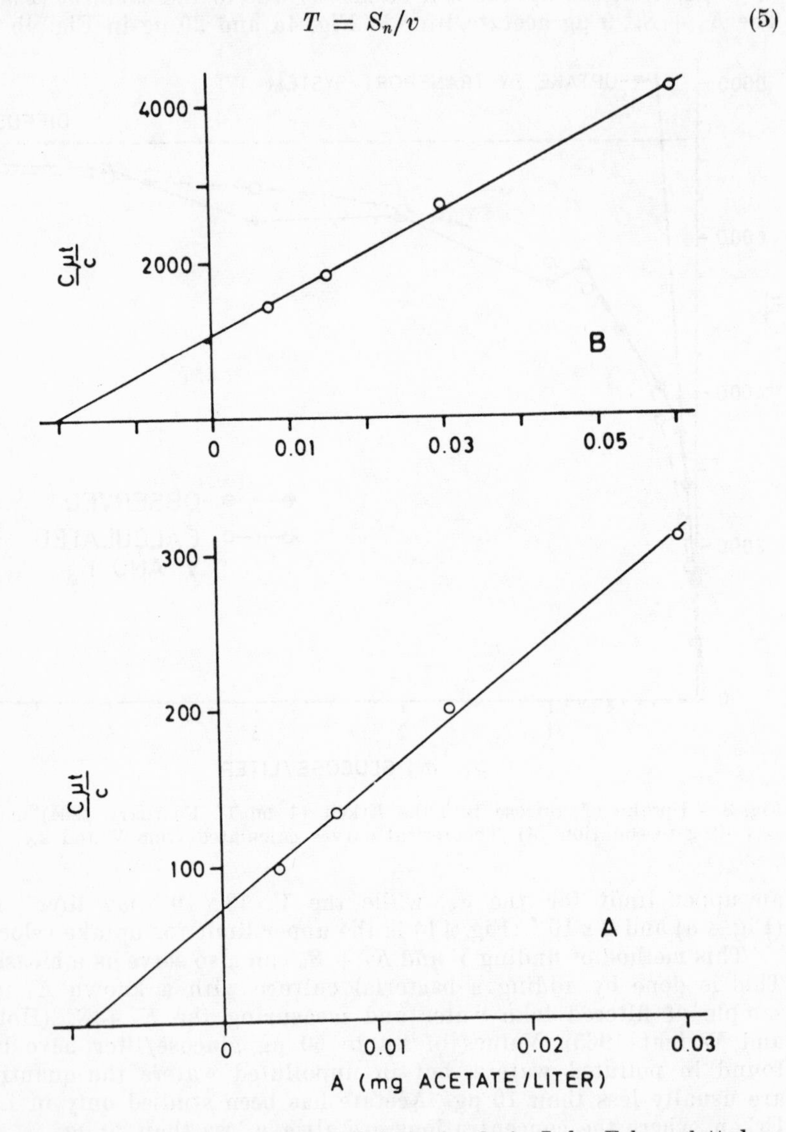

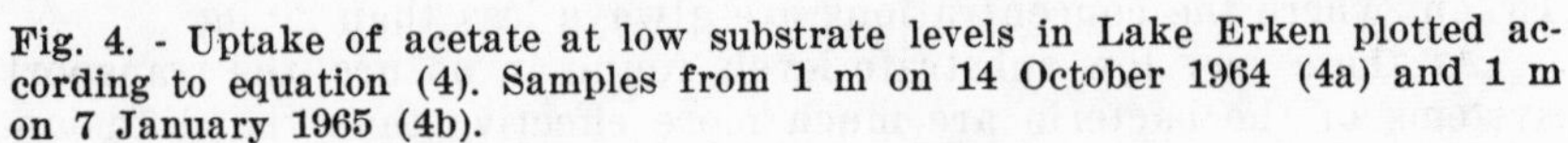

Fig. 4. - Uptake of acetate at low substrate levels in Lake Erken plotted according to equation (4). Samples from 1 m on 14 October 1964 (4a) and 1 m on 7 January 1965 (4b).

Assuming that the natural substrate level was 10 μg glucose/liter, the bacterial population in the laboratory experiment described above had a turnover time (T_t) of 130 hours, while the algal turnover time (T_d) was 1470 hours. This great disparity occurred in spite of the much greater biomass of the algae (250 times). Another way of approaching the problem of competition is to compare the time required to replace the cell carbon of the two populations. Again assuming 10 μg glucose/liter, it would take 22.1 years for the algae to replace their cell carbon and 100 hours for the bacteria. As the substrate concentration increases, the v_d increases while the v_t remains the same. At 10 mg glucose/liter, the cell carbon replacement time becomes 200 hours for the algae, and at 100 mg it is 20 hours. In nature, other compounds are available to the algae, and higher temperatures could increase the uptake rate. As a result, the findings reported here do not necessarily preclude algal heterotrophy in nature.

The turnover times in nature can be easily measured by these methods even though the actual substrate concentration (S_n) and the actual rate of uptake (v) are unknown. For algae, the T_d is $1/k_d$ (from equations (2) and (5)). For the bacteria, the ordinate is the T_t in hours when the data are plotted by equation (4). This intercept is $(K_t + S_n)/V$, and from equation (3) this equals S_n/v_t when $A = 0$. Thus, in Fig. 4 the T_t in October was 70 hours and in January, under the ice, over 1000 hours.

This line of reasoning can be carried one step further and the actual velocities of uptake by the two populations compared. From equation (5) the S_n is equal to $T(v)$. Therefore, the turnover times due to the algae and the bacteria may be compared as the S_n in a given water sample is equal in the equation for T_t and T_d.

$$v_d T_d = v_t T_t \quad \text{or} \quad v_d/v_t = T_t/T_d \tag{6}$$

Some results of these comparisons are given in Table 1. The algae always took up less than 8% of the amount the bacteria removed over the reported period, and in the summer the ratio was even lower. For a comparison, the ratio v_d/v_t in the laboratory experiment was 0.09.

The measurement of V and $K_t + S_n$ is dependent upon specific transport systems in the bacteria. We have studied the specificity of these systems for glucose, and found that of a large number of amino acids, fatty acids and sugars tested, only mannose interfered to any extent. In one experiment, the glucose uptake systems of plankton showed recovery of 8 μg/liter when 10 μg/liter of unlabeled glucose was added. When 100 μg mannose/liter was added, the uptake system « saw » it as 4 μg glucose/liter. In other words, this uptake system was 20 times more sensitive to glucose than to mannose. A thoretical discussion of these bacterial transport systems has been given by Kepes (1963), who indicated that 30-50 different specific systems might be present. Although we have tested only 3 compounds, it is likely that a number of carbohydrates, amino acids, and fatty acids could be studied with these methods.

Table 1. - Time (hr) required for the bacteria (T_t) and algae (T_d) to remove all of the substrate present in a natural sample. From these turnover times, the ratio (algae/bacteria) of the actual velocities of uptake has been calculated. Samples were from 11 m in Lake Erken, Sweden, 1964-65. Substrates were glucose (G) and acetate (A).

	16 Sept		14 Oct		18 Nov		16 Dec		7 Jan	
	G	A	G	A	G	A	G	A	G	A
T_t	16	17	17	37	60	200	100	270	70	430
T_d	715	1050	1000	1850	1280	3300	1800	3400	3125	6020
v_d/v_t	.022	.016	.017	.020	.047	.060	.055	.078	.022	.071

A number of conclusions can be drawn from these laboratory and field experiments, but the most basic one is that the bacteria are very effective at removing low concentrations of organic compounds from solutions. In contrast, the algae thus far studied appear to take up very little substrate at low concentrations, as their velocity of uptake is proportional to the substrate concentration. The bacteria take up glucose and acetate so well that they undoubtedly prevent any substantial accumulation of these compounds in most natural waters. Even in polluted waters, the $K_t + S_n$ values are very low, and the bacteria have responded to an increased rate of substrate formation by increasing their biomass. The general picture seems to be one of an almost constant amount of glucose and acetate present in the water the year round, but with a widely varying rate of uptake of these substances by bacteria.

Any competition between algae and bacteria for dissolved organic compounds appears to favor the bacteria, at least at the low substrate concentrations found in nature. However, a strong word of caution must be inserted here, as it is still possible that very small algae might have transport systems that are effective at low substrate concentrations. In the absence of any evidence for this, we have assumed that only bacteria have effective transport systems.

As we have studied only two organic compounds out of the many present in natural waters, the total picture of heterotrophy in lakes is still largely unknown. We do have some information on glycollic acid that can be mentioned here. The V for glycollate uptake is similar to that of glucose and acetate. The $K_t + S_n$, however, is 5 to 10 times higher than that for the other compounds.

If glucose and acetate are typical of the other organic compounds being actively used by the bacteria, then all of these compounds will be found in very low concentrations. Probably most of the dissolved organic matter in natural waters is not being rapidly used by the bacteria, but consists of large molecules which are but slowly broken down. Thus, the chief obstacle to algal heterotrophy would seem to be the bacteria, which keep the simple organic molecules at a very low level.

Acknowledgements

This investigation was supported in part by U.S. Public Health Service Postdoctoral Fellowship 15754 from the National Institute of General Medical Sciences (JEH) and by National Science Foundation Postdoctoral Fellowship 43096 (RTW). We are grateful to Prof. Wilhelm Rodhe for providing space and equipment in the Institute of Limnology, and for encouragement and suggestions in all phases of the work.

REFERENCES

Heukelekian, H. and N. C. Dondero. 1964. Research in aquatic microbiology: trends and needs, p. 441-452. *In* H. Heukelekian and N. C. Dondero (ed.), Principles and applications in aquatic microbiology. Wiley and Sons, New York, London, Sydney.

Hobbie, J. E. and R. T. Wright. 1965. Bioassay with bacterial uptake kinetics: glucose in fresh water. Limnol. Oceanog. *10*: 471-474.

Kepes, A. 1963. Permeases: identification and mechanism, p. 38-48. *In* N.E. Gibbons, (ed.), Recent progress in microbiology. 8th Internat. Congr. for Microbiol., Montreal, 1962. Univ. Toronto Press, Toronto.

Parsons, T. R. and J. D. H. Strickland. 1962. On the production of particulate organic carbon by heterotrophic processes in sea water. Deep-Sea Res. *8*: 211-222.

Wright, R. T. and J. E. Hobbie. 1965. The uptake of organic solutes in lake water. Limnol. Oceanog. *10*: 22-28.

ON THE TROPHIC ROLE OF CHEMOSYNTHESIS AND BACTERIAL BIOSYNTHESIS IN WATER BODIES

JU. I. SOROKIN
Institute ot Freshwater Biology « Borok », Ac. Sci. USSR
Jaroslav, Nekouz, Borok, USSR

For bibliographic citation of this paper, see page 10.

Abstract

Dark bacterial biosynthesis is in principle secondary production because it derives its energy from organic matter built during photosynthesis. It is, however, in some respects more nearly primary than secondary production because it proceeds mostly by the utilization of organic matter which would otherwise be practically lost for the direct feeding of animals. This is particularly true in the case of some inland waters and certain oligotrophic ocean waters where bacterial biosynthesis through the oxidation of allochthonous material is more important than photosynthesis. Experiments show that the trophic role of chemosynthesis is the utilization of energy bound in the end products of anaerobiosis, and that chemoautotrophes serve to connect the destructive processes in the sediments with biological productivity in water masses.

At the initial stage of the study of the biology and the chemistry of water bodies, the main purpose of microbiological research was to explain different aspects of the mineralization activity of water microflora, their participancy in the biogeochemical processes of turnover of substances, in the formation of chemical composition of the water media, and in the regeneration of nutrients. The results of that stage of investigations are discussed in the monographs by Kuznetzov (1959) and ZoBell (1946).

The next stage of the development of water microbiology is the study of the trophic role of bacteria. This is the process of bacterial mineralization of primary produced or allochthonous organic matter in the water mass or in bottom sediments which results in the biosynthesis of fresh particulate protein in the bodies of bacteria. The latter is an excellent food for the filter-feeding and detritus-eating animals. This was confirmed in the special studies by Rodina (1949) and others. It became clear that mineralizing activity of bacteria results in parallel biosynthesis of the fresh protein, and the latter is true not only for the single cell but also for the bacterial population as a whole. Therefore the final productivity of a water body is dependent on the rate of the initial food supply, and so it is also the function of the primary production and of the efficiency of utilization by bacteria of dissolved and particulate autochthonous and allochthonous organic material.

All kinds of dark bacterial biosynthesis (including chemosynthesis), in principle, are the secondary production processes, for their final energy source is the energy of organic matter, which was built during photosynthesis. However, the concrete ecological approach to the problem shows that bacterial biosynthesis in some respects is nearer to primary than to secondary production, for the basic process of the initial formation of bacteria in the water medium proceeds mostly by the utilization of organic matter which is practically lost for the direct utilization by animals. The quantitative studies of the rate of production of the bacterial biomass, the factors influencing it, and the utilization of bacteria in the food chain are at least as important for the characteristics of biological productivity of water basins as the measuring and investigation of primary photosynthetic production. This conclusion may be especially true in the case of inland waters and oligotrophic waters of seas and oceans, where the bacterial biosynthesis taking place through the oxidation of allochthonous organic matter plays the most important role.

The illustration of the importance of bacterial biosynthesis of the particulate food in comparison with the primary production is shown here on the example of the newly built Bratsk reservoir on the Angara river (Fig. 1). In this graph the results of the seasonal measuring of primary production (C^{14} method) are presented. Dark bottle *in situ* oxygen consumption (rate of destruction), production of bacteria (direct microscopic count of bacteria in bottles exposed *in situ,* and indirect calculation from the rate of destruction) are given. All the values presented on the graph are the mean daily values for all the water mass of the reservoir (83 km^3) as calculated from their monthly determinations at 13 stations in typical biotopes. The results show that the production of bacterial biomass is of the same range of values as the production of phytoplankton. The rate of the destruction under 1 m^2 of the surface is about 10 times more than the photosynthetic production, and so the destruction obviously proceeds through the oxidation of allochthonous as well as autochthonous organic material. Besides, when comparing the relative trophic importance of phytoplankton and bacteria in this reservoir it must be taken into account that the large part of the phytoplankton is represented by the big colonial forms of diatoms (*Melosira, Asterionella*) and the blue green algae (*Aphanizomenon*) which are not easily utilizable by the plankton crustaceans.

The initial decay of phytoplankton, phytobenthos and higher plants by aquatic microflora results in the vigorous synthesis of bacterial biomass usually in the surface layers of water and in the near-shore zone. The remains of the primary decay form water detritus and dissolved stable organic matter composed mainly of water humus. The significant part of the stable dissolved and particulate organic matter of the inland waters as well as of the seas and oceans is of terrestrial origin. The stable particulate and predominant dis-

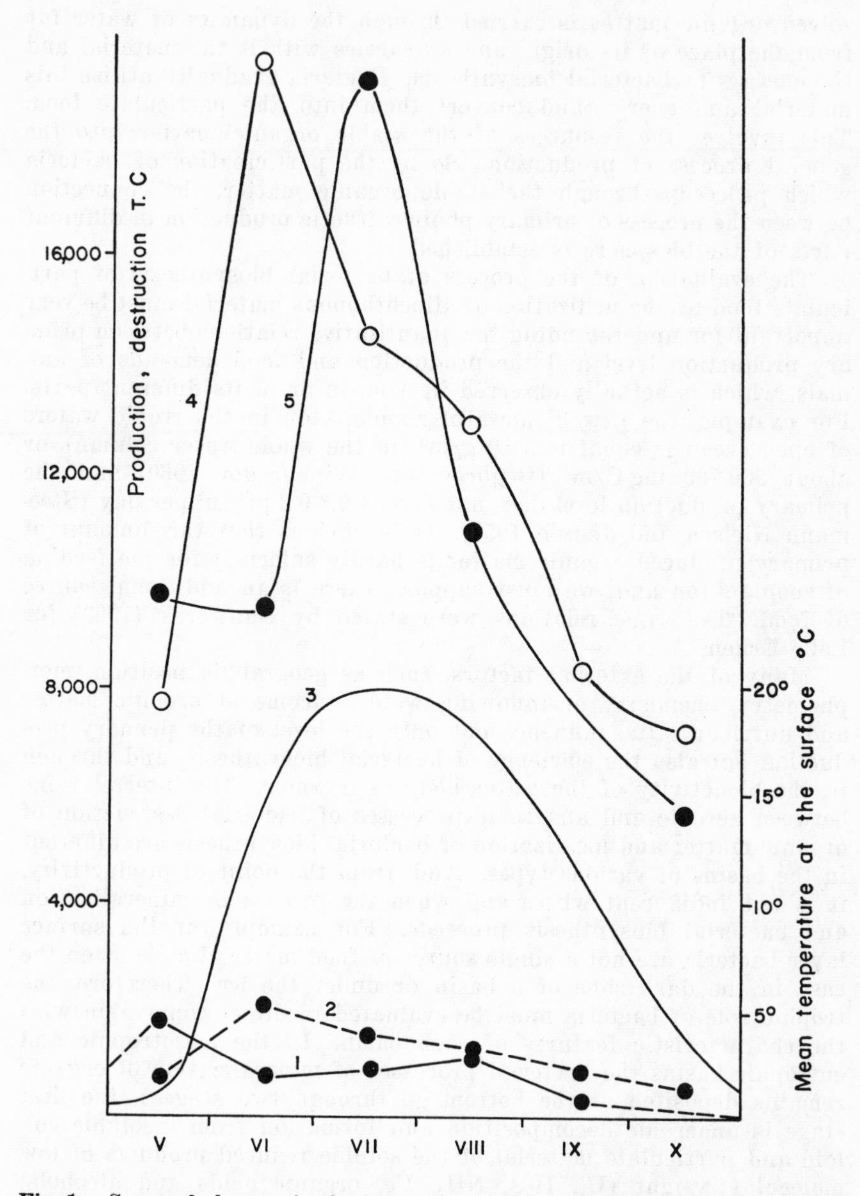

Fig. 1. - Seasonal changes in the rates of primary production of phytoplankton, destruction of organic matter, and production of bacteria in Bratsk reservoir in 1964 (summary of daily values for the whole water mass, expressed in tons of carbon). -1- primary production (C^{14}); -2- production of bacteria; -3- temperature; -4- destruction (dark bottle O_2 measurements); -5- destruction (calculated from the rate of reproduction of bacterial biomass).

solved organic matter is carried through the dynamics of water far from the place of its origin and so carries with it the material and the energy for bacterial biosynthesis. Bacteria gradually utilize this material and energy and convert them into the particulate food. This involves the resources of the stable organic matter into the general process of production. So by the participation of bacteria which proceeds through the stable organic matter, the connection between the process of primary photosynthetic production in different parts of the biosphere is established.

The evaluation of the process of bacterial biosynthesis of particulate food by the utilization of allochthonous material must be very important for understanding the quantitative relations between primary production level and the production and food demands of animals, which is actually observed in a basin or in its different parts. For example, the raw biomass of zooplankton in the tropic waters of open ocean is equal to 7-10 g/m^2 (in the whole water column), or about 500-700 mg C/m^2 (Bogorov and Winogradov 1960) and the primary production level does not exceed 0.1-0.2 gC/m^2 per day (Steemann Nielsen and Jensen 1957). It is obvious that this amount of primary produced organic matter is hardly sufficient for the feeding of zooplankton and, we must suppose, there is an additional source of food. The same relations were stated by Nauwerck (1963) for Lake Erken.

Many of the external factors, such as geographic position, morphometry, chemistry of inflowing water, income of organic matter and nutrient salts, influence not only the level of the primary production but also the efficiency of bacterial biosynthesis, and through it, the bioactivity of the water body as a whole. The interrelations between aerobic and anaerobic processes of bacterial destruction of organic matter and localization of bacterial biosynthesis are different in the basins of various types. And, from the point of productivity, it is not indifferent where and when the process of mineralization and bacterial biosynthesis proceeds. For example, at the surface layer bacteria are not a single source of food material as is often the case in the dark zone of a basin or under the ice. Therefore, the trophic role of bacteria must be evaluated in direct connection with the characteristic features of each basin. In the mesotrophic and eutrophic basins the bacterial processes of mineralization of organic remains deposited at the bottom go through two stages: the first stage is anaerobic decomposition and formation from insoluble colloid and particulate material of the soluble reduced products of low molecular weight (H_2, H_2S, NH_4, Fe, organic acids and alcohols) which contain most of the energy of the initial organic material. Molecules of CO_2, SO_4^{--}, NO_3^- and Fe^{++}-ions serve as carriers of the energy potential from silt to water because they operate as the principal hydrogen acceptors during the bacterial anaerobic decomposition of sedimented organic material (Fig. 2). The second stage of decomposition occurs at the boundary of the aerobic and anaerobic zone in the water mass or at the surface film of bottom sediments. Here the release of this bound energy is effected through the oxidation of these products by autotrophic and heterotrophic microflora, which utilize the energy for the biosynthesis of the particulate protein food. So the energy of the products of anaerobic decay by the participation of aerobic microflora is involved in the general process of the formation of particulate food.

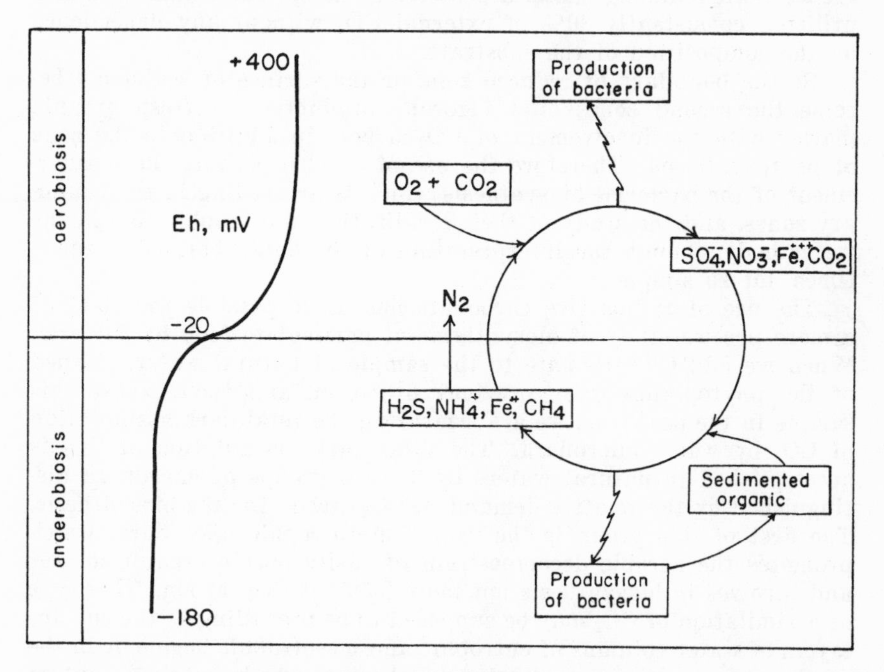

Fig. 2. - Scheme of the participation of SO_4, NO_3, Fe- ions and CO_2 in the transport of the reducing power and energy for bacterial biosynthesis from bottom sediments to the water column.

The special feature of the production of bacterial biomass through the oxidation of anaerobic products is the intensive involvement of CO_2-carbon during biosynthesis. The biomass of bacteria is formed in this case mostly by the participation of the true autotrophic bacteria, oxidizing H_2, NH_4, H_2S, S, $S_2O_3^{--}$, Fe^{+++}, and the bacteria, which oxidize C_1-compounds (methane, methanol, formic acid). Bacteria of the first group (chemautotrophes) are composed of CO_2-carbon. Bacteria of the second group are also characterized by high utilization of CO_2-carbon during the biosynthesis unlike other heterotrophes. Methane and methanol oxidizing bacteria utilize about 30-40% of needed carbon from the external CO_2. Bacteria which oxidize formic acid build their bodies almost completely of the carbon of external CO_2, without any significant reutilization of CO_2 formed during the respiration (Table 1).

The usual heterotrophes grown on the simple media containing C_2-C_6 organic substances utilize 6-10% of external CO_2 for biosynthesis. On more complex media and obviously under natural conditions this value is lowered to 2-4%. Among the other heterotrophes are sulphate-reducing bacteria possessing the specific metabolism and utilizing consistently 30% of external CO_2 without any dependence on the composition of the substrate.

So the boundary of reduced zone or the surface of sediments become the second zones of a vigorous production of fresh organic matter with the involvement of CO_2-carbon, in addition to the zone of photosynthesis. Therefore the use of C^{14}-CO_2 permits the measurement of the bacterial biosynthesis which is proceeding in the boundary zones, and the study of their localization and trophic role in the water body, though the interpretation of the data obtained is some times not so simple.

The use of radioactive carbonate has made possible the start of a more precise study of biosynthesis of particulate food by bacteria. When we add C^{14}-carbonate to the sample of natural water, cleaned of the phytoplankton by previous filtration, and then expose this sample in the darkness, we are measuring the total dark assimilation of CO_2 by water microflora. The total dark assimilation of CO_2 is accomplished in natural waters by 3 main groups of microflora distinguished by the relative demand in CO_2-carbon for the biosynthesis. The first of this group is the usual heterotrophic microflora, which promotes the aerobic decomposition of easily usable organic matter and involves in biosynthesis not more 3-5% of CO_2-carbon. This type of assimilation of CO_2 may be supposed to be prevailing at the surface layers of water columns of eutrophic and mesotrophic basins or in the whole water column of oligotrophic basins, which have no contact with the anaerobic zone.

The second group is the bacteria oxidizing simple reduced organic products of anaerobic decomposition (methane, methanol, formic acid). They use from 30% to 90% of carbon of external CO_2 for biosynthesis. This is the intermedial group between true autotrophes and true heterotrophes. When measuring C^{14}-O_2 assimilation by these bacteria, we cannot obtain the value of the whole biosynthesis, but only ½ to ⅓ of it. So the biosynthesis is underestimated, if we try to judge its value from the amount of assimilated CO_2. The assimilation of CO_2 by the last group may be supposed to be taking place mostly at the boundary of aerobic and anaerobic zones as the chemosynthesis does.

The third part of the total dark assimilation of C^{14}-O_2 may originate from consumption by true chemoautotrophic bacteria, which build

Table 1. - The use of CO_2-carbon for the biosynthesis by microorganisms of different types of metabolism.

Type of metabolism	Organism	Substrate	% of CO_2-C taken for biosynthesis
Usual hetero-trophes-aerobes	*Torula utilis*	glucose	1.76
	Actinomyces griseus	glucose	4.57
	Bacillus mesentericus	glucose	7.51
	» »	glucose+pepton	3.68
	Bacillus cereus	lactate	4.95
	» »	glucose	4.65
	» »	glucose+pepton	1.53
Usual hetero-trophes-anaerobes	*Bacillus macerans*	glucose	7.22
	Lactobacterium cereale	glucose	8.08
	Desulphovibrio desulphuricans	H_2(gas)+20 mg acetate	37.0
	» »	formic acid+20mg acetate	32.0
	» »	ethanol	33.1
	» »	lactate	31.8
	» »	formic acid + + pepton + yeast extract	30.6
Bacteria oxidizing C_1-compounds	Methanol-oxidizing bacterium isolated from reservoir	methane	28.5
	Methanol-oxidizing bacterium isolated from reservoir	methanol	28.8
	Bacterium formoxidans (n. sp., Sorokin 1961)	formic acid	97.7
	Bacterium formoxidans (n. sp., Sorokin 1961)	acetate	10.55
	Bacterium formoxidans (n. sp., Sorokin 1961)	glucose	3.82
	Bacterium formoxidans (n. sp., Sorokin 1961)	glucose + pepton	2.94
True autotrophes	*Hydrogenomonas flava*	H_2 (gas)	100.0
	» »	glucose	12.3
	» »	glucose + pepton	2.14

Bacteria were grown in the simple synthetic medium containing C^{14}-bicarbonate and strongly buffered with 5g/1 $H_3PO_4 + NaHCO_3$.
High concentration of CO_3 lowered the decrease of specific activity of $C^{14}O_3$ caused by the excretion of CO_2 by bacteria oxidizing nonlabelled substrate.
The values of relative assimilation of external CO_2 by bacteria were found by comparing the specific radioactivities of carbon in their cells and the average specific activity of CO_2-C in the media at the beginning and at the end of the experiment.

all the substances of the body from CO_2. At the present stage of using the measurement of dark C^{14}-O_2-assimilation as a tool in the study of bacterial biosynthesis by natural aquatic microflora, it is more convenient and possible, from the methodological point of view, to separate the true heterotrophic assimilation of CO_2 on the one side and the intermediate heterotrophic and chemosynthetic assimilation which is united under one term-chemosynthesis on the other side. When accepting such a rough division, however, it must be considered that, in this case, we could somewhat underestimate the real biosynthesis. This might be caused by the abovementioned underestimation of biosynthesis by bacteria of the intermediate type of metabolism (especially by methane oxidizing bacteria).

The total dark assimilation of CO_2 we have measured using the radiocarbon method of Steemann Nielsen in the following manner: the labelled carbonate ($2\text{-}4 \times 10^6$ c.p.m. of actual activity) is added to the 100 ml sample of water poured into the flask, or to the glass balloon from the plastic bottle through the funnel with a millipore filter which takes off the phytoplankton and lets the bacteria pass through (Sorokin 1958). After one (sometimes two) days of exposure in the total darkness at near the natural temperature, the water from the flask is filtered through the millipore filter with pore size less than 0.4 μ. The filters are dried, treated with weak solution of HCl, and the radioactivity of bacteria on it is counted. The total dark assimilation is calculated using the usual formula of Steemann Nielsen with a correction for isotopic discrimination (Sorokin 1964 b).

We have solved the problem of the separate determination of the usual heterotrophic assimilation of CO_2 and of the chemosynthesis in the following way. Besides the determination of the total dark assimilation of C^{14}-O_2 by bacterial flora in each sample, the possible value of the assimilation of CO_2 by heterotrophic microflora is measured by means of the estimation of the rate of production of bacterial biomass by direct microscopic count and the relative average amount of CO_2-carbon which is used by the usual heterotrophic microflora for biosynthesis (K). The calculations may be made accordingly using the following formula (Sorokin 1964 a)

$$H = \frac{24 \times K \times 0.07 \times V \times N}{G} \text{ mg C/l per day}$$

where: H is the possible assimilation of CO_2-carbon by heterotrophic microflora; the figures 0.07, V, N mean the biomass of microbial population in carbon units (V is the average volume of cells, N is the total initial number of bacteria in water counted directly under the microscope on the millipore filters); G is the value of the rate of reproduction of bacterial population. It means the time (in hours) needed for its duplication of the time of one generation. K is the average value of the relative heterotrophic assimilation of CO_2. The value of K was estimated using the radiocarbon method based on the comparison of the specific activities of CO_2 and the organic content of bacteria cells in pure cultures of heterotrophes, grown in a complex medium with a large amount of bicarbonates. This was stated to be about 0.02-0.03 (Sorokin 1961 a). Estimations of K have been made also with the use of microscopical measurement of the biomass of bacteria and gave the value 0.04. The latter can be used for the calculations of possibile heterotrophic assimilation in the natural waters. Romanenko (1964 a) accepts the value of K equal to 0.06.

Using this K value Romanenko has proposed a new method of calculation of the production of bacteria based on the measuring of heterotrophic assimilation of CO_2 (Romanenko 1964 a). He also found that the relation between the O_2-consumption, heterotrophic assimilation of CO_2 and production of bacteria in terms of carbon is near to 1000 : 6 : 100 (Romanenko 1964 b). The use of such a relation could help in the study of actual activity of metabolism and the growth of natural populations of aquatic microflora. The values of daily chemosynthesis production can be calculated by subtracting the possible value of heterotrophic assimilation from the total dark assimilation of CO_2.

The results of the estimations of chemosynthesis in the reservoirs of the Volga river are shown in Table 2.

Table 2. - Chemosynthesis in the Volga's reservoirs (in mg C/m³ per day).
S = surface layer, B = bottom layer

Biotope	Water layer	Reservoir, year				
		Rybinsk 1955	Gorky 1956	Kuibyshev 1957	Kuibyshev 1958	Kuibyshev 1959
Nonsilted bottom land	S	0.8	0	0	0	1.69
	B	1.8	1.0	11.0	0.4	2.3
River bed with slow current	S	2.1	0.9	4.0	0	0.6
	B	6.1	3.6	8.0	3.2	1.3
Silted bottom land	S	0.7-3.8	3.2	1.7	0	0.2
	B	11.0-37.3	12.1	21.9	9.5	5.6
Silty river bed, no current	S	4.0	5.0	1.0	0.2	1.0
	B	22.0	22.0	21.0-53.0	17.0-36.0	6.8

These data show that the process is actively proceeding in the boundary layers of aerobic zones and is dependent on the anaerobic decay, which is the source of the low molecular weight reduced substances. It was detected in the range of 5-30 mg C/m^3 per day down to the silty bottom in the deep parts of the reservoirs or at the boundary of anaerobic zones, which have occurred in relatively deep parts of reservoirs after the prolonged stratification (Fig. 3). During

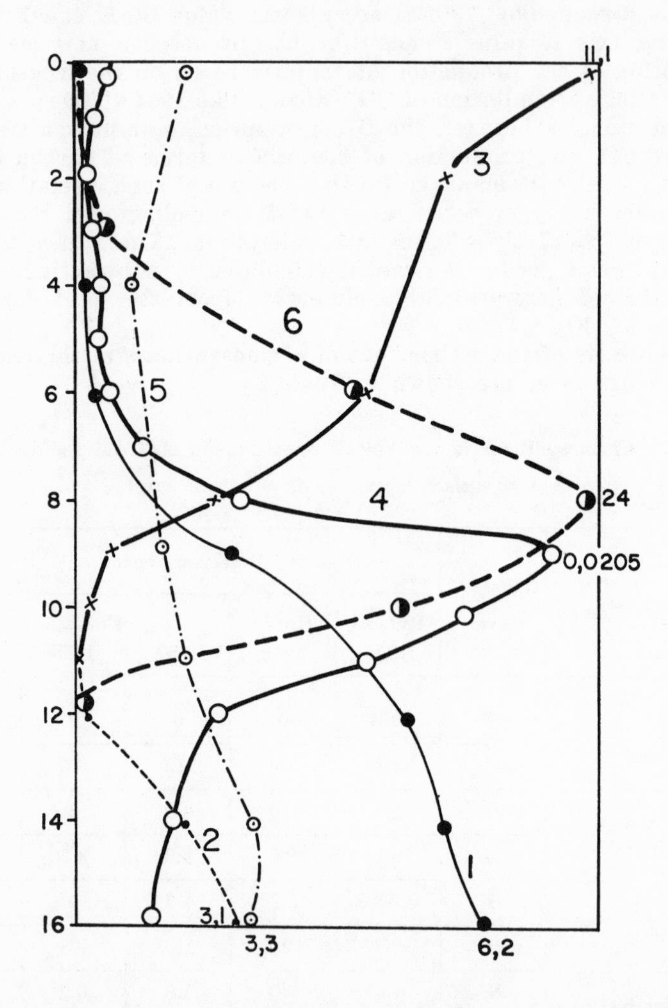

Fig. 3. - Chemosynthesis and other factors connected with it in Cheremshan bay of Kuybishev reservoir in August 1958. -1- CH_4 (dissolved), sm^3/l; -2- H_2S, mg/l; -3- O_2, mg/l; -4- Chemosynthesis, mg C/l; -5- total number of bacteria (microscopic count), 10^6/ml; -6- *Daphnia*, ind./l.

the constant homothermy and up through nonsilted bottom, chemosynthesis is practically absent and the total dark assimilation in this case is about equal to possible calculated assimilation of CO_2 by heterotrophic microflora (1-3 mg C/m^3 per day). During the process of stabilization of the chemical and biological regime of a reservoir and decreasing its trophical level, the intensity of chemosynthesis has also decreased (Table 2).

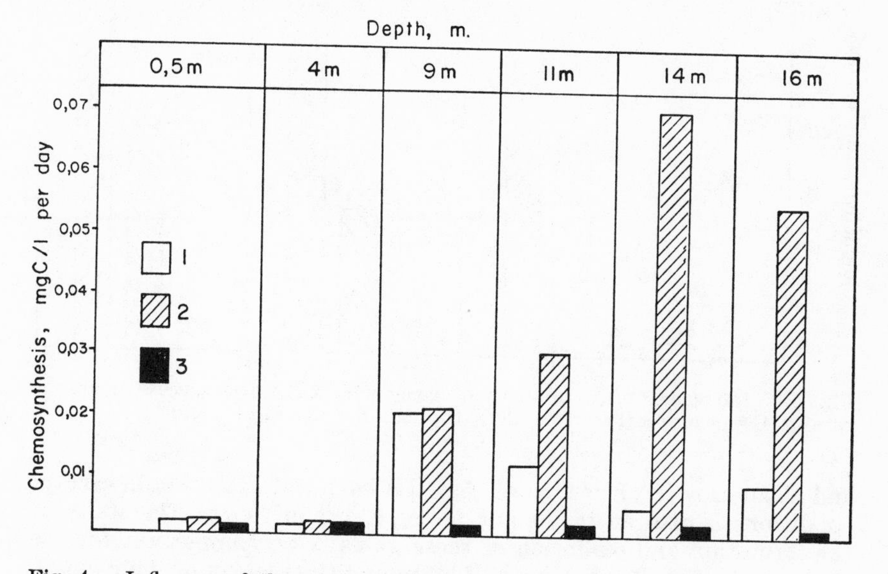

Fig. 4. - Influence of the addition of gases (2) and gas extraction (3) on the rate of *in situ* chemosynthesis (1) in Cheremshan Bay.

The main source of energy for chemosynthesis in the reservoirs is the gases H_2, H_2S, CH_4. The addition of these gases to the samples of water taken in the layer of intensive chemosynthesis increases the latter many times (Fig. 4). Extraction of the gases causes strong decrease of the rate of C^{14}-O_2-assimilation by microflora. At the zone of active chemosynthesis we have usually observed the accumulation of zooplankton, though often in these layers there was a strong deficit of oxygen (Fig. 3). The introduction of bubbles of the gases H_2 or CH_4 into the water samples greatly accelerate the incorporation of C^{14}-O_2-carbon into the bodies of filterfeeding invertebrates (Fig. 5). These experiments and observations prove the possibility of utilization of energy from the products of anaerobic decay in food chains through the participation of bacteria. Consequently the oxidation of these gases by bacteria is not only a serious factor influencing the dynamics of dissolved oxygen in the lakes, as was stated by Rossolino

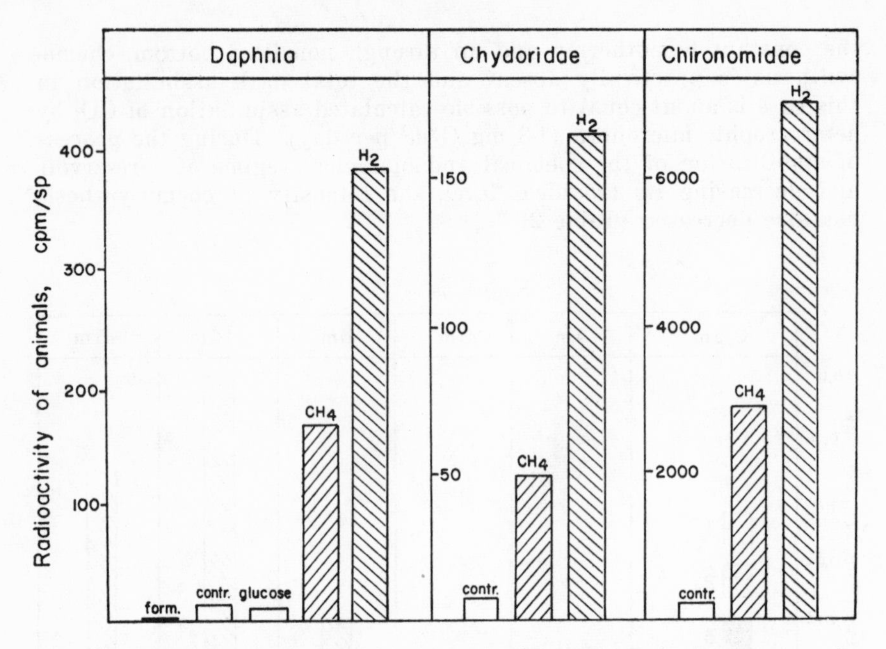

Fig. 5. - Influence of the addition of gases (H_2, CH_4) and glucose upon the rate of C^{14} incorporation from $C^{14}O_2$ through the bacterial chain.

and Kuznetzov (cf. Kuznetzov 1959). Its bacterial oxidation in certain conditions is also an important factor of productivity. The study of the formation and oxidation of these gases is very important for the characterization of bioactivity of lakes and the ecology of their population (Sorokin 1960).

Chemosynthesis plays an important role in the productivity of the organic matter in the meromictic lakes and in basins with a constant anaerobic zone. The examples of the latter are the meromictic lake Belovod and the Black Sea.

The lake Belovod is about 150 km east of Moscow (Kuznetzov 1959). Maximum depth is 24 m and below the 10 m depth the hydrogen sulphide zone begins. The concentration of H_2S strongly increases from the depth of 13 m where the limit of the penetration of light is, and increases up to 150 mg/l at the bottom. Bacterial oxidation of H_2S in the layer of 10-13 m occurs through the combined activity of chemoautotrophes (*Thiobacilli*) and photoautotrophes (*Chromatium*). The water from this layer is pink in color which is caused by the mass of *Chromatium*. The results of measuring the rate of CO_2-assimilation and other data are shown in Figs. 6 and 7. They show that in the intermedial layer the strong daily changes in H_2S and O_2-concentrations and in Eh occur as a consequence of the bacterial activity. In this layer abundant microflora is present. Here the

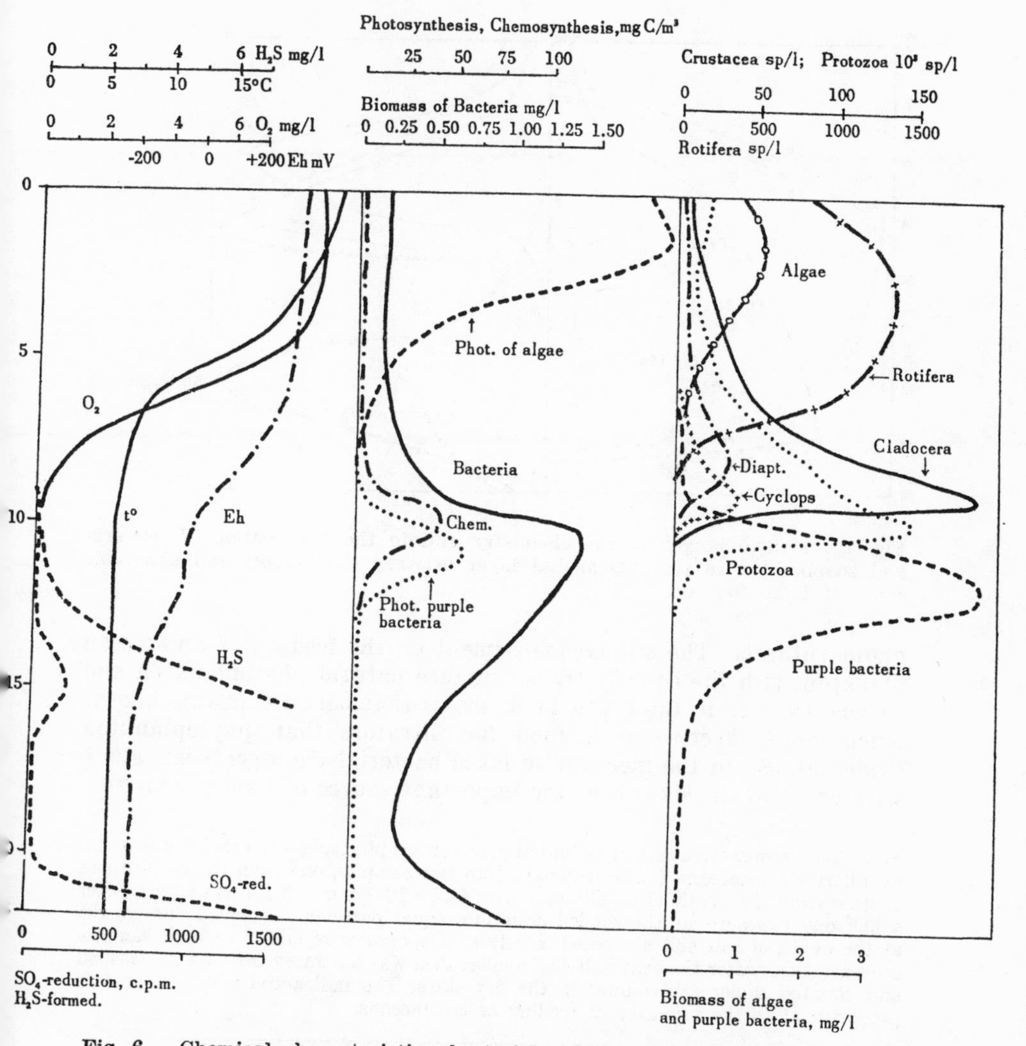

Fig. 6. - Chemical characteristics, bacterial processes, and zooplankton distribution in the water column of Lake Belovod, June 1964.

maximum rate of chemosynthesis was found and also photoautotrophic bacterial assimilation of CO_2 and maximum of bacterial biomass. The stained bacterial preparations from this zone revealed a picture more typical of pure culture than of the natural population. Maximum zooplankton was also found at the contact layer. It lowered with the lowering of the H_2S zone in day time. We have found crustaceans in midday in the layer populated with purple bacteria, when the H_2S had been completely oxidized during the bacterial

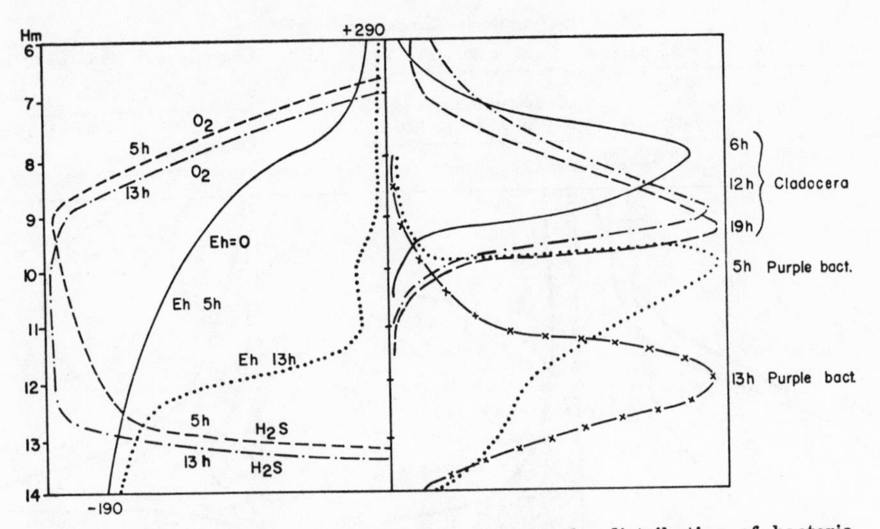

Fig. 7 - Daily changes in the chemistry and in the distribution of bacteria and zooplankton in the intermedial layer between the aerobic and anaerobic zones of Lake Belovod.

photosynthesis. The special experiment on the feeding of crustacean plankton with the equally traced surface natural phytoplankton and *Chromatium* from the depth 11 m shows that bacteria in this lake is much more effective as a food for filtrators than phytoplankton (Table 3). So in the meromictic lakes bacterial photosynthesis along with chemosynthesis can be the important source of food production.

Table 3. - Comparative rate of assimilation of surface phytoplankton and *Chromatium* by filtering crustaceans (Lake Belovod). Into two samples, one from the surface and another from the depth 11m., C^{14}-carbonate ($1.5 \cdot 10^6$ c.p.m. / 0.51) was added. After a half day exposure in the shaded aquarium equal portions of zooplankton caught at the depth of 8 m and composed mainly of *Daphnia* were added to both samples after aeration. After the next half day zooplankton was separated from water, washed and counted under the counter in the dry state. The radioactivity in this case is proportional to the intensity of feeding of crustaceans.

Depth of sampling	Radioactivity of zooplankton at the end of experiment, c.p.m. per 100 sp.
1 m. (phytoplankton)	260
11 m. (*Chromatium*)	1800

Table 4. - Total assimilation of CO_2, chemosynthesis, biomass of bacteria and activity of *Thiobacilli* in the open Black Sea (western Chalistase).

Depth, m	Eh, mv	Microflora (direct microscopy)		Assimilation of CO_2, mg C/m^3/day			Relative activity of thiobacilli (ass. of $C^{14}O_2$ in the presence of thiosulphate) (c. p. m/60 ml)	
		Total number of bacteria 10^3/ml	Raw biomass of bacteria mg/m^3	Total assimilation	Chemo-synthesis	Daily regeneration of biomass through the chemosynthesis, %	aerobic (+ O_2)	anaerobic + NO^3
0	+368	74	20.6	0.08	0	0	0	0
150	+268	71	19.8	0.42	0.33	22.3	110	0
165	+128	138	65	1.11	1.0	21.6	400	10
180	− 22	97	45	8.75	8.6	265.0	3160	26
190	− 62	79	37	0.77	0.68	25.2	22800	16700
225	−136	72	34	0.12	0.03	0.8	31700	12500
450	−178	9.4	4.4	0	0	0	212	260
1000	−182	2.0	0.9	0	0	0	0	0
2000	−190	5.5	2.6	0	0	0	0	0

In the Black Sea the active chemosynthesis has been revealed at the contact layer between the aerobic and anaerobic zones at the depths of 160-200 m, where the redox potential has the maximum gradient and changed from +50 to —100 mv of Eh (Sorokin 1964 c). Optimum rate of chemosynthesis was found at the Eh range —20 to —40 mv (rH_2 14). The daily production of chemosynthesis is at this point about 4-6 mg C/m^3. This production could provide the full regeneration of actual bacterial biomass in less than one day (Table 4).

A strong maxim of microflora (direct microscopic count) and dissolved organic matter was observed in this zone. The basic part of microflora is composed of thiobacilli which is very active in the contact zone. Microflora is represented here mostly by the chain forms which we have supposed to be thiobacilli. We have observed these forms in crude cultures of thiobacilli isolated from the contact zone of the Black Sea and have also found them in the anaerobic zone of lake Belovod in similar conditions. According to Lebedeva (1959) these chain forms are present in the stomachs of plankton crustaceans.

The experimental data show that the trophic role of chemosynthesis is in the utilization of the energy bound in the end products of anaerobiosis, for processes of production of particulate food from the CO_2 and C_1-compounds. Through the activity of chemoautotrophes the connection is accomplished between the processes of destruction in sediments and the process of biological productivity in water mass. It provides the continuous supply of definite water layers with fresh food, which is not dependent on fluctuations in primary production. The ecological peculiarity of chemosynthetic production of particulate food is demonstrated by the fact that this process can be found in those layers of the basin and in those times of the year where and when the primary photosynthetic production is low or absent.

The further study of chemosynthesis and heterotrophic carbon dioxide assimilation is only one part of an important problem of the bacterial biosynthesis of particulate food which plays not a less, but in some cases, a more important role in the regeneration of initial food resources in water basins than photosynthetic production. Progress in the study of this problem will still need special attention of the freshwater section of the International Biological Program for the development of the methods useful in carrying out further investigations.

REFERENCES

Bogorov, W. G. and M. E. Winogradov. 1960. Distribution of zooplankton biomass in the central Pacific ocean· Trans. Hydrobiol. Soc. of the USSR 10 (in Russ.).

Kuznetzov, S· I. 1959. Die Rolle der Microorganismen im Stoffkreislauf der Seen. VEB. Berlin.

Lebedeva, M. N. 1959. Filamentous bacteria from the depths of the Black Sea as a possible food for the filterfeeding zooplankton. Contr. Sta. Biol. Sevastopol *11*: 29-72 (in Russ.).

Nauwerck, A. 1963. Die Beziehungen zwischen Zooplankton und phytoplankton im See Erken. Symb. Bot. Upsalien. *17*: PP.

Rodina, A. G. 1949. Bacteria as a food of the aquatic animals. Priroda, *10*: 23 (in Russ.).

Romanenko, W. I. 1964a. Heterotrophic assimilation of CO_2 by the aquatic microflora. Microbiologia 33 (*4*): 679-683 (in Russ.).

— 1964 b. The dependence between the amounts of O_2 and CO_2 consumed by bacteria. Doklady Ac. Sci, USSR, 157 (*1*): 178-179 (in Russ.).

Sorokin, Ju. I. 1958. On the productivity of chemosynthesis in water column of the Rybinsk reservoir. Microbiologia 27 (*3*): 357 (in Russ.).

— 1960. Methan and hydrogen in the water of the Volga reservoirs. Contr. Inst. Biol. Reservoirs. Ac. Sci. USSR. 3 (*6*): 50 (in Russ.).

— 1961 a. Heterotrophic carbon dioxide assimilation by microorganisms. Jour. Obschei Biol. 22 (*4*): 265 (in Russ.).

— 1961 b. Autotrophic bacterium oxidizing formic acid. Microbiologia 30 (*3*): 385 in Russ.).

— 1964 a. The trophic role of the dark assimilation of CO_2 in water bodies. Microbiologia. 33 (5), p. 880-885 (in Russ.).

— 1964 b. On the trophic role of chemosynthesis in water bodies. Int. Rev. Ges. Hydrobiol. 49 (*2*): 307-324.

— 1964 c. On the primary production and bacterial activities in the Black Sea. J. du Conseil. 24 (*1*): 41-60.

Steemann Nielsen, E. and E. Aabye Jensen. 1957. Primary oceanic production. The autotrophic production of organic matter in the oceans. - Galathea Rep. *1*: 49.

Zo Bell, C. D. 1946. Marine Microbiology, Chronica Botanica, Waltham, Mass. U. S. .A.

IV

PRODUCTIVITY OF HIGHER AQUATIC PLANTS AND PERIPHYTON

SOME FACTORS INVOLVED IN THE USE OF DISSOLVED-OXYGEN DISTRIBUTIONS IN STREAMS TO DETERMINE PRODUCTIVITY

M. OWENS

Water Pollution Research Laboratory
Stevenage, England

For bibliographic citation of this paper, see page 10.

Abstract

Methods used to determine rates of reaeration, respiration and gross primary production in streams are described and possible sources of error discussed.

It has long been accepted that the dissolved-oxygen content of natural waters is an important sanitary index of their quality. The distribution of oxygen in a river gives a measure of the balance between the processes of supply and demand. Attempts are being made at the Water Pollution Research Laboratory, Stevenage, England, to determine the magnitude of some of these processes, and this paper describes some of the methods used to determine rates of reaeration, respiration, and gross primary production in streams. Methods used to determine net production have been described by Edwards and Owens (1960), and by Owens and Edwards (1961, 1962).

Efforts have been made to determine the metabolism of stream communities from analysis of diurnal curves of dissolved oxygen. Odum (1956, 1957) calculated the productivity of thermal springs from diurnal changes in the dissolved-oxygen concentration, employing corrections for the diffusion of oxygen through the surface. Similar methods have been used for temperate rivers by Hoskin (1959), Edwards and Owens (1962), and Gunnerson and Bailey (1963). McConnell and Sigler (1959) attempted to determine the productivity of a mountain stream.

In a stretch of river receiving no tributaries or run-off water, the change in the concentration of dissolved oxygen in the water per unit area of surface between an upstream and a downstream station can be expressed as follows:

$$X = P \pm D - R, \tag{1}$$

where X is the rate of gain or loss of oxygen per unit area of surface between the stations, P is the rate of production of oxygen (photosynthesis) per unit area, R is the rate of utilization of oxygen (respiration) per unit area, and D is either the rate of uptake of oxygen or its rate of loss by diffusion (depending upon whether the water is under-saturated or super-saturated with oxygen with respect to air).

X is calculated from the equation

$$X = (C_2 - C_1) \frac{F}{\Delta}, \tag{2}$$

where C_1 is the dissolved-oxygen concentration at the upstream station at time T_1, C_2 the dissolved-oxygen concentration at the downstream station at time T_2 such that T_2-T_1 is the average retention time between the stations, F is the flow (volume/time), and Δ the surface area between stations. Rates of flow can be determined radiochemically using Br^{82} (as ammonium bromide). The « total count » or « gulp » method described by Eden (1959) is generally employed. Lithium may also be used as a tracer for measuring rates of flow (Agg, Mitchell, and Eden 1961). If it is not possible to determine the flow rate of the stream, X may be calculated from the equation

$$X = \frac{(C_2 - C_1)\, H}{T_2 - T_1}, \tag{3}$$

where H is the average depth of the reach.

During the hours of darkness, the rate of change of the dissolved-oxygen content of the water is determined only by the rates of community respiration and of diffusion through the water surface:

$$X = D - R. \tag{4}$$

The rate of diffusion, D, depends upon the degree of saturation of the water:

$$D = f\,(\overline{C}_s - \overline{C}), \tag{5}$$

where f is the exchange coefficient of the reach, $\overline{C}_s$ is the average saturation concentration within the reach, and $\overline{C}$ the average dissolved-oxygen concentration. If the oxygen deficit is large, and the difference between the deficits at the two stations is relatively small, then the average driving force causing oxygen absorption is very nearly equal to the arithmetic mean of the deficits at the two stations. On theoretical grounds it would be more correct to use the geometric mean (Edwards, Owens, and Gibbs 1961). Thus during the hours of darkness

$$X = f\,(\overline{C}_s - \overline{C}) - R. \tag{6}$$

It is possible to calculate the exchange coefficient (f) and the respiration rate (R) from Equation 6 by measuring the rate of change of oxygen (X) at different times during the hours of darkness at different saturation deficits ($\overline{C}_s - \overline{C}$) if it is assumed that the community respiration during this time does not vary (Odum 1956) or that it varies in a predictable manner with temperature and oxygen changes (Edwards, Owens, and Gibbs 1961; Edwards and Owens 1962; Odum and Wilson 1962; Beyers 1963). This method may be used only when there is a large change in the saturation deficit during the hours of darkness. Where no such large changes occur, the concentration of oxygen can be reduced by the controlled addition of sodium sulphite and a cobalt catalyst (Gameson, Truesdale, and Downing 1955; Edwards, Owens, and Gibbs 1961), and the rate of change of oxygen content between stations measured before and during the passage of the partially deoxygenated water. Profiles of the dissolved-oxygen content during an experiment in which the oxygen concentrations were reduced to two successive plateaus by the addition of sodium sulphite are shown in Fig. 1.

In small productive streams, such as the River Ivel, where there is a pronounced diurnal rhythm in the dissolved-oxygen concentration, an alternative method has been used to determine the exchange coefficient. An opaque plastic sheet was suspended on a nylon net about 60 cm above the river surface and fixed to the river banks. Measurements with chemical actinometers placed underneath the sheet showed that it effectively darkened the reach and prevented photosynthesis. The change in oxygen concentration down the covered reach was measured throughout 24 hours, and thus the naturally occurring changes in oxygen concentration, brought about by photosynthesis upstream, were utilized in determining the exchange coefficient. The daily range in oxygen content of the water flowing into the darkened reach was about 6 mg/l, similar to changes usually obtained by the addition of sulphite. Good agreement was achieved between the results of the « opaque-sheet » and « sulphite » methods in the Ivel (Edwards 1962), but they were not compared at the same time or under identical flow conditions. Odum et al. (1963) have also covered ponds with plastic sheets to determine diffusion, but in this instance a translucent plastic was used to prevent diffusion occurring, the sheet being laid on the water surface. Estimates of the rate of diffusion were made by determining the difference between the diurnal oxygen curves obtained in the covered pond and those in an uncovered pond.

Calculations of reaeration or exchange coefficients from field data of this kind involve assumptions concerning the effects on oxygen consumption of changes in temperature and in dissolved-oxygen concentration which may occur throughout the period of observation. For two lowland streams where the effects of these environmental parameters on the respiration rates of plants (Owens and Maris 1964) and mud (Edwards and Rolley 1965) had been determined, calculated exchange coefficients were reduced by nearly 30 per cent when cor-

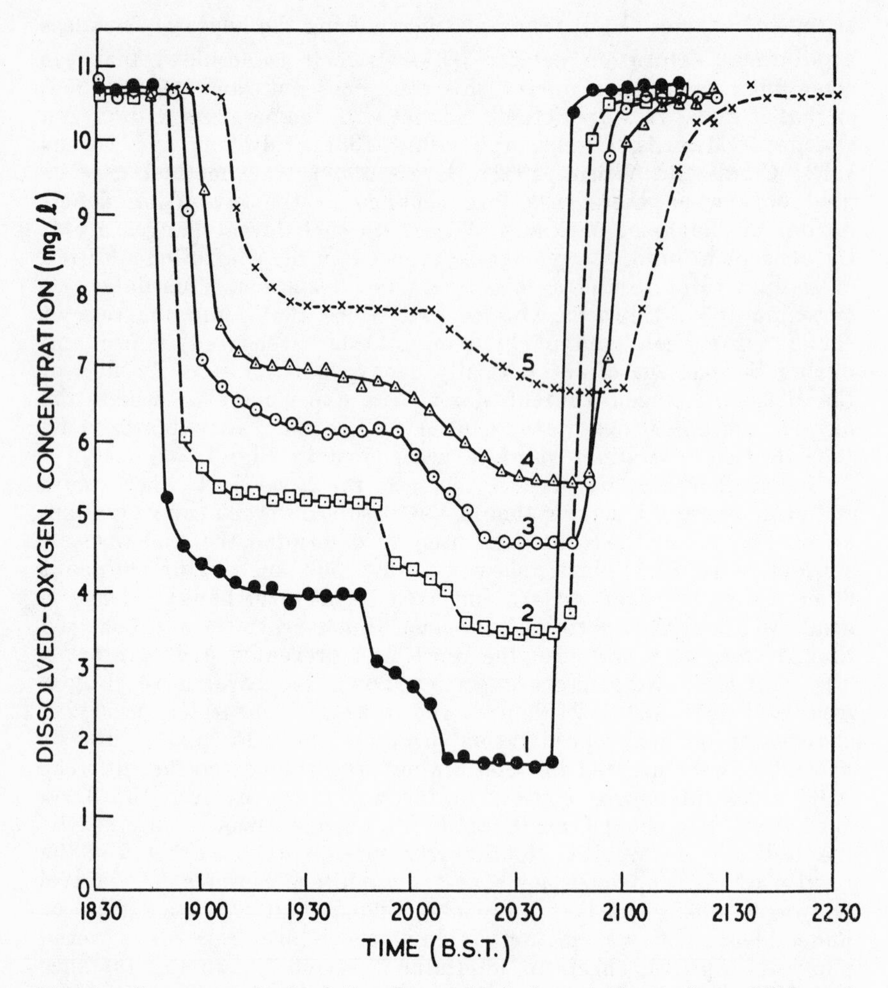

Fig. 1. - Oxygen content of the Derwent at five stations after addition of sulphite.

rections for the effects of changing temperatures and oxygen concentration on community respiration were applied (Edwards, Owens, and Gibbs 1961).

Rates of diffusion may also be determined by enclosing part of the water surface either under a free-floating polythene tent, supported by metal frames and floats (Department of Scientific and Industrial Research 1964), or under a clear plastic dome (Copeland and Duffer 1964) and measuring the amount of oxygen taken up by the water from air or oxygen enclosed by the tent. Although such tents or domes affect turbulence within their confines, comparative experiments using this and other methods have given similar results.

Attempts have been made to develop a method for predicting, from hydraulic and water-quality parameters, the rate at which deoxygenated stream water will absorb oxygen from the air (Streeter and Phelps 1925; O'Connor and Dobbins 1956; Churchill, Elmore, and Buckingham 1962; Dobbins 1964). Data collected by the W.P.R.L. and by Churchill, Elmore, and Buckingham (1962) on reaeration rates in streams have been combined and analysed using multiple regression procedures (Owens, Edwards, and Gibbs 1964), and constants have been derived for the equation relating the exchange coefficient to the average velocity and depth of the stream:

$$f\,(20\,°\mathrm{C}) = 27.5\,U^{0.67}\,H^{-0.85}, \tag{7}$$

where f is the exchange coefficient (cm/h) at 20°C, U is the average velocity (ft/s), and H is the mean depth (ft). A comparison of computed and observed values of the exchange coefficient is shown in Fig. 2. This equation makes it possible to predict, with a reasonable degree of accuracy, the reaeration rate which would be expected in rivers from their mean velocities and depths, provided these are within the experimentally observed ranges (velocity, 0.1 — 5.0 ft/s; depth, 0.4 — 11.0 ft). These computed values are for clean water only and have to be corrected when applied to polluted water, for pollutants reduce reaeration rates (Downing, Melbourne, and Bruce 1957). The effect of temperature on the exchange coefficient has been described by the Committee on Sanitary Engineering Research (1961) of the American Society of Civil Engineers; the rate of reaeration was shown to increase by 2.4 per cent per degC throughout a temperature range of 5-30°C.

Before rates of oxygen production by photosynthesis (P) can be determined for daylight hours from Equation 1, the oxygen consumption (R) must be calculated over this period. Odum (1956, 1957), Odum and Hoskin (1958), and Hoskin (1959) assumed that respiration was constant throughout 24-hour periods of observation. Varying respiration rates have been applied by Odum and Wilson (1962) and Odum et al. (1963). When pre-sunrise and post-sunset values of respiration rate differed in their experiments, a constant rate of change in respiration during the intervening period was assumed in estimating photosynthesis. When pronounced changes occur in the temperature and oxygen concentration it is probably better to apply correction factors to the average of the night-time respiration values calculated from Equation 6, for although the oxygen consumption of aerobic bacteria generally appears to be independent of the oxygen concentration of the medium, provided it is above 1 mg/l (ZoBell and Stadler 1940), this is not so for plants, animals, and aquatic muds. Measurements of oxygen consumption of aquatic macrophytes

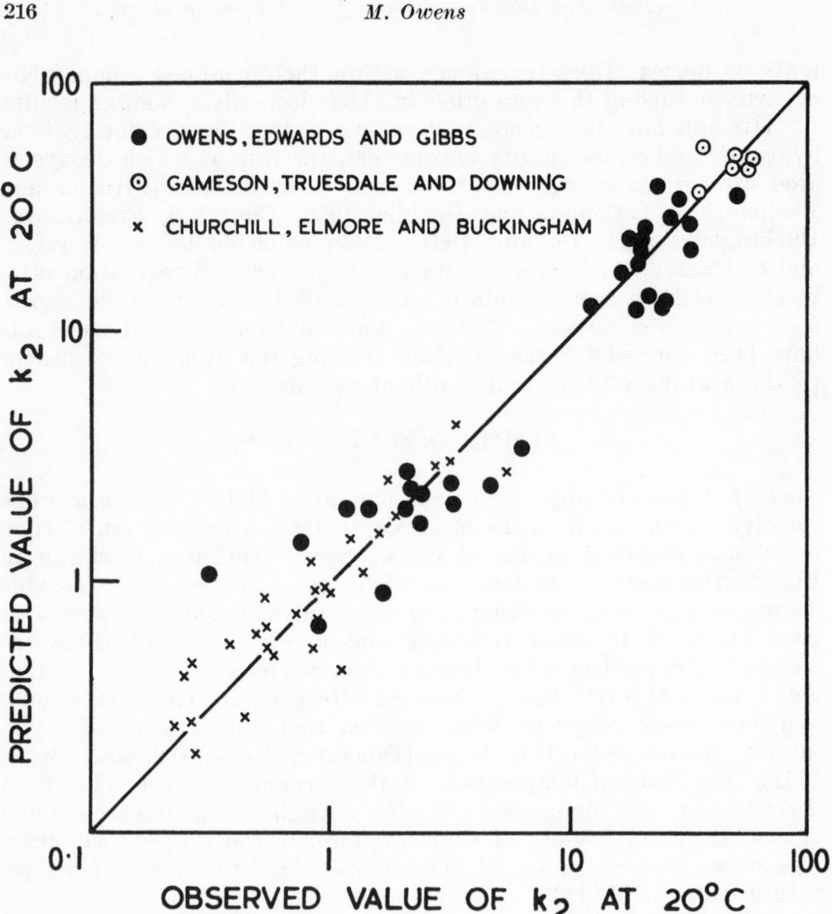

Fig. 2. - Comparison of values of the reaeration coefficient (k_2) predicted from the equation $f = 27.5\ U^{0.67}\ H^{-0.85}$ with those observed ($f = 2.92\ Hk_2$).

in relation to oxygen concentrations and temperature have been made by Gessner and Pannier (1958) and Owens and Maris (1964). In both investigations the rate of oxygen consumption of the species studied varied with the oxygen concentration. In the latter study, a relation between oxygen consumption and oxygen concentration (for a range of 1.5-16.0 mg dissolved oxygen per liter) of the form $R' = aCb$, where R' is the rate of respiration per unit dry weight of weed, C is the oxygen concentration, and a and b are constants, was determined. A similar relation between the oxygen consumption of muds and oxygen concentration was found by Edwards and Rolley (1965). Temperature coefficients (Q_{10}) of 2.7 (*Callitriche*), 2.3 (*Ranunculus*), 2.2 (*Berula*), 1.5 (*Hippuris*), and 2.0 (mud) were obtained between 10 and 20°C.

From laboratory studies of this kind, and from the relative abundance of these plant species in the River Ivel (determined by cropping studies), it has been possible to estimate the effect of oxygen concentration on the oxygen consumption of the aquatic community.

Fig. 3 shows typical curves of oxygen concentration at two stations in the Ivel, an unpolluted chalk stream, over 24 hours in May 1964. The results of analyses of these diurnal curves are given in Table 1. The value of the exchange coefficient used in these analyses was calculated from Equation 7. Two analyses have been made: (a) assuming that the respiration rate is constant throughout the 24 hours, and (b) assuming that the respiration rate varies with changes in oxygen concentration and temperature. Values of P and R calculated by method (b) are greater (by 14 and 17 per cent respectively) than those calculated by method (a).

The values of P and R obtained by method (b) are in the ranges 3.2 to 17.6 and 6.7 to 15.4 g oxygen/m^2 day found for gross primary production and community respiration respectively in a previous study of the Ivel (Edwards and Owens 1962).

Odum (1956) has proposed a simplification of the upstream-downstream oxygen method in which diurnal rhythms at a single station are utilized to calculate diffusion, respiratory, and photosynthetic rates. This method may be employed only where the whole of a reach of river is « experiencing a simultaneous rise and fall of oxygen », and under these conditions « a second station would reveal a curve iden-

Table 1. - Photosynthesis (P) and Respiration (R) in a reach of the River Ivel (May 1964)

Method of analysis	Photosynthesis (P) g/m^2/day	Respiration (R) g/m^2/day
Upstream-downstream method		
Constant R	10.9	8.0
Variable R	12.4	9.4
Single-curve method (Constant R)		
Station 1	9.7	9.2
Station 2	15.1	14.0

Values of P and R have been calculated using the upstream-downstream method with constant respiration and with variable respiration (corrections were applied for the effect of changes in oxygen concentration and temperature). Values of P and R calculated from analysis of the single curves at Stations 1 and 2 are also included. All values are in g oxygen/m^2/day.

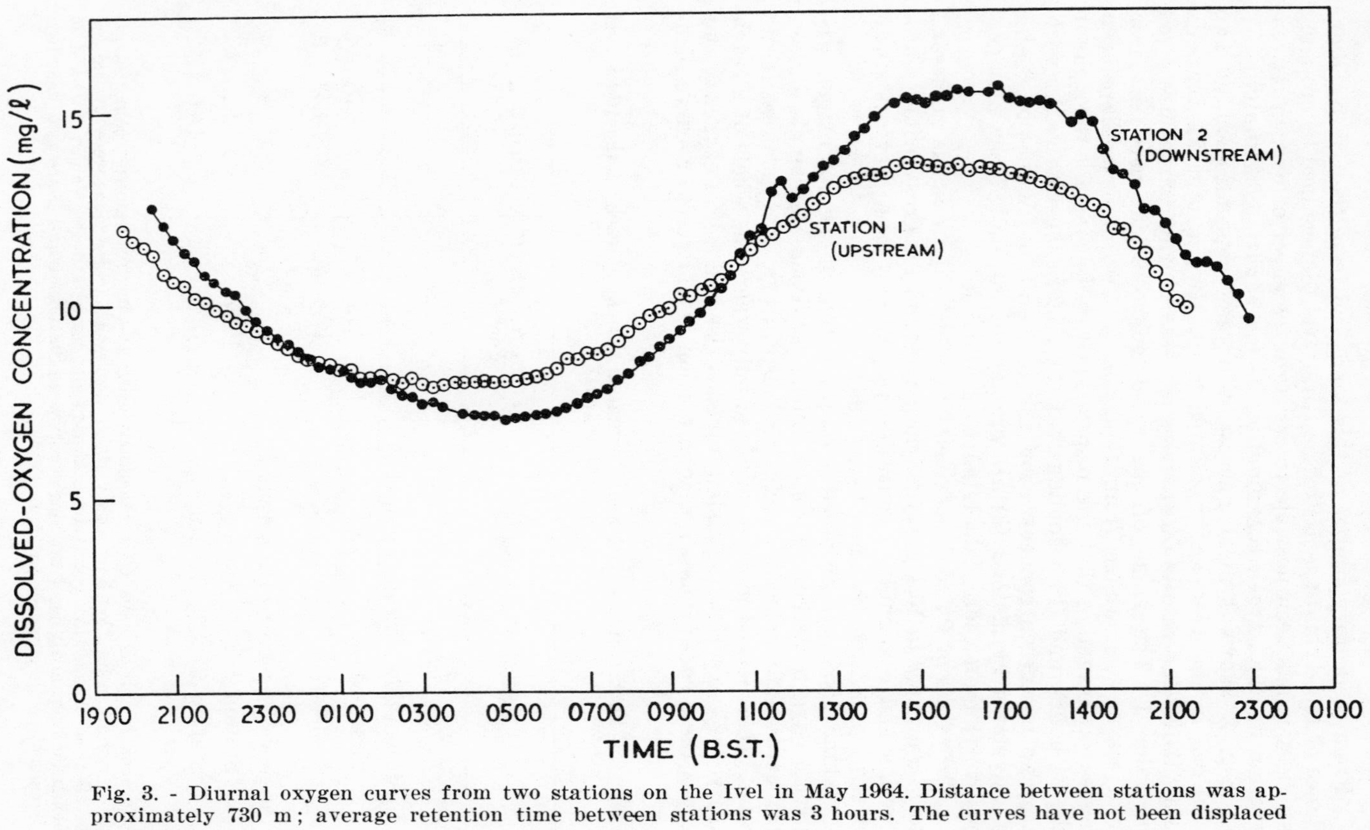

Fig. 3. - Diurnal oxygen curves from two stations on the Ivel in May 1964. Distance between stations was approximately 730 m; average retention time between stations was 3 hours. The curves have not been displaced for time.

tical with that of the first station ». Though the diurnal curves of oxygen shown in Fig. 3 almost meet this requirement, the photosynthetic and respiratory rates calculated from these curves (Table 1) are not in agreement with each other or with the results obtained by the upstream-downstream method. The same exchange coefficient was used in both analyses. Edwards and Owens (1962) showed more pronounced differences between the results obtained from single-curve analysis and those obtained from analysis of the change between stations. Single-curve analysis gave values for photosynthetic and respiratory rates of 2.6 and 2.0 g/m^2/ day respectively at the upstream station, and 10.1 and 6.6 g/m^2/ day at the downstream station; these stations were only 550 m apart. Analyses by the twin-curve method gave 11.8 and 7.7 g oxygen/m^2/day. Odum (1956) employed the single-curve technique to analyse diurnal oxygen curves from various sources, but there was no evidence to show that the rivers were spatially homogeneous with respect to oxygen distribution.

In the Ivel, where average water velocities rarely exceed 6 cm/sec, the exchange of oxygen through the water surface is relatively unimportant; the oxygen contributed by diffusion in the curves shown for May, 1964 amounted to a loss of 1 g/m^2/ day. In a previous study (Edwards and Owens 1962), the oxygen contributed by diffusion ranged from a loss of 1 g/m^2/ day, when the water was predominantly supersaturated, to a gain of 4 g/m^2/ day. In shallow, swiftly flowing streams the rates of diffusion can be so great that the effects of stream metabolism on the distribution of dissolved oxygen are generally masked (McConnell and Sigler 1959; Hoskin 1959).

The extent to which variation in the exchange coefficient could affect the calculated rates of community respiration and photosynthesis given in Table 1 is shown in Fig. 4. Assuming extreme values of 2 and 10 cm/h for the exchange coefficient, the daily respiration and photosynthesis vary between 7.8 and 11.6, and 9.9 and 15.6 g oxygen/m^2/day respectively; the observed rates determined, using an exchange coefficient of 5.5 cm/h, were 9.4 and 12.4 g oxygen/m^2/ day. Adoption of these extreme values of the exchange coefficient could lead to errors in the estimation of respiration and photosynthesis of about 25 per cent. Where diurnal rhythms in the oxygen concentration are less pronounced, or where there are lower exchange coefficients, estimations of the gross primary productivity and respiration will be less affected by errors in the determination of the exchange coefficient.

A possible source of error in the estimation of gross primary production is the loss of oxygen to the atmosphere in the form of bubbles. Odum (1957), from observations in Silver Springs, Florida, where the dissolved oxygen concentration did not exceed the air-saturation value, found that the volume of gas bubbles released was 224 ml/m^2/ day; even assuming these to be composed of pure oxygen, less than 1 per cent of the oxygen produced was lost in this way.

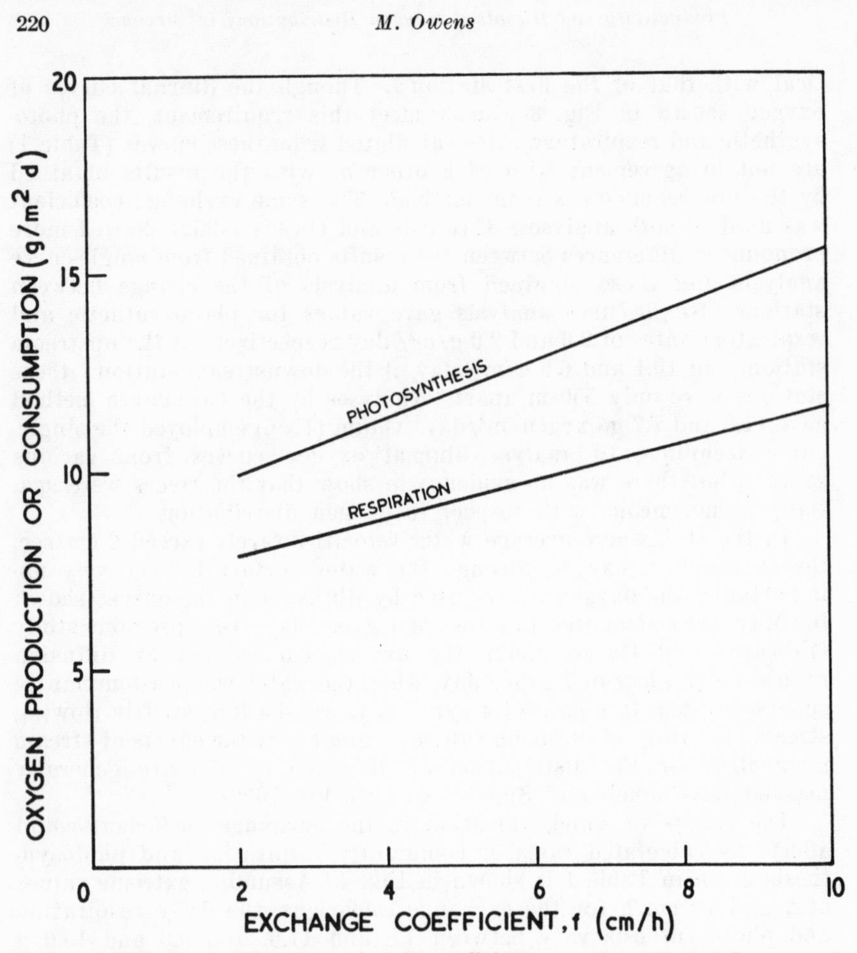

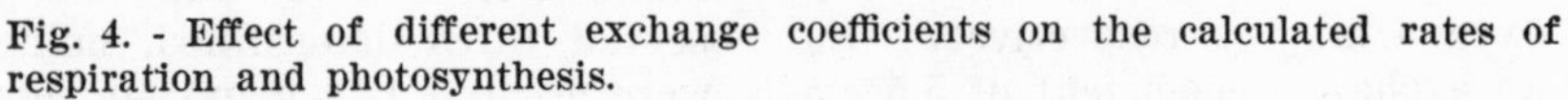

Fig. 4. - Effect of different exchange coefficients on the calculated rates of respiration and photosynthesis.

During the summers of 1962 and 1963, bubbles of gas released from weedy areas of the Ivel were collected in glass funnels. Fig. 5(a) shows the bubble loss from beds of *Hippuris vulgaris* and *Berula erecta* on 20th June, 1962, together with surface light intensities and dissolved-oxygen concentrations in the overlying water. The total loss of gas was about 1000 ml/m^2/day from each weed-bed, and, as there is some evidence that when these bubbles reach the surface of the water they are in equilibrium or slightly enriched with oxygen in respect to the water flowing past the plant, this loss is equivalent to about 6.5% of the total oxygen production. Fig. 5(b) shows the results of similar experiments carried out in August 1963, when the total losses of gas from *Hippuris* and *Berula* beds were equivalent to 4% of the total oxygen production.

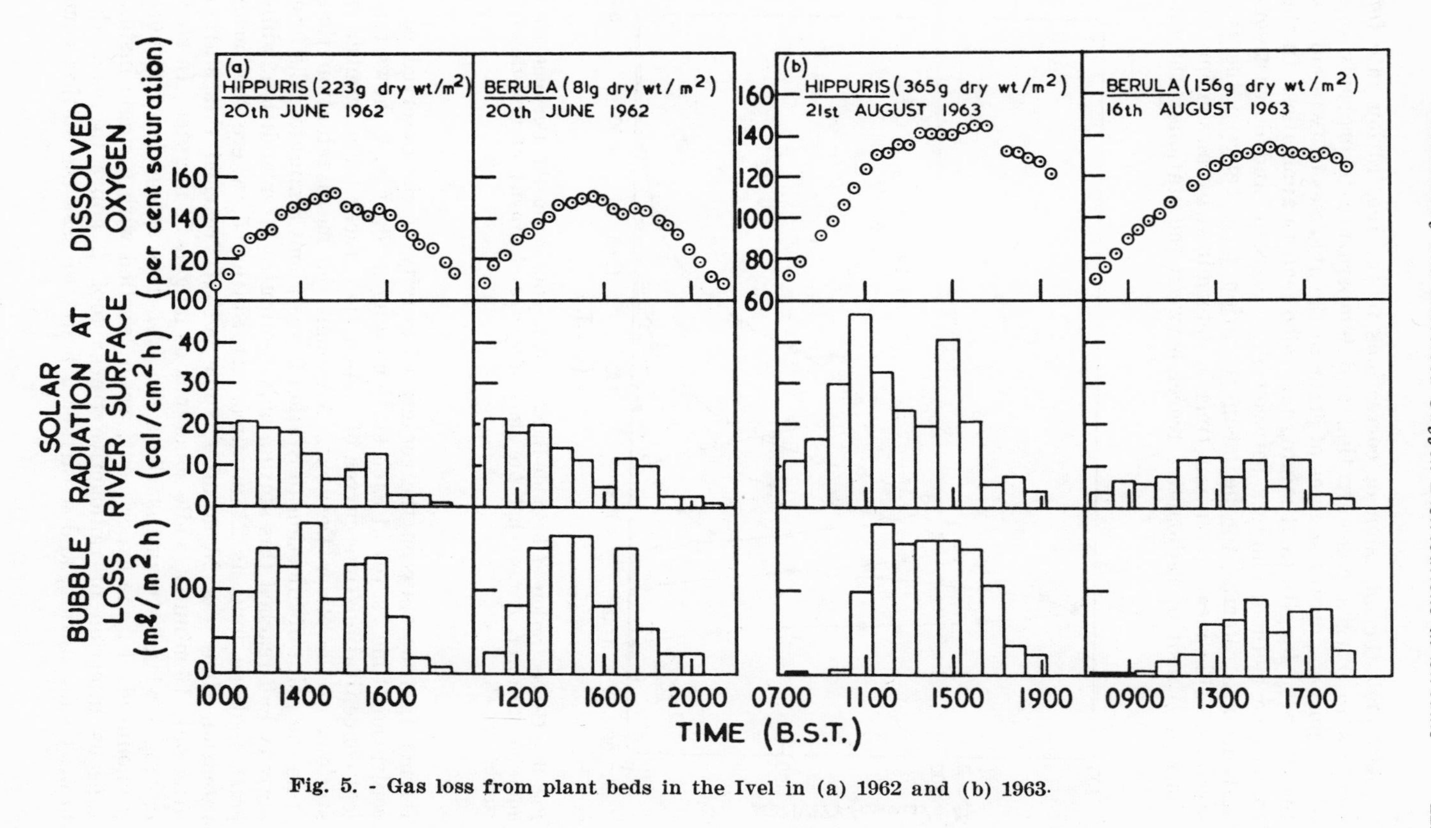

Fig. 5. - Gas loss from plant beds in the Ivel in (a) 1962 and (b) 1963.

Although the omission of corrections to the respiration rate for changes in oxygen concentration and temperature is probably one of the most important sources of error in the analyses of twin curves, the effects of longitudinal mixing may also lead to anomalies. In the twin-curve method, changes between the upstream and downstream stations are referred to the mean retention time between the two stations. This mean retention time is determined by measuring the times of transit of radioactive tracer between the stations; the dif-

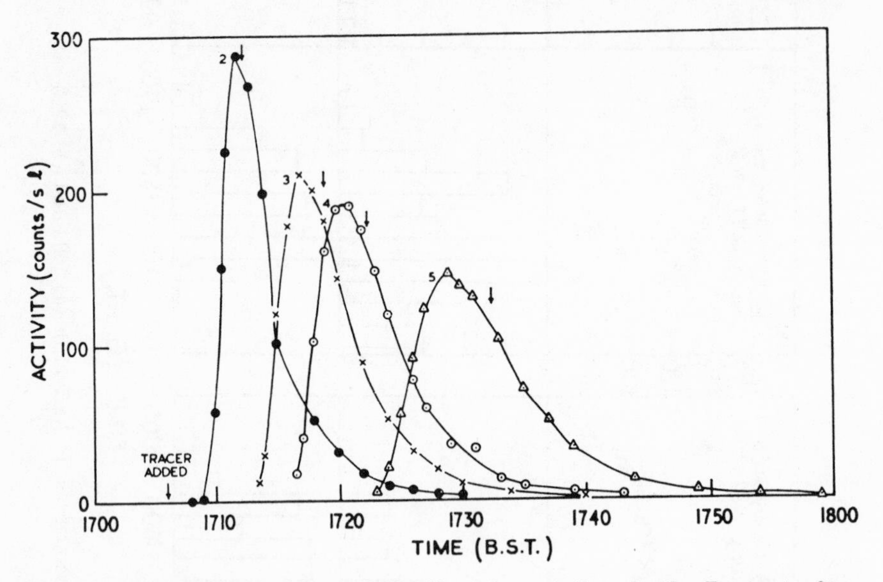

Fig. 6. - Concentration of bromide-Br 82 at four stations in the Derwent after addition of tracer. Arrows indicate time of centre of gravity of flow-through curve.

ference in time between the centers of gravity of the observed concentration/time curves at each station being used. Fig. 6 shows typical concentration/time curves of radioactive tracer for a number of stations in the River Derwent. Although the mean retention time from the point of injection to Station 2 was about 8 minutes, the first activity had reached this station only 3 minutes after addition, while some took as long as 24 minutes. At Station 5, where the mean retention time was 26 minutes, some activity reached this station after only 16 minutes, while some took as long as 54 minutes. In slow moving lowland streams, such as the Ivel, this spread is even more pronounced. In one experiment, water labelled with a single instantaneous injection of radioactive tracer took 2 hours to reach the downstream station, and a further 3 hours to pass the station completely, while the mean retention time from the point of addition was 3¼ hours.

Attempts must be made to embody mixing corrections, based on the distribution of flow tracers, into the analyses of changes in the oxygen concentration to see if the rates of gross primary production and community respiration so obtained will differ markedly from those obtained without the application of this elaborate refinement. Attempts to determine values of the longitudinal mixing coefficients, employing methods described by Thomas (1958) and Krenkel and Orlob (1962), have so far proved unsuccessful (Owens, Edwards and Gibbs 1964). Further work is in progress to develop a method for determining these coefficients.

REFERENCES

Agg, A. R., N. T. Mitchell, and G. E. Eden. 1961. The use of lithium as a tracer for measuring rates of flow of water or sewage. - J. Proc. Inst. Sew. Purif. 240-245.

Beyers, R. J. 1963. A characteristic diurnal metabolic pattern in balanced microcosms. - Publ. Inst. Mar. Sci. Univ. Tex. *9* : 19-27.

Churchill, M. A., H. L. Elmore, and R. A. Buckingham. 1962. The prediction of stream reaeration rates. - Int. J. Air Wat. Pollut. *6* : 467-504.

Committee on Sanitary Engineering Research. 1961. Thirty-first Progress Report. Effect of water temperature on stream reaeration. - J. Sanit. Eng. Div. Am. Soc. Civ. Engrs. *87*, SA6 : 59-71.

Copeland, B. J. and W. R. Duffer. 1964. Use of a clear plastic dome to measure gaseous diffusion rates in natural waters. - Limnol. Oceanog. *9* : 494-499.

Department of Scientific and Industrial Research. 1964. Effects of Polluting Discharges on the Thames Estuary. - H. M. Stationery Office, London. pp. 354-360.

Dobbins, W. E. 1964. BOD and oxygen relationships in streams. - J. Sanit. Eng. Div. Am. Soc. Civ. Engrs. *90*, SA3 : 53-78

Downing, A. L., K. V. Melbourne, and A. M. Bruce. 1957. The effect of contaminants on the rate of aeration of water. - J. Appl. Chem., London *7* : 590-596.

Eden, G. E. 1959. Some uses of radioisotopes in the study of sewage treatment processes. - J. Proc. Inst. Sew. Purif. 522-538.

Edwards, R. W. 1962. Some effects of plants and animals on the conditions in freshwater streams with particular reference to their oxygen balance. Int. J. Air Wat. Pollut. *6* : 505-520.

— and M. Owens. 1960. The effects of plants on river conditions I. Summer crops and estimates of net productivity of macrophytes in a chalk stream. J. Ecol. *48* : 151-160.

— and M. Owens. 1962. The effects of plants on river conditions IV. The oxygen balance of a chalk stream. - J. Ecol. *50* : 207-220.

— M. Owens, and J. W. Gibbs. 1961. Estimates of surface aeration in two streams. - J. Inst. Wat. Engrs. *15* : 395-405.

— and H. L. J. Rolley. 1965. Oxygen consumption of river muds. - J. Ecol. *53* : 1-19.

Gameson, A. L. H., G. A. Truesdale, and A. L. Downing. 1955. Reaeration studies in a lakeland beck. - J. Inst. Wat. Engrs. *9* : 57-94.

Gessner, F. and F. Pannier. 1958. Der Sauerstoffverbrauch der Wasserpflanzen bei verschiedenen Sauerstoffspannungen - Hydrobiologia. *10* : 325-351.

Gunnerson, C. G. and T. E. Bailey. 1963. Oxygen relationships in the Sacramento River. - J. Sanit. Eng. Div. Am. Soc. Civ. Engrs. *89*, SA4 : 95-124.
Hoskin, C. M. 1959. Studies of oxygen metabolism of streams in North Carolina. Publ. Inst. Mar. Sci. Univ. Tex. *6* : 186-192.
Krenkel, P. A. and G. T. Orlob. 1962. Turbulent diffusion and the reaeration coefficient. - J. Sanit. Eng. Div. Am. Soc. Civ. Engrs. *88*, SA2 : 53-83.
McConnell, W. J. and W. F. Sigler. 1959. Chlorophyll and productivity in a mountain river. - Limnol. Oceanog. *4* :335-351.
O'Connor, D. J. and W. E. Dobbins. 1956. Mechanism of reaeration in natural streams. - J. Sanit. Eng. Div. Am. Soc. Civ. Engrs. *82*, SA6 ; Paper 1115.
Odum, H. T. 1956 Primary production in flowing waters. - Limnol. Oceanog. *1* : 103-117.
— 1957. Trophic structure and productivity of Silver Springs, Florida. - Ecol. Monogr. *27* : 55-112.
— and C. M. Hoskin. 1958. Comparative studies on the metabolism of marine waters. - Publ. Inst. Mar. Sci. Univ. Tex. *5* : 15-46.
— W. L. Siler, R. J. Beyers, and N. Armstrong. 1963. Experiments with engineering of marine ecosystems. - Publ. Inst. Mar. Sci. Univ. Tex. *9* : 373-403.
— and R. F. Wilson. 1962. Further studies on reaeration and metabolism of Texas Bays 1958-1960. - Publ. Inst. Mar. Sci. Tex. *8* : 23-55.
Owens, M. and R. W. Edwards. 1961. The effects of plants on river conditions II. Further crop studies and estimates of net productivity of macrophytes in a chalk stream. - J. Ecol. *49* : 119-126.
— — 1962. The effects of plants on river conditions III. Crop studies and estimates of net productivity of macrophytes in four streams in Southern England. - J. Ecol. *50* : 157-162.
— — and J. W. Gibbs. 1964. Some reaeration studies in streams. - Int. J. Air Wat. Pollut. *8* : 469-486.
— and P. J. Maris. 1964. Some factors affecting the respiration of some aquatic plants. - Hydrobiologia *23* : 533-543.
Streeter, H. W. and E. B. Phelps. 1925. A study of the pollution and natural purification of the Ohio River III. - U. S. Public Health Bull., Wash., No. 146. 75 pp. (Reprinted 1958).
Thomas, H. A. 1958. Mixing and diffusion of wastes in streams. In 'Oxygen Relationships in Streams'. Robert A. Taft Sanitary Engineering Center, Tech. Rep. W58-2 : 97-106.
ZoBell, C. E. and J. Stadler. 1940. The effect of oxygen tension on the oxygen uptake of lake bacteria. - J. Bact. *39* :307-322.

SUBLITTORAL BENTHIC PRODUCTION ON A SANDY BEACH

J. H. STEELE
Marine Laboratory
Aberdeen, Scotland

For bibliographic citation of this paper, see page 10.

Abstract

Preliminary results from a year's sampling on an exposed sandy beach on the west coast of Scotland show the following results:

(*a*) Chlorophyll *a* content in the sand is low at low water mark and increases to a depth of water of 15 m below low water. There is little sign of seasonal cycles.

(*b*) Chlorophyll *a* is constant to a depth in the sand of 10-15 cm at low water and 3-10 cm at 15 m with intermediate values at intermediate depths.

(*c*) Only 3-5% of the organic matter is unattached to sand grains and there is a good relation between chlorophyll *a* and attached organic matter.

(*d*) Using C^{14}, estimates of organic carbon production by photosynthesis suggest low yearly uptake values, probably about 10 g C/m^2 year. The depth of the « euphotic zone » in sand is about 3 mm.

The major problem raised by these preliminary data concerns the ability of the benthic plant populations to show a positive net production based solely on photosynthesis.

SOME BASIC DATA FOR INVESTIGATIONS OF THE PRODUCTIVITY OF AQUATIC MACROPHYTES

D. F. WESTLAKE

River Laboratory

Freshwater Biological Association, Wareham, England

For bibliographic citation of this paper, see page 10.

Abstract

The biomass and annual production of different types of community are briefly compared. In designing productivity experiments some initial simple but detailed information is often required. Data on the volume, dry weight, ash, carbon, and energy contents of aquatic plants are reviewed and methods of analysis are discussed when special problems arise. A survey of the proportion of the biomass that is to be found underground shows that it is often very important to sample the roots and rhizomes, especially with emergent plants. The relations between biomass and production are discussed, illustrated by some relevant investigations. Information on losses by death, damage, disease, and grazing, and on the age of parts persisting for more than a year, is reviewed.

INTRODUCTION

Ecological investigations are often hampered by a lack of what Lund has called « basic data » (1964). That is to say quite simple, but detailed, information on the chemical composition, morphology, and phenology of the organisms present. This paper attempts to summarize such basic data as are available for aquatic macrophytes, particularly those relevant to plans for studies of the productivity of freshwater communities.

THE BIOMASS AND PRODUCTIVITY OF AQUATIC COMMUNITIES

In general, macrophytic communities are more productive per unit area than phytoplankton communities under comparable conditions (Westlake 1963). On the other hand, in lakes which are generally deeper than some 10 m (depending on the transparency), oceans, and deep, turbid rivers, the macrophytes are confined to the littoral zone and their total production will normally be less than that of the phytoplankton (Gessner 1959, Westlake 1960, Straskraba in press). Many communities of emergent plants are more productive than the adjacent submerged communities, but this is very dependent on local conditions such as exposure to wave action. Typical values for the net production of fertile sites are given in Table 1; the gross production is probably 1.5-3 times greater.

Table 1. - The annual net primary production of aquatic communities on fertile sites (Westlake 1963)

(m. t dry organic matter/ha)

Marine phytoplankton	1 - 4.5
Lake phytoplankton	1 - 9
Freshwater submerged macrophytes (water weeds)	4 - 20
Marine submerged macrophytes (seaweeds)	25 - 40
Marine emergent macrophytes (salt marsh)	25 - 85?
Freshwater emergent macrophytes (reedswamp)	30 - 85

The biomass of phytoplankton is often only 10-100 g dry weight/m^2, whereas the biomass of macrophytes is between 0.2 and 10 kg dry wt/m^2 on fertile sites (e.g. Westlake 1963). In most phytoplankton communities the seasonal maximum biomass is much less than the annual production, but in many freshwater communities of submerged macrophytes the two are similar, and in many communities of submerged marine macrophytes and emergent macrophytes the biomass is considerably greater (Westlake 1963). The annual turnover of macrophytes is discussed in more detail later.

The biomass of luxuriant periphyton is probably about 100-500 g dry wt/m^2 (from Rickett 1924, Odum 1957, Edwards and Owens 1965). In many waters such populations, and corresponding productivities, are only maintained for short periods, but the possibility that the periphyton makes an important contribution must always be considered.

CRITERIA OF BIOMASS AND PRODUCTIVITY

Possible criteria, which have been used in various investigations, include volume, fresh weight, fresh organic weight, dry weight, dry organic weight, carbon content, oxygen or carbon dioxide exchange, energy content, and chlorophyll content. Other analyses which may be useful include the ash, protein, carbohydrate and fat contents. To compare investigations using different criteria it is necessary to know their inter-relations (Westlake 1965). Data on phytoplankton have been reviewed (Strickland 1960) and the early information on marine organisms is assembled by Vinogradov (1953). More recent determinations on many marine macrophytes have been published in papers from the Institute of Seaweed Research, Inveresk, and the Norsk Institut for Tang-og-Tare-forskning. Here I propose to attempt a survey of data on freshwater macrophytes partly derived from the literature and partly from my own observations.

Volume. This is a poor measure because the density of most freshwater macrophytes is mainly a function of the proportion of internal air-spaces, so the relations between volume and any weight criterion are very variable. Also it is particularly difficult to determine accurately with submerged plants. The best procedure is to remove adherent water in a spin-drier for a minute, determine the volume of the plant by displacement of water and repeat this sequence several times. A smaller volume is recorded after each operation and if volume is plotted against total time of spinning a curve is obtained which may be extrapolated to zero to give the original volume. For example the volume of a sample of *Potamogeton pectinatus* by extrapolation was 1.60 cm^3 per 1.0 g fresh weight and after spin-drying four times, one minute each time, the observed volume was 1.20 cm^3/g. Forsberg (1960) gives values for the relation between dry weight and fresh volume for several species of submerged macrophytes, ranging between 11 and 17 cm^3/g for non-calcareous plants but falling to 6 cm^3/g for the calcareous *Chara fragilis.* By assuming appropriate water contents it can be estimated that the volume of 1.0 g of fresh weight ranges from about 1.3 cm^3 for *Myriophyllum verticillatum* to about 1.06 cm^3 for *Nitella mucronata* and *C. fragilis.* Hejny (1960) reports values of 1.5-2.3 ml/g fresh weight for *Chara foetida* but these seem rather high for a plant without air spaces. He also gives data for underground parts, green shoots and whole plants of several emergent species including *Cyperus fuscus* (c. 1.1 cm^3/g fresh weight), *Butomus umbellatus* (1.3-1.7 cm^3/g) and *Scirpus lacustris* (2.6-3.7 cm^3/g whole plant, 3.5-5.6 cm^3/g green shoots). The others lie within this range. The shoots usually had a higher volume than the underground parts, especially with *S. lacustris.* Seidel (1955) has found aerial shoots of this plant with 3.1-4.7 cm^3/g fresh weight. The volume of most parts of the floating plant, *Eichhornia crassipes,* is about 1.3 cm^3/g but the floats occupy over 7 cm^3/g (Penfound and Earle 1948).

Fresh weight. This is the weight of the living plant without any adherent water. In practice the wet weight is determined, i.e. some residual adherent water remains, but with suitable techniques there is little difference between the two. This is often the form in which all field determinations are made and the other criteria are estimated from it by factors determined from analyses of sub-samples. Hence it is important to determine fresh weight as accurately as possible. The best method is to use a spin-drier, which gives consistent results rapidly (Gortner 1934, Edwards and Owens 1960). The latter compared spin-drying with air-draining on racks. In the absence of a spin-drier the weed may be hung on lines and turned frequently until water almost ceases to drip from it and the outer-most fronds are just beginning to shrivel (Westlake 1961). The time taken may be from 30 minutes to over 3 hours, depending on the weather. Small pieces of weed falling down should be collected on damp sacks.

A check on spin-drying in air has been made by centrifuging small samples of weed in air and under petroleum ether, determining both volumes and weights and comparing densities. Petroleum ether displaces water added to cotton wool quantitatively and it does not displace water from land plants or spin-dried water plants. A Behren's test showed that the probability of obtaining the observed difference by chance was much greater than 1 in 20.

Comparisons of spin-drying and air-draining on lines out-of-doors have been made with samples of *Potamogeton pectinatus*. The air-drained weight fell rapidly at first but the curve gradually flattened out; the final slope and the general level, in relation to the spin-dried weight, depended on weather conditions. On warm or hot days, with a breeze blowing, the spin-dried weight was reached after about 30 minutes. Air-drained and spin-dried weights were very similar, within the 5 per cent experimental error. About 5 per cent of the air-drained weight was adherent water but this was presumably balanced by losses of internal water. All adherent water was gone after 2-3 hours, or even earlier on hot days. The final rate of loss of internal water was about 2 per cent/hour on a hot day. On still, humid days, however, 10-30 per cent of the air-drained weight was adherent water even after 3 hours.

Dry weight. This should be determined by drying to constant weight in an oven at 105°C but sometimes air-dry weights are determined after leaving the fresh sample to come to equilibrium with atmospheric moisture. Air-dry weights, and the weights of oven-dried plant materials allowed to cool, or stored, in contact with air, are 5-10 per cent higher than the true oven-dry weight. Some generalizations about the dry matter content of fresh material may be made, but are not to be relied on for any particular sample. The dry weight of many of the submerged higher plants, and the softer emergent plants, ranges between 5 and 20 per cent of the fresh weight, but if calcareous deposits are present it may approach 25 per cent. Plants with floating or emergent leaves tend to contain more dry matter (c. 15 per cent) than those which are totally submerged (c. 10 per cent). The macrophytic algae and mosses are similar, ranging from *Nitella,* which may contain as little as 4 per cent, to *Cladophora* with about 15 per cent and calcareous populations of *Chara* with over 20 per cent. The larger reeds, rushes, and sedges contain more dry matter, and, when they are mature, it forms from 14 to over 40 per cent of the fresh weight. The very high values here may include some dead material, and the average for living plants is probably about 25 per cent. There is little sign of any seasonal variation in submerged macrophytes. The emergent plant *Glyceria maxima* shows an increase of about 10 per cent between spring and summer in the dry matter content of the tops. The underground parts of *G. maxima* contain slightly less dry matter than the tops in the spring and the increase is only about 3 per cent

(Table 2). When woody rhizomes are present (e.g. *Scirpus lacustris, Typha angustifolia*), it has been found that the average dry matter content of the underground material is very similar to that of the tops in late summer, but that the rhizomes contain more and the roots less dry matter.

The chief practical problems in the determination of ash and carbon contents arise from the presence of calcareous materials, but if suitable methods are used with adequate precautions there are no insuperable difficulties. The ash content is usually determined as the residue of an ignition at 550°C. This needs to be performed carefully to avoid contamination by uncombusted carbon, losses by « spitting », and losses by the decomposition of calcium carbonate which starts above 550°C. If magnesium carbonate, which decomposes above 350°C, is present to a significant extent the carbonates should be decomposed by acetic acid, the displaced carbonate determined and added to the weight of the residue after ignition. This will also ensure the inclusion of bicarbonate in the ash value.

Organic carbon is best determined by dry combustion of the material (e.g. Belcher and Ingram 1950). Carbonate and bicarbonate should be removed before the combustion, by treatment with sulphur dioxide. Wet oxidations are to be avoided (see Westlake 1963).

In submerged higher plants the ash is usually between 15 and 25 per cent of the dry weight, but it may exceed 50 per cent in some very calcareous material and be as low as 5-15 per cent in a few species, most of which have floating leaves. The macrophytic algae are similar except that calcareous *Characeae* may contain as much as 70 per cent as ash. Aquatic mosses appear to contain only 3-7 per cent ash in soft waters but possibly up to 40 per cent in harder waters.

The softer emergent plants tend to have less ash than the submerged macrophytes (usually 10-20 per cent) though a few, in particular *Eichhornia crassipes,* can reach over 30 per cent in mineral- rich waters. The harder emergent reeds, rushes and sedges contain markedly less ash, ranging from 2 to 12 per cent in their green shoots.

As the great majority of aquatic macrophytes are not woody, and do not store great amounts of sugar or fat, their organic carbon contents have a very limited range when expressed in terms of the ash-free dry weight. When data on rhizomes are available, the range may be extended. Few have less than 43 per cent or more than 48 per cent, and the mean is about 46.5 per cent. Some published values of over 50 per cent are probably due to errors in the methods used for the carbon or the ash analysis. A lower value of about 43 per cent for a *Nitella* sp. seems well established and may be due to the reducing sugars in the cell sap. Occasional samples with contents between 48 and 50 per cent may be rich in protein. The common practice of assuming the organic carbon to be 45 per cent of the dry weight is very unsound because of the relatively high

Table 2. - Dry weight, ash and carbon content of some freshwater macrophytes.

		Dry matter % fresh wt	Ash % dry wt	Carbon % ash-free wt
(1)	*Berula erecta*. R. Ivel, 6-9/1958	6-7	24-27	41-46
(1 & 4)	*Callitriche stagnalis*. R. Ivel, 6-9/1958, 7/1960	7- 9	11-23	44-48
(1)	*Cladophora* sp., R. Lea, 6/1958	16	25 (sand?)	48
(2)	*Eichhornia crassipes*. Marion Co., Florida, 8/1962	6	—	—
(3)	*Elodea canadensis*. Windermere, 7/1962	9	—	—
(3)	*Fontinalis antipyretica*. R. Eamont, 10/1959	12	7	49
(2)	*Glyceria maxima*. 5/1964 (underground) R. Frome	14	9	—
	5/1964 (green shoots) R. Frome	18	9	—
	7/1964 (underground) R. Frome	17	8	—
	7/1964 (green shoots) R. Frome	28	8	—
(4)	*Hippuris vulgaris*. R. Ivel, 5 & 7/1960	11	15	47
(3)	*Littorella uniflora*. Dubbs Reservoir, 9/1959	9-10	24	46
(1)	*Mentha aquatica*. R. Ivel, 9/1958	9	16	43
(1)	*Myosotis palustris*. R. Ivel, 9/1958	6	20	45
(3)	*Myriophyllum alterniflorum*. Windermere, 10/1959	6	25	48
(1)	*Nasturtium officinale*. R. Ivel, 9/1958	5	23	44
(5)	*Nitella opaca*. Windermere, 2/1955	—	12	43
(3)	*Nitella* sp. Windermere, 10/1959	4	17	43
(4)	*Potamogeton densus*. R. Chess, 7/1960	13	12	47-48
(4)	*P. lucens*. R. Yare, 6/1960	21	32	43
(1)	*P. pectinatus*. R. Lea, 6/1958	14	19	48
(3)	*P. praelongus*. Grasmere, 10/1959	8	14	47
(3)	*Ranunculus heterophyllus*. R. Lowther, 10/1959	8	10	45
(4)	*R. pseudofluitans*. R. Ivel, 7/1960	12-13	12	47
(1)	*Sparganium* sp. R. Ivel, 9/1958	9	18	42
(2)	*Scirpus lacustris*. Velky Pálenec, Czechoslovakia, 10/1963 (underground)	26	—	—
	(green shoots)	26	8	—
(2)	*Typha angustifolia* Velky Pálenec, Czechoslovakia, 10/1963 (underground)	22	—	—
	(green shoots)	23	9	—

(1) Dry weights and carbon/dry weight from Edwards and Owens (1960), ash values by personal communication.
(2) D. F. Westlake.
(3) Material collected and dried by D. F. Westlake, ash and carbon by Mr. F. J. H. Mackereth and Mr. J. Heron.
(4) Material collected and dried by Owens and Edwards (1962), ash and carbon by Mr. F. J. H. Mackereth and Mr. J. Heron.
(5) Material collected by Dr. J. W. G. Lund, analyses by Mr. F. J. H. Mackereth and Mr. J. Heron.

ash contents. In calcareous material, the true factor may be less than 30 per cent, and even in normal submerged aquatics it is about 35-40 per cent.

There is little evidence of seasonal variations in ash or carbon contents of freshwater plants apart from some indications that calcareous deposits may increase during periods of rapid growth.

On the basis of the organic carbon content and proximate analyses the energy content of most aquatic macrophytes will lie between 4.3 and 4.8 kcal/g organic matter, with a mean value of 4.6. In the harder emergent plants the calorific value is generally below the average. Most of the data on proximate analyses is discussed by Straskraba (in press). Bray (1962) has found 4.34 kcal/g in hybrid *Typha*, by bomb calorimetry, which compares with calculated values of 4.21-4.35 for *T. latifolia.* I have found 4.5-4.8 kcal/g in *G. maxima.*

Most of the published data on the composition of freshwater macrophytes is reviewed in Straskraba (in press), where values for individual species and references may be found. Some data, most of which have not been published elsewhere, are collected in Table 2.

UNDERGROUND PARTS

It is often considered that the roots of submerged macrophytes can be neglected (Edwards and Owens 1960, Forsberg 1960, Shcherbakov 1950). Many have very few roots or rhizoids (e.g. *Ceratophyllum,* many seaweeds), and in others the roots are less than 10 per cent of the biomass (e.g. 2.6 per cent for *Elodea canadensis,* Borutskii 1950). However it cannot be assumed that roots are always a small proportion for they can form 46-55 per cent of the biomass of *Littorella uniflora,* and Wilson (1935) has suggested that this is characteristic of rosette plants. Investigations of plants such as *Potamogeton pectinatus,* which has underground storage organs, and *Ranunculus pseudofluitans,* which has roots for anchorage in strong currents, are needed. The marine angiosperm *Thalassia testudinum* may have 75-85 per cent of the biomass in the sub-stratum (Burkholder, Burkholder and Rivero 1959).

Many perennial emergent plants have a large proportion of their biomass buried, even at the time of the seasonal maximum standing crop of green shoots (see Table 3, also many other species photographed by Hejny (1960). There is general agreement that the underground parts of *Phragmites communis* are at least equal to the green shoots, but the evidence on their relative proportions is confused (Aario 1933, Pallis 1916, Seidel 1956, Ranwell, personal communication). Hurlimann (1951) found some influence of habitat conditions on the ratio of root to shoot in *P. communis,* and the roots of healthy seedlings, two months to two years old, formed between 16 and 38 per cent of the biomass. These results probably refer only to the true roots.

Seidel states that the rhizomes of *Scirpus lacustris* are 95 per cent of the whole plant, but this seems an extremely high proportion if the samples were taken in summer. Mature plants of two annual emergent species, *Zizania aquatica* and *Cyperus fuscus*, have only 7-8 per cent underground (Rogosin 1958, Hejny 1960).

In August, plants of *Eichhornia crassipes* from a wide ditch in Marion County, Florida had 48 per cent of their biomass as roots when growing in open water, but only 21 per cent near to the bank. In Louisiana, Penfound and Earle (1948) found that the roots and rhizomes were 29 and 23 per cent respectively in April, and about 20 per cent and 10 per cent in June.

Table 3. - Relative weights of parts of perennial reedswamp plants (% biomass)

			Green shoots		Underground	
			Tops	Stubble	Rhizomes	Roots
(1)	*Alisma plantago-aquatica*, July-August, fresh		60		40	
(2)	*Butomus umbellatus*. August, fresh		<64		>36	
(2)	*Carex riparia*. August, fresh		<70		>30	
(1)	*Equisetum fluviatile*. July-August, fresh		17		83	
(3)	*Glyceria maxima*. July,	fresh	49±15		51	
		dry	61	(2.4)	39	
	September,	fresh	33±16		67	
		dry	45	(3.6)	55	
(1)	*Phragmites communis*. July-August, fresh		17		83	
(2)	*Scirpus lacustris*. August, fresh		<54		>46	
(1)	*S. lacustris*. July-August, fresh		10		90	
(4)	*S. lacustris*. October,	fresh	17	8	47	28
		dry	25		75	
(2)	*Sparganium erectum*. August, fresh		<70		>30	
(1)	*Sparganium ramosum*. July-August, fresh		33		66	
(2)	*Typha angustifolia*. August, fresh		<48		>52	
(4)	*T. angustifolia*. October,	fresh	20	21	34	25
		dry	43		57	
(5)	*T. hybrid*. September, dry		37		64	
(2)	*T. latifolia*. August, fresh		<54		>46	
(1)	*T. latifolia*. July-August, fresh		50		50	

(1) Aario (1933). Finnish lake.
(2) Hejny (1960). Data for undamaged plants grown from small pieces planted in ricefields in Czechoslovakia and Hungary during July. Proportion underground is likely to be higher for mature plants.
(3) Biomass samples taken by River Frome, Kingston Maurward. Stubble included in underground. ± mean range.
(4) Plants sampled at Velky Pálenec, Czechoslovakia. Stubble included in dry green shoots.
(5) Bray *et al* (1959). Minnesota swamp.

The bases of the green shoots (« stubble ») may be a significant part of the biomass, especially if no special care is taken when cutting the tops of the plant off (Table 3). The stubble of *Glyceria maxima* cut at the soil (litter) surface was between 2 and 4 per cent of the biomass (4-8 per cent of green shoots) during the growing season. When *Scirpus lacustris* was cut at about 15 cm above the soil, 8 per cent of the biomass (32 per cent of the green shoots) was stubble, and for *Typha angustifolia* cut at about 25 cm, just over 20 per cent was stubble (50 per cent of the green shoots). The total length of the average green shoot was approximately 80 cm for the *Glyceria* and 170 cm for the other two. This may be important when comparing data on the standing group of tops with biomass determinations.

THE ESTIMATION OF PRODUCTIVITY FROM BIOMASS CHANGES

If there are no losses of plant material except by respiration between two sampling times, the net production, by definition, is equal to the observed change in biomass. The biomass changes may be determined from random samples in a community, or from plants of known weight introduced into a community and subsequently recovered (the « box » technique).

In agricultural and horticultural research, methods fundamentally similar to the box technique are commonly used, and elaborate procedures have been developed for growing large numbers of introduced plants in outdoor and greenhouse communities under different conditions. Underwater planting and recovery is more difficult, and water movements prevent the maintenance of different temperature and chemical conditions, so only a few, relatively simple, experiments have been made. Also, many aquatic plants are perennial, and the growth of the first year after planting out may not be representative. Borutskii (1950) and Odum (1957) have planted out submerged macrophytes in boxes or cages, respectively. Penfound and Earle (1948), Hitchcock et al. (1949), and Earle (1956) have placed plants of *Eichhornia crassipes* in ponds, but most of these results were in terms of increase in numbers. Hejny (1960) has determined the growth of pieces of several reedswamp plants in rice fields for short periods. Hurlimann (1951) has grown *Phragmites communis* plants, mostly from seed, in boxes and experimental plots. None of these investigations have given good data for the annual productivity of natural stands; the chief objection being that the planted individuals were not placed among a natural population and were thus relatively free from grazing, damage, and competition for nutrients and light.

For annuals the initial biomass (seed weight) is negligible (cf. Filzer 1951). If a series of determinations of biomass are made the typical growth curve is sigmoid, but finally the biomass starts to

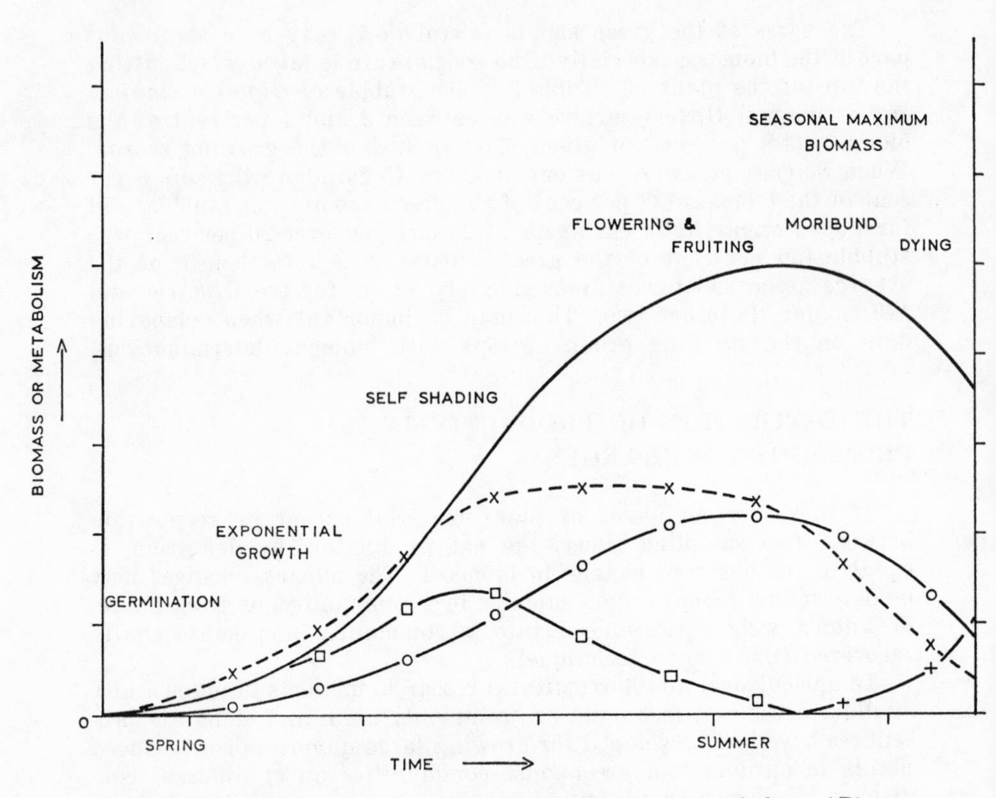

Fig. 1. - Hypothetical growth and metabolism curves for an annual plant. / Biomass. X Current gross productivity. □ Current net productivity. ○ Current respiration rate. + Death losses.

decrease (Fig. 1). It seems probable that gross productivity reaches a plateau and later declines, especially in the older tissues, but that respiration continues to increase with biomass, so that the net productivity decreases and then becomes negative. In many plants this seems to occur before losses by death of tissues become important. The maximum biomass is reached when the current daily net productivity becomes zero, and this biomass is thus the maximum cumulative net production. Subsequently both the biomass and cumulative net production decrease, and the terminal net production is the material that has not been respired by the time the whole of the plant is dead. The actual annual net production available to the consumers of a community depends on the time at which the plant is exploited. For general comparisons, it is best to assume that the consumer population is ideally adapted to exploit the maximum cumulative net production (Westlake 1963).

Seasonal changes in biomass (or standing crop of green shoots) have been followed in a number of temperate and sub-tropical freshwater macrophyte communities. In general there is a rapid increase in the spring and a summer maximum, similar to the idealised curve for an annual, but sometimes a significant initial biomass is present (Bray, Lawrence and Pearson 1959, Borutskii 1950, Edwards and Owens 1962, Elster and Vollenweider 1961, Forsberg 1959, 1960, Gröntved 1958, Lindemann 1941, Odum 1957, Owens and Edwards 1961, Penfound 1956, Peterson 1913, Wetzel 1964, Zaki 1960). It should be noted that there are some arithmetical errors in the time intervals in Table 5 of Penfound's paper, which affect the data shown in his Fig. 1. Many of the curves observed are irregular, compared with the theoretical curves, because of changes in environmental conditions. Less typical patterns of growth are often found for seaweed communities (e.g. Printz 1950, Walker and Richardson 1955).

Many authors have assumed that the maximum biomass occurs about the time of flowering, usually July or August in the Northern hemisphere. However examination of the data in the above papers shows that, although these assumptions are often true, there is frequently a difference of a month or so between different species, different places, or different years, and in special cases the maximum may be at quite a different time, or there may be two maxima.

If the initial biomass is negligible, or if it dies before the seasonal maximum is reached, and losses are negligible, the seasonal maximum biomass is equal to the maximum cumulative net production (Fig. 2, curves A and B). This may be called annual regrowth, which is manifest when the plants die down to a small quantity of perennating organs, but less obvious when much of the plant survives until the next spring. Many submerged freshwater macrophytes in temperate climates appear to be plants of annual regrowth, but this should always be checked. Some species always die down (e.g. *Potamogeton natans* and *pectinatus*), others persist (e.g. *Ranunculus pseudofluitans*, *Elodea canadensis*), and some vary with local conditions (e.g. *Callitriche stagnalis*, *Myriophyllum spp.*) (see Butcher 1933). If the initial biomass persists without losses until after the seasonal maximum, the annual net production may be obtained from the difference between the final and initial biomass (Fig. 2, curve C), but, more commonly, a variable proportion of the initial biomass is lost (curve D, e.g. *Chara*, *Laminaria cloustoni*). The annual net production is then often 50-80 per cent of the maximum biomass (e.g. Forsberg 1960, Gröntved 1958, Walker 1954, Walker and Richardson 1955). Losses of the current year's production may occur, which will also complicate the estimation of production from biomass changes (curve E).

Some authors have criticised the use of biomass data for estimations of productivity (Forsberg 1959, Penfound 1956, Wetzel 1964), but these criticisms should not be taken as general condemnations

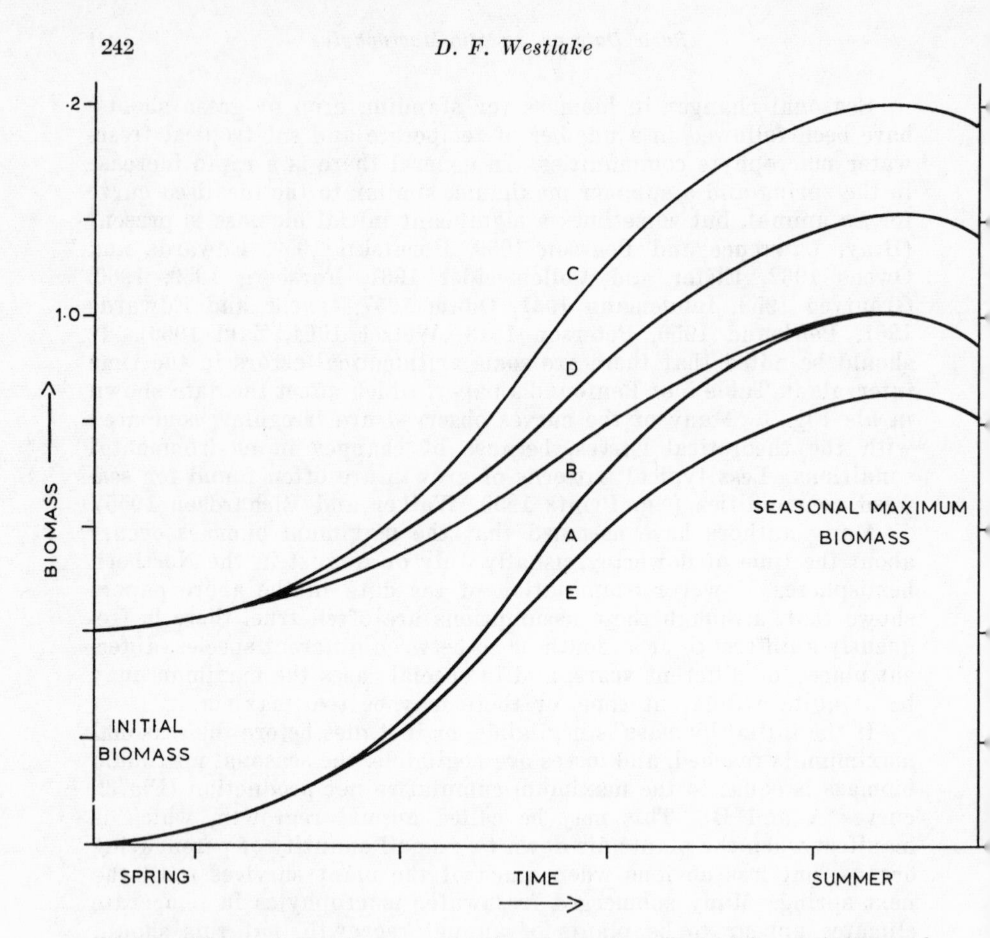

Fig. 2. - Types of growth curves. A. True annual or plant with manifest annual regrowth. B. Plant with obscured annual regrowth. C. Plant with spring biomass persisting until seasonal maximum. D. Plant with only part of spring biomass persisting until seasonal maximum. E. Plant with annual regrowth, losses from current year's biomass before seasonal maximum.

of appropriate methods. A single biomass determination should not be expected to be a measure of the current productivity or production capacity (cf. Fig. 1), but (in the absence of losses) the biomass difference over a short period is a valid measure of current net productivity. The biomass difference over a long period (in the absence of losses) is a measure of the average or cumulative productivity. The «terminal» biomass, which is usually sampled at the end of summer, later than the seasonal maximum, is not a good measure of average productivity because losses have become important (see Penfound). If early losses are important then the biomass methods are difficult to use. In some marine and many tropical communities growth and losses may be almost continuous, so the biomass remains relatively constant and the annual production may be 2-3 times larger. Such situations require special investigations of the losses (e.g. Petersen 1913) or determinations of metabolic rates (e.g. Odum 1957, 1963, Odum, Burkholder and Rivero 1959). In phytoplankton investigations these approaches are nearly always essential.

Very few studies of biomass changes in relation to losses and metabolic changes have been made, and it is to be hoped that the IBP will stimulate such complete investigations. Losses by death, damage, or grazing, from temperate zone plants before the seasonal maximum biomass, have often been considered small. Harper (1918) and Borutskii (1949) estimated the losses from the current year's crop of green shoots of a number of freshwater macrophytes to be 2-10 per cent of the final biomass (possibly post-maximal in some cases). In a *Glyceria maxima* community dead leaves have been found to be 10-20 per cent of the maximum biomass, but some of these may have persisted from the previous year before dying. Losses from seaweeds by wave-action may be very great (Walker and Richardson 1955). Evidence from terrestrial plants suggests that communities not showing obvious signs of losses by death, disease or grazing are unlikely to have lost more than 15 per cent of the current year's production by the time of the maximum biomass (Westlake 1963). Only the phytoplankton communities (and the exploited forest) among the communities compared by Macfadyen (1964) show important grazing losses. The high losses from a *Typha latifolia* community reported by Penfound (1956) either occurred after the maximum or were the result of exceptionally bad frost and drought.

A remarkably detailed investigation of the dynamics of the biomass of *Elodea canadensis* in Lake Beloie has been made by Borutskii (1950). This includes biomass samples throughout the year, details of the proportions and ages of different parts of the plants, and determinations of growth by plants in boxes. He concludes that the annual production could be over five times the seasonal maximum biomass, but there are some errors in the calculations and the data may be interpreted in other ways. It seems probable that most of the material present in the autumn persists through the winter, although it is redistributed among fewer shoots, and eventually dies during the summer, causing a summer minimum. The autumnal maximum biomass is then all material of the current year, and in undisturbed sites, is probably close to the annual production. In the majority of the lake, which is subject to intense human interference, there are probably high losses by damage, and the production there greatly exceeds the biomass. He made a similar study of the green shoots of *Phragmites communis* and estimated that the annual crop was only about 2 per cent greater than the maximum standing crop.

There are two investigations which permit some comparisons to be made between the biomass of plants of annual regrowth and their productivity estimated from observations of metabolic rates. From the oxygen changes observed in a stretch of the River Ivel (Edwards and Owens 1962), and by making some assumptions (Westlake 1963), the annual production may be estimated as about 0.85 kg dry matter/ m^2 in 1959. In 1958 the maximum biomass of part of this stretch was 0.52 kg/m^2, mostly *Berula erecta* (Edwards and Owens 1960). Remembering that the metabolic observations will include some algal production (Edwards and Owens 1965), that neither the area or the year are the same, and that there is some uncertainty in the assumptions made, there is no definite evidence that the maximum biomass is really much less than the annual production.

Wetzel (1964) measured productivity as fixation of radiocarbon by *Ruppia maritima* in Borax Lake during the growing season, and made concurrent determinations of biomass. Assuming that the ash content of R. *maritima* is 20 per cent dry wt, that the carbon content is 47 per cent organic wt, that the radiocarbon fixation is approximately equal to the daytime net productivity and deducting about 10 per cent for night-time respiration losses, the annual net production estimated from his radiocarbon experiments by graphical integration is between 50 and 70 g dry wt/m^2, which is close to the maximum biomass of 64 g/m^2. Again, these values are not known precisely, but losses are probably insignificant for practical purposes.

Even if the green shoots are renewed annually, underground parts may persist for a number of years, and when these form an important part of the biomass it is necessary to determine their age, or to observe their annual increments directly, before the annual production can be estimated. Very little is known about this. Aario (1933) considered that the underground parts were usually several years old. Approximately half the underground material of *Typha* was produced in a year while the annual increments for *Scirpus lacustris* were a much smaller fraction.

Hejny's data on the relative proportions of the annual increments of green shoots and underground parts (1960) can be combined with the proportions in Table 3 to estimate ages. For *S. lacustris* the underground increment was about 70 per cent of the increment of the green shoots and if there was five times as much underground as above, then the underground parts would be about 7 years' accumulation. From similar data the accumulation period underground would be about 1.5 years for *T. latifolia* and just over a year for *Sparganium erectum.*

Observations by Pallis (1916) and Hurlimann (1951) on *Phragmites communis* show that the rhizomes persist for at least three years, and from Merinotti (1941) and Perdue (1958) it appears that the rhizomes of the related *Arundo donax* lasts some four years.

Preliminary observations on marked rhizomes of *S. lacustris* have shown that they persist for at least 3 years, probably more. Similar experiments have indicated the life of the rhizomes of other reed-swamp plants: *Glyceria maxima,* rather less than 2 years; *P. com-*

munis, 2 to more than 3 years; *Sparganium erectum*, about 18 months; *T. latifolia*, less than 2 years. Some indications that soil conditions may influence the length of life were noticed. The roots of most of these plants probably lived for less than a year.

When there is some accumulation underground, and some green shoots die, while others are growing, the determination of production from biomass becomes complicated. Preliminary analyses of data on *Glyceria maxima* suggest that there are several distinct « waves » of the growth of new green shoots. Many initiated in the summer continue growing after the maximum biomass is reached, while the mature shoots are dying. Many of these summer shoots, and also many shoots initiated in the autumn, die during the winter, but some of their production is probably translocated to the rhizomes. In such communities, particularly when the production is largely consumed after death, it is probably best to try to determine the losses of biomass; since the total losses over the year, in a stable community, are equal to the production (cf. Borutskii 1950).

Many aquatic macrophyte communities are scarcely subject to grazing and their production is consumed by bacteria, fungi and detritus feeders. Many species of animals will eat live macrophytes but few rely on this as their sole source of food or cause much damage (Frohne 1938, McGaha 1952, Gajevskaja 1958, Hynes 1960). Most herbivores, such as molluscs, living among submerged weeds appear to browse on the periphyton and dead material (Smirnov 1958, 1959 a, b). Emergent plants are more likely to be grazed but losses are often very small. In a *Glyceria maxima* reedswamp consumption by molluscs and leaf-miners could hardly be measured, and was less than 1 per cent of the maximum biomass. Smirnov (1961) found from 0.4 to 7.5 per cent of the leaves and inflorescences of various emergent communities was consumed per day. At the lowest level, less than 5 per cent of the maximum biomass had been consumed, but only stems remained at the highest level. Large vertebrates, such as grass carp, turtles, swans, manatee, and coypu, may be the commonest cause of extensive grazing losses, but are often most active in shallow water and emergent communities (e.g. Allsop 1960, Fasset 1957, Glazner 1958 a and b, Hickling 1962, Hotchkiss 1941, Odum 1957). Cattle graze some bank-side plants intensively, especially *Glyceria maxima*. Grazing may stimulate productivity, so when it is important the exclusion of grazers from an experimental site may give underestimates of the productivity.

Acknowledgements

I wish to thank Dr. R. W. Edwards, Mr. M. Owens, and the Director of the Water Pollution Research Laboratory for permission to include some unpublished data, and Mr. F. J. H. Mackereth and Mr. J. Heron of the Freshwater Biological Association for performing ash and carbon analyses for me. I am very grateful to the Directors and staffs of the University of Wisconsin Hydrobiological Laboratory, the Hydrobiological Laboratory of the Czechoslovak Academy of Sciences, and the Ross Allen Reptile Institute, Florida, for providing facilities and assistance for collecting and drying plant samples. Mrs. T. Thompson, Mr. J. J. Holmes, and Mr. K. E. Marshall prepared translations of Russian papers. Dr. M. Straskraba provided valuable information and comments.

REFERENCES

Aario, L. 1933. Vegetation and postglaziale geschichte des Nurmijärvi-Sees. Suomal. elain-ja kasvit. Seur. nan Julk. *3* (2): vi+132 p. and map.

Allsopp, W. H. L. 1960. The manatee: ecology and use for weed control. - Nature *188*: 762.

Belcher, R. and S. Ingram. 1950. A rapid micro-combustion method for the determination of carbon and hydrogen. - Analytica Chim. Acta *4*: 118-129.

Borutskii, E. V. 1949. (Changes in the growth of the macrophytes in Lake Beloie at Kossino from 1888-1938. In Russian). - Trudy vses. gidrobiol. Obshch. *1*: 44-56.

— 1950. (Data on the dynamics of the biomass of the macrophytes of lakes. In Russian). - Trudy vses. gidrobiol. Obshch. *2*: 43-68.

Bray, J. R. 1962. Estimates of energy budgets for a *Typha* (cattail) marsh. Science *136*: 1119-1120.

— D. B. Lawrence, and L. C. Pearson. 1959. Primary production in some Minnesota terrestrial communities for 1957. - Oikos *10*: 38-49.

Burkholder, P. R., L. M. Burkholder, and J. A. Rivero. 1959. Some chemical constituents of turtle grass, *Thalassia testudinum*. - Bull. Torrey Bot. Club *86*: 88-93.

Butcher, R. W. 1933. Studies of the ecology of rivers I. On the distribution of macrophytic vegetation in the rivers of Britain. - J. Ecol. *21*: 58-91.

Earle, T. T. 1956. In Penfound (1956) p. 99.

Edwards, R. W. and M. Owens. 1960. The effects of plants on river conditions I. Summer crops and estimates of net productivity of macrophytes in a chalk stream. - J. Ecol. *48*: 151-160.

— and M. Owens. 1962. The effects of plants on river conditions IV. The oxygen balance of a chalk stream. - J. Ecol. *50*: 207-220.

— and M. Owens. 1965. The oxygen balance of streams. In Ecology and the industrial society, ed. R. W. Edwards and J. M. Lambert. - Brit. Ecol. Soc. Symp. *5*: 149-172.

Ekzertsev, V. A. 1958. (The production of littoral vegetation of the Ivankovskoyo Reservoir. In Russian.) - Byll. Inst. Biol. Vodokhran. *1*: 19-21.

Elster, H. J. and R. Vollenweider. 1961. Beiträge zur Limnologie ägyptens. Archiv Hydrobiol. *57*: 241-343.

Fasset, N. C. 1957. A manual of aquatic plants. - Univ. Wisconsin Press, Madison, Wisc. x + 405 p.

Filzer, P. 1951. Die naturlichen Grundlagen des Planzenertrags in Mitteleuropa. Schweizerbartsche Verlags. Stuttgart. 198 p.

Forsberg, C. 1959. Quantitative sampling of subaquatic vegetation. - Oikos *10*: 233-240.

— 1960. Subaquatic vegetation in Ösbysjön, Djursholm. - Oikos *11*: 183-199.

Frohne, W. C. 1938. Contribution to knowledge of the limnological role of the higher aquatic plants. - Trans. Am. Microsc. Soc. *57*: 256-268.

Gajevskaja, N. S. 1958. Le role de groupes principaux de la flore aquatique dans les cycles trophiques des differents bassins d'eau douce. - Verh. int. Ver. Limnol. *13*: 350-362.

Gessner, F. 1959. Hydrobotanik: die physiologischen Grundlagen der pflanzenverbreitung im Wasser, 2. Stoffhaushalt. - VEB Deutscher Verlag Wissenschaften, Berlin. viii+701 p. and 8 plates.
Glazner, H. 1958 a. First report on the growth of nutria in fish ponds and their influence on destruction of coarse vegetation. - Bamidgeh *10*: 32-35.
— 1958 b. Supplement to the report on nutria culture in fish ponds at Kfar Rupin. - Bamidgeh *10*: 54-59.
Gorham, E. 1953. Chemical studies on the soils and vegetation of waterlogged habitats in the English Lake District. - J. Ecol. *41*: 345-360.
Gortner, R. A. 1934. Lake vegetation as a possible source of forage. - Science *80*: 531-533.
Gröntved, J. 1958. Underwater macrovegetation in shallow coastal waters. J. Cons. int. Explor. Mer *24*: 32-42.
Harper, R. M. 1918. Some dynamic studies of Long Island vegetation. - Plant World *21*: 38-46.
Hejny, S. 1960. Ökologische Charakteristik der Wasser - und Sumpfpflanzen in den slowakischen Tiefebenen (Donau - und Theissgebiet). (In German; Czech summary). - Vdyavatel' stvo Slovenskej Akadémie Vied. Bratislava. 492 p.
Hickling, C. F. 1962. Fish culture. - Faber and Faber Ltd., London. 295 p. and 36 plates.
Hitchcock, A. E., P. W. Zimmerman, H. Kirkpatrick, and T. T. Earle. 1949. Water hyacinth; its growth, reproduction and practical control by 2,4-D. Contr. Boyce Thompson Inst. Pl. Res. *15*: 363-401.
Hotchkiss, N. 1941. The limnological role of the higher plants. *In* A symposium on hydrobiology: 152-162. - Univ. Wisconsin Press, Madison, Wisc.
Hurlimann, H. 1951. Zur Lebensgeschichte des Schilfs am den Ufern der Schweizer Seen. - Beitr. geobot. Landes-aufn. Schweiz *30*: 1-232.
Hynes, H. B. N. 1960. The biology of polluted waters. - Liverpool Univ. Press. Liverpool. xiv+202 p., 2 plates and map.
Lindeman, R. L. 1941. Seasonal food cycle dynamics in a senescent lake. Am. Midl. Nat. *26*: 636-673.
Lund, J. W. G. 1964. Primary production and periodicity of phytoplankton. Verh. int. Ver. Limnol. *15*: 37-56.
Macfadyen, A. 1964. Energy flow in terrestrial and marine environments. *In* Grazing in terrestrial and marine environments, ed. D. J. Crisp. - Brit. Ecol. Soc. Symp., *4*: 3-20. Blackwell. Oxford.
McGaha, Y. J. 1952. The limnological relations of insects to certain aquatic flowering plants. - Trans. Am. Microc. Soc. *71*: 355-381.
Merinotti, F. 1941. L'utilizzazione della canna gentile *Arundo donax* per la produzione autarchica di cellulosa nobile per raion. - Chimica, Milano *8*: 349-355.
Nelson, I. W., H. V. Lindstrom, L. S. Palmer, W. M. Sandstrom, and A. N. Wick. 1939. Nutritive value and chemical composition of certain freshwater plants of Minnesota. - Tech. Bull. Univ. Minn. Agric. Exp. Stn. *136*: 1-47.
Odum, H. T. 1957. Trophic structure and productivity of Silver Springs, Fla. Ecol. Monogr. *27*: 55-112.
— 1963. Productivity measurements in Texas turtle grass and the effects of dredging an intracoastal channel. - Publs. Inst. Mar. Sci. Univ. Tex., *9*: 48-58.
— P. R. Burkholder and J. Rivero. 1959. Measurements of productivity of turtle grass flats, reefs and the Bahia Fosforescente of Southern Puerto Rico. - Publs. Inst. Mar. Sci. Univ. Tex. *9*: 159-170.
Owens, M. and R. W. Edwards. 1961. The effects of plants on river conditions II. Further crop studies and estimates of net productivity of macrophytes in a chalk stream. - J. Ecol. *49*: 119-126.
— and R. W. Edwards. 1962. The effects of plants on river conditions III. Crop studies and estimates of net productivity of macrophytes in four streams in Southern England. - J. Ecol. *50*: 157-162.

Pallis, M. 1916. The structure and history of plav: the floating fen of the delta of the Danube. - J. Linn. Soc. (Bot.) *43*: 233-290.
Penfound, W. T. 1956. Primary production of vascular aquatic plants. - Limnol. Oceanog. *1*: 92-101.
— and T. T. Earle. 1948. The biology of the water hyacinth. - Ecol. Monogr. *18*: 447-472.
Perdue, R. E. 1958. *Arundo donax* - source of musical reeds and industrial cellulose. - Econ. Bot. *12*: 368-404.
Petersen, C. G. J. 1913. Om Baedeltangens (*Zostera marina*) Aarsproduktion i de danske Farvande (In Danish, English summary.) - Mindeskr. Japetus Steenstrups Föds., *9*.
Printz, H. 1950. Seasonal growth and production of dry matter in *Ascophyllum nodosum* (L.) Le Jol. - Avh. norske Vidensk. Akad. Oslo *4*: 15 p.
Rickett, H. W. 1924. A quantitative study of the larger aquatic plants of Green Lake, Wisconsin. Trans. Wis. Acad. Sci. Arts Lett. *21*: 381-414.
Rogosin A. 1958. Wild rice (*Zizania aquatica*) in Northern Minnesota. - M. S. Thesis, Univ. Minnesota. (Quoted in Bray *et al.*, 1959).
Seidel, K. 1955. Die Flechtbinse. - Binnengewässer *21*: xv + 216 p.
Seidel, K. 1956. *Scirpus lacustris* im eutrophen See. Z. - Fisch. 7-8, 553-567.
Shcherbakov, A. P. 1950. (Productivity of the littoral macrovegetation of Lake Glubokoie. In Russian.) - Trudy vses. gidrobiol. Obshch. *2*: 69-78.
Smirnov, N. N. 1958. Some data about food consumption of plant production of bogs and fens by animals. - Verh. int. Ver. Limnol. *13*: 363-368.
— 1959 *a*. (Approximately quantitative investigation of food of aquatic invertebrates by dissections. In Russian.) - Byull. Inst. Biol. Vodokhran. *5*:43-47.
— 1959 *b*. (Role of higher plants in the nutrition of animal populations of bogs and fens. In Russian.) - Trudy mosk. tekhnol. Inst. ryb. Prom. Khoz. *10*: 75-87.
— 1961. Consumption of emergent plants by insects. - Verh. int. Ver. Limnol. *14*: 232-236.
Straskraba, M. 1963. The share of the littoral region in the productivity of two ponds in southern Bohemia. - Rozpr. csl. Akad. Ved. (mat. prírod Véd.) *73*, (13):1-63.
— (In press). Der Anteil der höheren Pflanzen an der Produktion der Gewässer. - Mitt. int. Ver. Limnol. No. 14, Stoffhaushalt der Binnengewässer: Chemie und Mikrobiologie.
Strickland, J. D. H. 1960. Measuring the production of marine phytoplankton. Bull. Fish. Res. Bd Can. No. *122*: 172 p.
Vinogradov, A. P. 1953. Translation by J. Efron and J. K. Setelow. The elementary composition of marine organisms. - Mem. Sears. Fdn. Mar. Res. No. *2*: 647 p.
Walker, F. T. 1954. Distribution of *Laminariaceae* around Scotland. - J. Cons. Int. Explor. Mer *20*: 160-166.
— and W. D. Richardson. 1955. An ecological investigation of *Laminaria cloustoni*. - J. Ecol. *43*: 26-38.
Westlake, D. F. 1960. Water-weed and water management. - Inst. publ. Hlth. Engrs. J. *59*: 148-60.
— 1961. Aquatic macrophytes and the oxygen balance of running water. - Verh. int. Ver. Limnol. *14*: 499-503.
— 1963. Comparisons of plant productivity. - Biol. Rev. *38*: 385-425.
— 1965. Theoretical aspects of the comparability of productivity data. Mem. Ist. Ital. Idrobiol., *18* Suppl.: 313-321.
Wetzel, R. G. 1964. A comparative study of the primary productivity of higher aquatic plants, periphyton, and phytoplankton in a large shallow lake. Int. Revue ges. Hydrobiol. Hydrogr. *49*: 1-61.
Wilson, L. R. 1935. Lake development and plant succession in Vilas Co. Wisconsin. - Ecol. Monogr. *5*: 207-247.
Zaki, S. 1960. Density distribution of rooted hydrophytes in Nozha Hydrodrome. Notes Mem. hydrobiol. Dep. U.A.R. No *48*: 28 p.

TECHNIQUES AND PROBLEMS OF PRIMARY PRODUCTIVITY MEASUREMENTS IN HIGHER AQUATIC PLANTS AND PERIPHYTON [1]

ROBERT G. WETZEL [2]

Department of Zoology
Indiana University, Bloomington, Indiana

(1) Contribution No. 770, Department of Zoology, Indiana University, Bloomington, Indiana.

(2) Present address: Department of Botany and Plant Pathology, W. K. Kellogg Biological Station, Michigan State University, Hickory Corners, Michigan.

For bibliographic citation of this paper, see page 10.

Abstract

A brief summary is presented of the techniques for measuring biomass and the *in situ* production rates of aquatic macrophytes and periphyton. Problems of interpretation, intercalibration of methods, and adaptations of techniques to various habitats are emphasized.

Evidence indicates that the *in situ* oxygen light and dark chamber technique should not be used for estimates of production rates of vascular aquatic plants. Oxygen evolved during photosynthesis accumulates, is stored in the internal lacunae, and is utilized to various degrees during periods of darkness. As a result, dissolved oxygen of the surrounding water is not proportional to the production of oxygen and photosynthetic rates. The *in situ* C^{14} technique of determining production rates of macrophytes is discussed in view of recent findings that indicate diurnal variations in photosynthesis and the efficiency of light utilization.

Major problems of estimating biomass of periphyton include the placement of substrata for colonization, heterogeneity of distribution, and the rate of population turnover. The advantages and problems of various *in situ* methods of measuring production rates of periphyton are considered.

The literature on quantitative and semi-quantitative investigations of higher aquatic plants and periphyton is indeed voluminous. Much of the literature on vascular aquatic plants has been reviewed in detail by Gessner (1955 a, 1959), Penfound (1956), Raspopov (1963), and Wetzel (1964 a). The numerous reports of investigations of sessile algae have been variously assembled and analyzed in many reviews (Cooke 1956, Blum 1956, Castenholz 1960, Sládecková 1962 *et seqq.*, Round 1964, Wetzel 1964 a). A majority of the works cited in the reviews are concerned with biomass determinations, some of which are extended on a time basis with various interpretive problems. There is probably little disagreement that rates of growth, preferably *in situ* measurements of photosynthetic productivity, are far superior to biomass estimates and calculated rates of productivity. I will attempt to review these works only in general terms, except for the addition of recent studies, with emphasis on possible interrelationships of techniques for the International Biological Program.

AQUATIC MACROPHYTES

In discussing the problems associated with the sampling of aquatic macrophytes for estimates of production, a distinction is necessary between standing crop and biomass (Westlake 1963). Standing crop refers generally to the weight of plant material sampled by conventional methods at one time from a given area. Biomass, on the other hand, is similar to standing crop in that it includes the weight of plant material of a unit area at a given time, but all parts of all plants are included. The latter distinction is an important one and the root systems and organs of storage cannot be neglected. The biomass of the underground organs varies widely in comparison to that of the submerged or emergent portions of aquatic and semi-aquatic plants. The biomass of these organs can vary from 0-10% in submergents to over 50% among other aquatic plants (summarized in Westlake 1963). Underground stems, roots, and especially the tuberous stems of perennial aquatic plants usually are not included or only incompletely, or simply estimated. In certain cases, as among perennial plants, the annual increment of growth of the rootstocks is unknown for the most part, and exceedingly difficult to determine. Some satisfactory estimate should be determined for each species before reasonable and comparable estimates of biomass can be made. *In situ* measurements of carbon fixation would indicate the importance of growth contributions to storage organs and circumvent this problem. However, technical problems of such analyses would be great among the large plants.

Oven dry weight (105°C) is the most straightforward method of quantifying biomass on an areal basis. Approximate percentage factors for converting wet weights to dry weights must be avoided. Water content varies greatly among species, seasonal stages of development, and environmental conditions (Edwards and Owens 1960, Owens and Edwards 1961, 1962, Westlake 1961, Wetzel 1964 a). Determination of the organic dry weight by subtraction of the ash weight is to be recommended strongly. While the ash weight is generally small in comparison to the organic weight, it can also be highly variable among species and environmental conditions (Westlake 1963), especially when carbonate encrustation is excessive (Wetzel 1960). Inasmuch as calcium carbonate decomposes at temperatures much above 550°C, ignition temperatures should not exceed this value. Errors resulting from the loss of magnesium carbonate above 350°C are probably small among calcareous plants. Although the proportion of $MgCO_3$ in marl deposits is variable and occurs as high as 6.5%, usually the amounts are less than 3 per cent (Blatchley and Ashley 1900). Organic carbon is highly variable as a percentage of dry weight, less so when related to organic weight (see discussion and approximate conversion factors of Westlake 1963). It appears that the dry organic weight, determined from dry weight and ash weight, is the most generally acceptable and widely applicable of the determinations of biomass for aquatic macrophytes.

Applications of quantitative methods of pigment concentrations to higher aquatic plants have been few (see review in Wetzel 1964 a). Moreover, the techniques of extraction of pigments and instruments for analysis of extracts have been highly variable. Wetzel (1964 a, 1964 b) found only rough correlations between concentrations of pigments and biomass, and highly varying correlations between pigments and productivity. Since momentary concentrations of pigments are influenced by a host of dynamic variables (*e.g.*, quality and quantity of light, temperature, nitrogen, antimetabolites, salinity, phosphorous, magnesium, iron, and others), they are not considered a good estimate of biomass in absolute, and often even relative, terms.

The biomass of epiphytic periphyton of submergent or the aquatic portions of emergent macrophytes is generally small in comparison to the biomass of the larger plants. In certain cases where large accumulations may be easily removed without introducing other errors, *e.g.*, aquatic portions of bulrushes such as *Scirpus*, special attempts at removal may be desirable. In general, however, periphytic errors are probably negligible in comparison with those of sampling within the heterogeneous macrophytic community.

Estimates of the annual productivity by the increase in biomass are applicable, with some reservations, to aquatic macrophytes in that a majority of the macrophyte communities demonstrate marked annual fluctuations. Generally few losses in biomass occur during the period of growth in comparison to phytoplankton populations. In certain submerged communities, however, losses could be heavy by faunal consumption or destruction. Practically no quantitative evidence is available on such a relationship. If the initial biomass is negligible, as is often the case among submergents, the maximum biomass gives an approximate estimate of net production. Loss of biomass by grazing, disease and death can be significant, somewhat negating the value of such a simple biomass estimate. Since the maximal terminal biomass can vary from year to year and in spatial distribution (*cf.* Westlake 1963), determination of the maximal biomass can be made only by frequent measurements of seasonal biomass in a natural situation. Wetzel (1964 a, 1964 b) found conflicting results in the simultaneous comparisons between biomass measurements with calculated productivity and actual measurements of growth rates by C^{14} techniques of a saline submergent *Ruppia maritima*. Carbon biomass per time interval of the season as well as accumulative dry weight were completely in opposition to measured productivity, since maximal biomass occurred at the point of very low carbon fixation in the terminal phases of flowering. Cumulative carbon biomass per period (biomass divided by the days from the beginning of the growing season) was less misleading in this regard but still exhibited large, opposing deviations from actual measurements of productivity. While estimates of the total seasonal net productivity of macrophytes can be approximated from biomass measurements, static appraisal of growth by biomass can be erroneous in intraseasonal comparisons of the productivities of several producer components.

Measurements of the primary productivity of submerged vascular aquatic plants and macroalgae *in situ* have been meager (reviewed in Wetzel 1964 a). Techniques involving the changes in oxygen production of the water surrounding enclosed plants should be discarded or, at best, used with extreme caution. Errors can be considerable as a result of oxygen metabolism of many other components of the community enclosed by the experimental vessel. The suspicions of Wetzel (1964 b) that large errors may be involved in the oxygen method because of storage of oxygen in the lacunal systems of vascular hydrophytes have been confirmed in the studies of Hartman and Brown (1966). The latter work indicates a rapid diurnal accumulation of oxygen in the internal atmosphere and a slow diffusion into the surrounding water. Dissolved oxygen concentrations of the surrounding water were not proportional to production of internal oxygen and lagged considerably (several hours). Relative diffusion rates of oxygen into the water are not directly correlated to the intensity of photosynthesis (Górski 1935). The greater the intensity of assimilation, the smaller the amounts of oxygen that diffuse into the water and the more that is accumulated in intercellular lacunae. Use of stored internal oxygen during periods of darkness (Górski 1929, Bourn 1932) for respiration can occur without effect on the concentrations of dissolved oxygen of the medium. Apparently lacunal reserves of oxygen also are used during brief periods of anaerobsis of the surrounding water as under ice and snow cover in shallow situations (Hartman and Brown 1966). Simultaneous experiments comparing photosynthesis by C^{14} uptake to changes in the oxygen concentrations of the surrounding water demonstrated large inconsistencies (Wetzel 1964 b, Hartman and Brown 1966).

Diurnal fluctuations in lacunal concentrations of carbon dioxide occurred also in opposition to those of oxygen. A slight lag, similar to that for oxygen, was observed between minimal carbon dioxide values of the water and the internal atmosphere of several submergents (Hartman and Brown 1965). Carbon dioxide diffuses readily in comparison to oxygen, however, and was not observed to accumulate internally to the extent observed with oxygen (Górski 1929). It is likely that CO_2 is assimilated before it reaches the intercellular spaces (Górski 1935). If carbon dioxide accumulates internally and is used preferentially to that from external sources, results of the few studies involving the change in pH of the medium (Verduin 1952, Whitwer 1955), and the C^{14} techniques discussed below are weakened. More study is needed to determine fully the extent of these problems.

The C^{14} techniques for phytoplankton were extended and developed for the measurement of the *in situ* productivity of submerged aquatic plants in 1961 (Wetzel 1964 a, 1964 b). The methods are similar in principle to those for photosynthetic measurements of phytoplankton. Transparent and opaque Plexiglas chambers of various sizes (Fig. 1) were placed surrounding the plants for *in situ* incubation. C^{14} was injected into the chambers, mixed by small internal propeller-like blades, and incubated for a 4-5-hour period. After incubation the entire plant, including the root system, was removed and quick-frozen in the field. Self-absorption problems of C^{14} radiation by the plant tissues were circumvented by Van Slyke chemical combustion techniques and conversion of organic carbon to CO_2 and radioassay in gas-phase. Productivity was calculated in a proportional manner as in the techniques for phytoplankton. Results were expressed on a mg carbon fixed per dry biomass weight per unit time and extended to mg C/m^2/time from biomass distributional analyses for the individual plant species.

Prior to oxidation and conversion to CO_2 plants were exposed to fumes of concentrated HCl for the removal of extracellular contamination of C^{14} precipitated as carbonates. Such decontamination is necessary for phytoplankton (Wetzel 1965) and evidence has accumulated that rinsing with dilute solutions of acid should not be employed because of cellular loss. Brief exposure (10 minutes) to fumes of concentrated HCl is a superior technique even though it is less convenient. The necessity for removal of extracellular C^{14} contamination has not been shown directly for submerged macrophytes, nor that dilute HCl rinsing causes cellular loss. Yet until experimental results are available, exposure to fuming HCl seems desirable as a precautionary measure, especially in studies on macrophytes from calcareous waters.

In the original studies Wetzel (1964a) kept the incubation to a short mid-day period of four hours to avoid as much as possible the deleterious effects of the containers that occur with prolonged incubation (*e.g.,* Zobell and Anderson 1936). In order to calculate daily values, the productivity of the mid-day four-hour increment was expanded by a factor determined from the integrated area of the incubation period on the light curve as fraction of the total integrated area of the daily photoperiod. This method was a practical measure where sampling through the littoral zone necessitated a large number of samples and it was not possible to determine diurnal productivity measurements in every case. However, recent evidence (Hartman and Brown 1966) indicates that this technique of expansion to daily productivity values must be used with caution. By attenuating the amount of incident illumination with neutral density filters diurnal patterns of photosynthesis shifted somewhat (Fig. 2). Photosynthetic rates were maximal in the late morning under full illumination and in mid-afternoon at 75% illumination. At greatly reduced illumination of 12.5%, highest rates occurred in

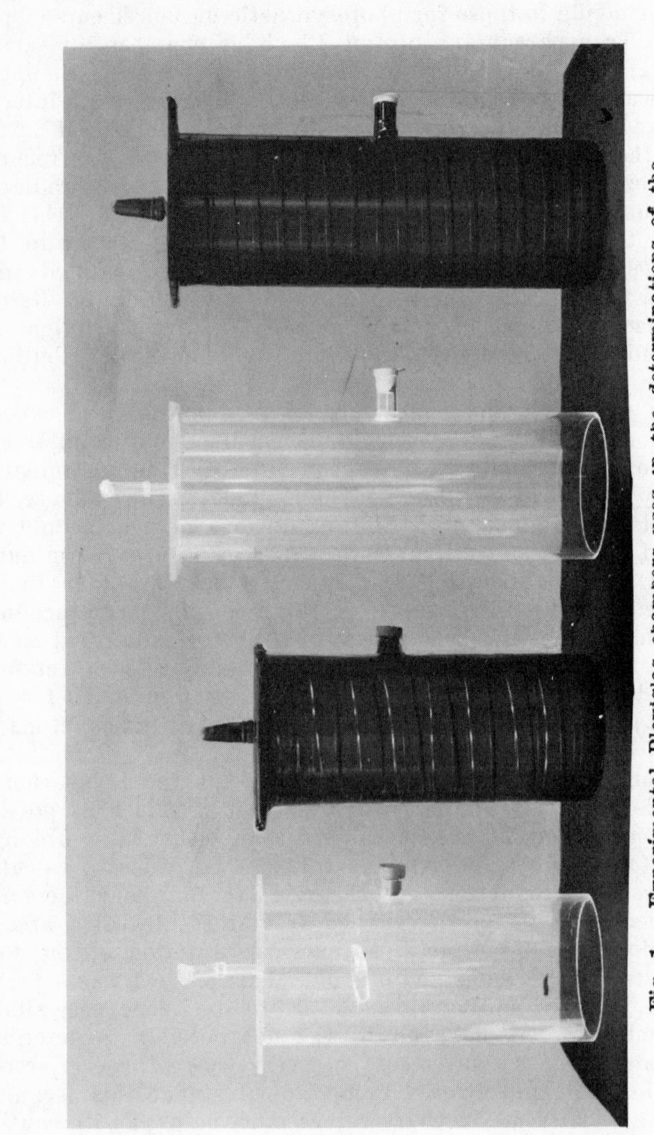

Fig. 1. - Experimental Plexiglas chambers used in the determinations of the productivity of macrophytes.

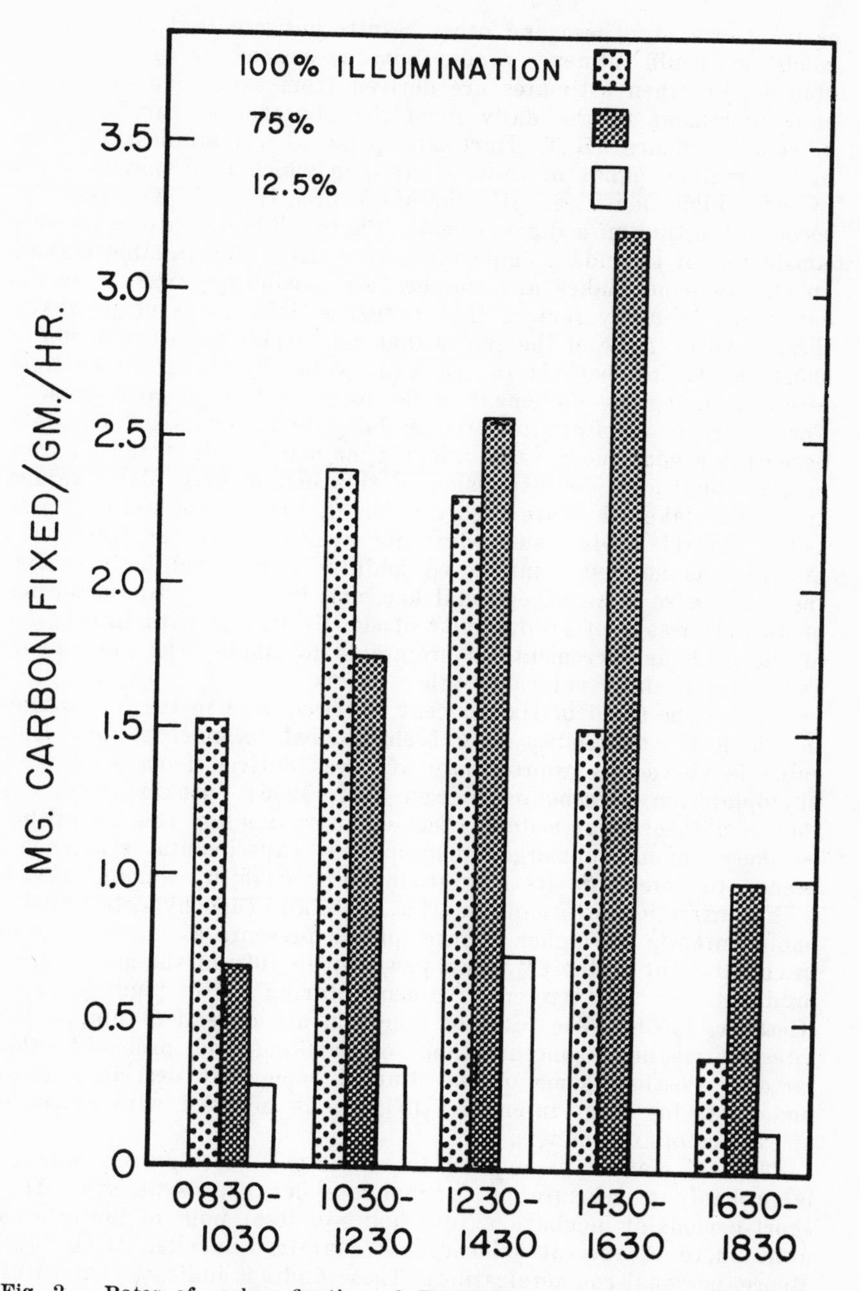

Fig. 2. - Rates of carbon fixation of *Elodea canadensis* exposed to varying degrees of incident illumination, October 20, 1963 (redrawn from Hartman and Brown, 1966).

early afternoon. These and other results indicate that atmospheric conditions could influence productivity measurements to a considerable degree when estimates are derived from expansion of a four-hour increment to the daily from the photoperiod curves. There is some evidence (R. T. Hartman, personal communication) that in the shallow ponds of these studies, in which macrophytes occur in very high densities, pH fluctuates greatly and available CO_2 becomes limiting on a diurnal basis. Photo-inhibitory processes were considered of secondary importance to sources of available carbon. In shallow ponds, lakes, and marshes, the possibility exists of diurnal variations in many factors that influence daily rates of photosynthesis. An example of the errors that may result from expansion of short incubation periods to the daily value by the ratio of light received during the increment to the total light is given in Table 1. The evidence suggests that the mid-day increment (*e.g.* 1000-1400 period) is a good mean value under a majority of light and environmental conditions, for expansion of this increment to daily values. In certain lakes that are highly buffered, such as the saline Borax Lake (Wetzel 1964a), such diurnal variations are negligible and growth was no doubt influenced mainly by seasonal factors. But the latter case is an exceptional one, and in most instances the use of diurnal assays of productivity of short duration with integration of the 3-4-hour increments is preferable to single brief assays and expansion to daily values by other means.

An isotope effect of six per cent has been used in the few studies in which C^{14} techniques have been applied to macrophytes. This value is viewed as approximate and was derived from studies on phytoplankton (Steemann Nielsen 1952, 1955; Sorokin 1959). Although a theoretical isotope effect of approximately this magnitude is likely among submerged macrophytes, experimental studies are needed to determine its magnitude and variability among species.

The excretion of organic matter, especially carbohydrates, during photosynthesis by higher aquatic plants presents a source of error in employment of the C^{14} techniques. Up to 10% of the assimilation products may be excreted as glucose during active photosynthesis (Gessner, 1955b). The rates of secretion are dependent upon light, internal medium, concentrations of calcium, and probably other factors. Possibly some of the diurnal variability described above under varying light intensities (Fig. 2) is involved with excretion of extracellular products.

The C^{14} method as applied to submerged vascular macrophytes is probably a measure of approximate net photosynthesis. After short periods of incubation (four hours or less), none of the internal atmosphere of several submergents contained labelled $C^{14}O_2$ (J. R. Moore, personal communication). These findings indicate that within this period labelled C^{14} taken up was not respired, and that intracellular respiratory CO_2 was derived from unlabelled sources. A similar situation has been found for phytoplankton during the initial 5 to 10 hours of incubation (reviewed by Strickland 1960; see also Thomas 1964). When plants occur in shallow situations where pH and sources of available carbon shift drastically during a diurnal period, respiration of cellular and internal atmospheric carbon probably increases considerably over conditions of exponential growth. Hence, there may exist diurnal as well as species differences in the measurement of net photosynthesis. It is likely, however, that the measurements with C^{14} technique for submergents represent values close to net photosynthesis when short incubation are used and environmental factors are near optimal.

PERIPHYTON

Measurements of the productivity of periphyton have been almost entirely limited to biomass. Quantitative investigations of sessile periphytic algae in lakes, and less frequently in lotic and intertidal marine situations, have been numerous. The oldest and most widely used approach is to provide an artificial, uniform substratum for colonization, submersed for a given period of incubation. In such studies the colonized organisms are counted per unit area, or wet, air dry, dry (105°C), or ash-free dry (organic) weight estimates of biomass are determined. The chlorophyll content of the periphyton of artificial substrates has also been used as an estimate of biomass. The historical development of the techniques of estimating the rates of colonization of periphyton on artificial substrata has been reviewed in detail alsewhere (Cooke 1956, Blum 1956, Castenholz 1960a, 1961, Sládecková 1960, 1962, Round 1964, Wetzel 1964a).

The individual techniques for the determination of biomass of periphyton have been described and critically considered in the reviews cited above, especially by Sládecková. The artificial substrate methods of determining the magnitude of periphytic colonization have numerous serious sources of error (*e.g.*, selectivity of attached organisms, variable turnover rates, mechanical and manipulative losses, etc.). Assuming that a majority of these problems can be corrected or accounted for in careful analyses on a seasonal or annual basis, indirect biomass estimates of productivity may be determined. As discussed earlier concerning the macrophytes, such estimates represent net production (here mean biomass times turnover rate, Sládecek and Sládecková 1964). Such measures represent cumulative biomass after losses from respiration, predation, death, and decomposition.

Two serious problems of the artificial substrate methods of determining periphytic productivity seem especially pertinent to the present author. Firstly, in a great majority of the studies on periphyton the substrata are suspended in the pelagic regions of standing bodies of water or the main flow areas in lotic situations. Natural periphytic substrates are primarily benthic and macrophytic in nature. Any substrates that occur for any length of time in the open water are strictly fortuitous and completely insignificant to the productivity of an ecosystem. Only rarely are natural substrata studied (*e.g.,* Young 1945, Blum 1956, Douglas 1958, Round 1964), or are artificial substrata placed in an ecologically realistic position (*e.g.,* Castenholz 1960a, 1960b). Therefore, many of the estimates of periphytic production represent only colonization rates of certain of the phytoplankton and may be, and I suspect usually are, entirely unrelated to true productivity by sessile producers. This erroneous relationship is especially true in reservoirs, where turbidity severely limits macrophytic and benthic areas to relatively small regions of the immediate littoral zone.

Secondly, the rate of population turnover is difficult to determine. Turnover rate of the periphytic population represents an average of several variable species rates. The rates of turnover vary markedly with season, extent of colonization of the substrate, type and position of the substrate, environmental parameters, and other factors. Moreover, subjective errors occur in determinations of the population turnover that are dependent on frequency of sampling and estimation of what degree of colonization represents a climax stage of the community (Sládecek and Sládecková 1964). Higher turnover rates are determined with increased frequency of sampling than with monthly or longer sampling intervals as is commonly the case.

Attempts to assay the rates of productivity of periphyton *in situ* have been few. Technical problems are many and revolve around the difficulties of heterogeneous distribution of periphyton on natural substrates. Perhaps the first quantitative estimate of the productivity of epipelic periphyton was made by the change in oxygen concentrations in glass aquaria inverted over the sediments of small, shallow (0.4 m), dystrophic Lake Piavochnoye in 1937 (Bervald 1939). These brief studies indicated productivity of the epipelic periphyton to exceed twice that of the plankton per square meter. An excellent variant of the oxygen techniques was made in the *in situ* assays of the productivity of epiphytic periphyton of the horsetail (*Equisetum*) (Assman 1951, 1953). External casings, transparent and opaqued, of Liebig condensers were placed directly around the emergent portions of the plant from which changes in oxygen of the water were determined. The annual productivity of the periphyton constituted approximately 35% of that of the macrophytes determined by other methods. Several of the studies on community metabolism have included periphyton in the estimate of productivity of the benthic producers by oxygen difference techniques (Odum 1957, Park, Hood, and Odum 1958). Pomeroy (1959a, 1959b) estimated photosynthetic rates of salt marsh algae by light and dark bell jar methods. In this instance the jars were filled with boiled, filtered water, a technique that should be avoided for obvious reasons of alteration of the medium. Productivity of epilithic periphyton on tile plates of pools has been estimated by oxygen methods (Kurasawa 1959). Felföldy (1961a, 1961b) determined the photosynthetic rates of natural periphytic communities under laboratory conditions with manometric techniques. Studied were the effects of temperature on growth as well as relationships between photosynthesis and several measures of biomass (chlorophyll, nitrogen, and dry weight). Measures of biomass were poorly correlated with photosynthetic rates. Similar findings between chlorophyll biomass and *in situ* production rates were evident in epilithic periphyton populations (Wetzel 1963). The importance of estimating the « functional chlorophyll » by conversion of chlorophylls to phaeophytins by simple spectrophotometric techniques (Wetzel 1964 a) is emphasized. The use of the light and dark bottle oxygen methodology for the estimation of photosynthetic rates of periphyton in flowing water (Odum 1957, McConnell and Sigler 1959, Kobayasi 1961) must be viewed with reservation. The restriction of water movement has been shown to affect metabolism greatly in both rheophilic periphyton (Whitford 1960, Whitford and Schumacher 1961) and macrophytes (Gessner 1937, Barth 1957).

C^{14} techniques have been applied to periphyton in only a few cases. Vollenweider and Samaan (1958) suspended glass rods of known surface area near or below the water surface vertically among emergent macrophytes for simulated colonization (3-4 weeks) of periphyton epiphytic on the larger plants. The rods colonized with algae were then carefully placed into bottles of pre-filtered lake water, innoculated with C^{14}, and incubated at the depth from which they were taken. After a brief incubation period, the periphyton was removed from the rods, homogenized in a small aliquot of water, and a portion filtered onto membrane filters for radioassay and analysis similar to that of phytoplankton. An extension of this technique could include chemical combustion of the algal matter and analysis of the $C^{14}O_2$ in gas phase (Wetzel 1963, 1964a), eliminating self-absorption problems.

In the detailed studies of the productivity of epipelic periphyton of a Danish fjord, Grontved (1960, 1962) devised a sampling technique whereby a suspended subsample of the benthic algae and some of the littoral phytoplankton were incubated in bottles with C^{14}. By a complicated but sound series of manipulations the fraction of the total assayed carbon fixation was estimated for benthic producers. Self-absorption of radiation by the sediment particles was estimated by several techniques of radioassay of filters before and after removal from the larger inorganic particles. The resultant values were expressed as the potential rate of production in that the experiments were not performed *in situ*. The actual rate of production was suggested tentatively as one-half of potential rate when light intensity was optimal. In studies on light inhibition and injury effects on photo-

Table 1. - Values of measured diurnal productivity of several macrophytes and daily productivity calculated by expansion of the values of the increments from the ratio of the fraction of the light received during the incubation to the total daily radiation. Conneaut Lake, western Pennsylvania, 1964. (From unpublished data of Drs. R. T. Hartman, J. R. Moore, and D. L. Brown, Department of Biology, University of Pittsburgh).

Date	Time of increment	% of daily radiation	*Elodea canadensis*		*Myriophyllum exalbescens*		*Najas flexilis*		*Ceratophyllum demersum*	
			Measured mg C/g/period	Calculated mg C/g/day from increment	Measured mg C/g/period	Calculated mg C/g/day from increment	Measured mg C/g/period	Calculated mg C/g/day from increment	Measured mg C/g/period	Calculated mg C/g/day from increment
23 April	0600-1000	12.2	3.51	28.76	2.04	16.73	0.91	7.42	0.96	7.86
	1000-1400	51.2	1.27	2.48	2.10	4.10	1.10	2.14	0.76	1.49
	1400-1800	31.7	2.90	9.15	3.96	12.49	1.22	3.85	0.58	1.81
	1800-2000	4.9	0.21	4.21	0.14	2.84	0.11	2.25	0.14	2.83
	Total	100.0	7.89		8.24		3.34		2.44	
4 June	0600-0700	0.24	0.21	85.66	0.13	55.37	0.20	82.70	0.17	72.66
	0700-1100	23.11	3.52	15.22	2.19	9.50	6.36	27.52	3.26	14.11
	1100-1515	49.14	5.37	10.93	3.68	7.50	5.11	10.39	4.51	9.18
	1515-2100	26.03	2.31	8.85	2.07	7.97	1.38	5.30	4.53	17.14
	2100-2300	1.45	0.23	15.62	0.53	3.66	0.31	21.28	0.45	3.11
	Total	99.97	11.64		8.60		13.36		12.92	
2 Sept.	0640-1100	24.85	4.08	16.44	4.57	18.38	6.92	27.83	4.59	18.49
	1100-1500	47.66	4.09	8.59	5.34	11.20	6.47	13.57	4.52	9.49
	1500-2145	27.49	2.87	10.44	1.29	4.69	1.58	57.65	2.66	9.68
	Total	100.00	11.04		11.20		14.97		11.77	

synthesis of phytoplankton in Antarctic ponds (Goldman, Mason, and Wood 1963), a few assays of carbon fixation were made of the epilithic algal mat by suspension of small disks of the algal material in bottles. Following brief incubation the disks were dried, combusted to CO_2, and counted in gas-phase. Excellent, nearly complete replication of natural stream populations of epilithic periphyton was found on polyethylene film substrates exposed for 30 days positioned among natural substrates (Backhaus 1965). Selective species colonization occurred on both sides of the 0.2 mm sheets in relation to light distribution. Disks were removed and exposed to C^{14} in test tubes, containing a bubble of air, that were suspended freely to be agitated by the current. After a four-hour incubation period the samples were removed, dried, and assayed.

The only assays of C^{14} measured productivity of benthic periphyton *in situ* known to the author are the studies of Wetzel (1963, 1964a) in a shallow, saline lake in the coastal mountains of northern California. Plexiglas chambers, similar to those employed in measurements of the productivity of macrophytes (Fig. 1), were worked by short rotation movements into the sediments along transects perpendicular to the shoreline. C^{14} innoculum of known assay was injected underwater and the chambers sealed. Samples were incubated for a four-hour period. After incubation the chambers were removed by working a stainless steel plate under the open end which permitted removal of the entire sample, effectively an undisturbed short core of the superficial sedimentary material and the overlying water. The water was then removed by a large syringe and the upper centimeter of sediments was frozen under desiccation. Just prior to radioassay the sedimentary material encrusted with periphyton was exposed to flowing fumes of HCl to remove possible extracellular precipitated C^{14}. The organic material of the samples was then oxidized to CO_2 by Van Slyke combustion for radioassay in gas phase. While these techniques are tedious, they circumvent the problems of self-absorption of the weak β-radiation. Other difficulties are encountered using these C^{14} methods. Technical restrictions of the equipment limit the maximum sample size to 400 mg total carbon that may be effectively converted to CO_2; with modification the carbon capacity can be increased to 1,720 mg. Hence, small sample sizes are necessary and this is especially true when the sediments are calcareous in nature.

The relative lack of homogeneity of the distribution of periphyton in natural situations reduces considerably the validity that can be assigned to measurements of periphtic productivity by present techniques. The homogeneous littoral covering of small, angular pebbles of Borax Lake, California, was ideal for application of the C^{14} techniques (Wetzel 1964a). This condition was an exception, however, which minimized errors of expansion of experimentally measured areas to square meter values. Perhaps the placement of a large number (sufficient for a detailed annual study) of uniform, simulated substrates in a natural situation in realistic position, combined with C^{14} techniques of measurements, would yield an accurate estimate of the productivity of periphyton. The problem of heterogeneous distribution is difficult to circumvent and will remain a continual detriment to the significance that can be attained in any method of estimating biomass or productivity of periphyton.

Once the hurdle of reasonably accurate and reproducible methods of assaying growth rates of periphyton in nature is surmounted, experimental approaches will be possible to investigate physical and biological parameters regulating growth. We can only speculate on these relationships at the present time (*vid.* discussion of Round 1964).

Acknowledgments

Numerous discussions among many people have contributed to the development of the work on the productivity of macrophytes and periphyton. In particular to be mentioned is the kindness of Drs. R. T. Hartman, D. L. Brown, and J. R. Moore, all of the Department of Biology, University of Pittsburgh. The enthusiasm and encouragement of Dr. C. R. Goldman, University of California, Davis, and Dr. D. G. Frey, Indiana University, is greatly appreciated. The assistance of the National Science Foundation (Grant GB-1452) to the author is gratefully acknowledged.

REFERENCES

Assman, A. V. 1951. Rol vodoroslevykh obrastanii v obrazovanii organicheskogo veshchestva v vodoeme. (The role of algal periphyton in the production of organic matter in a body of water). - Doklady Akad. Nauk, SSSR *76*: 905-908. (In Russian)

— 1953. Rol vodoroslevykh obrastanii v obrazovanii organicheskogo veshchestva v Glubokom ozere. (The role of algal periphyton in the production of organic matter in Lake Glubokoje). - Trudy Vsesoyuz. Gidrobiol. Obchestva *5*: 138-157. (In Russian).

Backhaus, D. 1965. Ökologische und experimentelle Untersuchungen an den Aufwuchsalgen der Donauquellflüsse Breg und Brigach und der obersten Donau bis zur Versickerung bei Immendingen. Doctoral dissertation, Universität Freiburg.

Barth, H. 1957. Aufnahme und Abgabe von CO_2 und O_2 bei submersen Wasserpflanzen. - Gewässer Abwässer *4* (17/18): 18-81.

Bervald, E. A. 1939. Opyt izucheniya prevashcheniya organicheskikh veshchestv v presnovodnom vodoeme. (Experiment of the study of conversion of organic substances in a body of fresh water). - Sb. nauchn. stud. rabot MGU, biologiya, *4*. (In Russian; unable to obtain original; cited in Vinberg, 1960).

Blatchley, W. S. and G. H. Ashley. 1900. The lakes of northern Indiana and their associated marl deposits. - Rep. Ind. Dep. Geol. *25*: 31-321.

Blum, J. L. 1956. The ecology of river algae. Bot. Rev. *22*: 291-341.

Bourn, W. S. 1932. Ecological and physiological studies on certain aquatic angiosperms. - Contr. Boyce Thompson Inst. Plant Res. *4*: 425-496.

Castenholz, R. W. 1960a. Seasonal changes in the attached algae of freshwater and saline lakes in the Lower Grand Coulee. Washington. - Limnol. Oceanogr. *5*: 1-28.

— 1960 b. The algae of saline and freshwater lakes in the Lower Grand Coulee, Washington. - Res. Studies *28*: 125-155.

— 1961. An evaluation of a submerged glass method of estimating production of attached algae. - Verh. int. Ver. Limnol. *14*: 155-159.

Cooke, W. B. 1956. Colonization of artificial bare areas by microorganisms. Bot. Rev. *22*: 613-638.

Douglas. B. 1958. The ecology of the attached diatoms and other algae in a small stony stream. - J. Ecol. *46*: 295-322.

Edwards, R. W. and M. Owens. 1960. The effects of plants on river conditions. I. Summer crops and estimates of net productivity of macrophytes in a chalk stream. - J. Ecol. *48*: 151-160.

Felföldy, L. J. M. 1961a. Effect of temperature on the photosynthesis of a natural diatom population. - Ann. Biol. Tihany *28*: 95-98.

— 1961b. On the chlorophyll content and biological productivity of periphytic diatom communities on the stony shores of Lake Balaton. - Ann. Biol. Tihany *28*: 99-104.

Gessner, F. 1937. Untersuchungen über Assimilation und Atmung submerser Wasserpflanzen. - Jb. wiss. Bot. *85*: 267-328.

— 1955a. Hydrobotanik. Die physiologischen Grundlagen der Pflanzenverbreitung im Wasser. I. Energiehaushalt. - Berlin, VEB Deutscher Verlag der Wissenschaften. 517 pp.

— 1955b. Discussion. In: Fogg, G. E. and D. F. Westlake. The importance of extracellular products of algae in freshwater. - Verh. int. Ver. Limnol. *12*: 219-232.

— 1959. Hydrobotanik. Die physiologischen Grundlagen der Pflanzenverbreitung im Wasser. II. Stoffhaushalt. Berlin, VEB Deutscher Verlag der Wissenschaften. 701 pp.

Goldman, C. R., D. T. Mason, and B. J. B. Wood. 1963. Light injury and inhibition in Antarctic freshwater phytoplankton. - Limnol. Oceanog. *8*: 313-322.

Górski, F. 1929. Recherches sur les méthodes de mesure de photosynthèse chez les plantes aquatiques submergées. - Acta Soc. Bot. Poloniae *6*: 1-29.

— 1935. Wymiana gazów u roslin wodnych podckas asymilacji. Gas interchange in aquatic plants during photosynthesis. - Bull Int. Acad. Polon. Sci. Lett., Cl. Sci. Math. Nat., Ser. B, *1935*: 177-198. (In English).

Grontved. J. 1960. On the productivity of microbenthos and phytoplankton in some Danish fjords. - Medd. Danmarks Fiskeri- og Havundersogelser N. S. *3*: 55 - 92.

— 1962. Preliminary report on the productivity of microbenthos and phytoplankton in the Danish Wadden Sea. - Medd. Danmarks Fiskeri- og Havundersogelser N. S. *3*: 347-378.

Hartman, R. T. and D. L. Brown. 1966. Internal atmosphere of submersed vascular hydrophytes in relation to photosynthesis. - Ecology (In press).

Kobayasi, H. 1961. Productivity in sessile algal community of Japanese mountain river (*sic*). - Bot. Mag. Tokyo *74*: 331-341.

Kurasawa, H. 1959. Studies on the biological production of fire pools in Tokyo. XII. The seasonal changes in the amount of algae attached on the wall of pools. - Misc. Rep. Res. Inst. Nat. Resources *51*: 15-21.

McConnell, W. J. and W. F. Sigler. 1959. Chlorophyll and productivity in a mountain river. Limnol. Oceanog. *4*: 335-351.

Odum, H. T. 1957. Trophic structure and productivity of Silver Springs, Florida. Ecol. Monogr. *27*: 55-112.

Owens, M. and R. W. Edwards. 1961. The effects of plants on river conditions. II. Further crop studies and estimates of net productivity of macrophytes in a chalk stream. - J. Ecol. *49* : 119-126.
— 1962. The effects of plants on river conditions. III. Crop studies and estimates of net productivity of macrophytes in four streams in southern England. - J. Ecol. *50* : 157-162.
Park, K., D. W. Hood, and H. T. Odum. 1958. Diurnal pH variation in Texas bays, and its application to primary production estimation. - Publ. Inst. Mar. Sci. (Texas) *5* : 47-64.
Penfound, W. T. 1956. Primary production of vascular aquatic plants. - Limnol. Oceanog. *1* : 92-101.
Pomeroy, L. R. 1959a. Algal productivity in salt marshes of Georgia. - Limnol. Oceanog. *4* : 386-398.
— 1959b. Productivity of algae in salt marshes. - Proc. Salt Marsh Conf. *1958* : 88-95
Raspopov, I. M. 1963. Ob osnovnykh ponyatiyakh i napravleniyakh gidrobotaniki v Sovetskom Soiuze. (On the basic concepts and directions of hydrobotany in the Soviet Union). - Uspekhi Sovremennoi Biol. *55* : 453-464. (In Russian).
Round, F. E. 1964. The ecology of benthic algae. *In*: Jackson, D. F. (Ed.). Algae and man. New York, Plenum Press. pp. 138-184.
Sládecek, V. and A. Sládecková. 1964. Determination of periphyton production by means of the glass slide method. - Hydrobiologia *23* : 125-158.
Sládecková, A. 1960. Limnological study of the Reservoir Sedlice near Zeliv. XI. Periphyton stratification during the first year-long period (June 1957 - June 1958). - Sci. Pap. Inst. Chem. Techn., Fac. Techn. Fuel Water *4* :143-261.
— 1962. Limnological investigation methods for the periphyton («Aufwuchs») community. - Bot. Rev. *28* : 286-350.
Sorokin, J. I. 1959. Opredelenie velichin izotopicheskogo effekta pri fotosinteze v kulturakh *Scenedesmus quadricauda*. (Determination of the isotopic discrimination by photosynthesis in cultures of *Scenedesmus quadricauda*). Bull. Inst. Biol. Vodokhranilisch *4* : 7-9. (In Russian).
Steemann Nielsen, E. 1952. The use of radio-active carbon (C^{14}) for measuring organic production in the sea. - J. Cons. int. Expl. Mer *18* : 117-140.
— 1955. The interaction of photosynthesis and respiration and its importance for the determination of C^{14}-discrimination in photosynthesis. - Physiol. Plant. *8* : 945-953.
Strickland, J. D. H. 1960. Measuring the production of marine phytoplankton. Bull. Fish. Res. Bd. Can. *122* : 172 pp.
Thomas, W. H. 1964. An experimental evaluation of the C^{14} method for measuring phytoplankton production, using cultures of *Dunaliella primolecta* Butcher. - Fish. Bull. U.S.F.W.S. *63* : 273-292.
Verduin, J. 1952. The volume-based photosynthetic rates of aquatic plants. Amer. J. Bot. *39* : 157-159.
Vinberg, G. G. 1960. Pervichnaia produktsiia voedoemov. (Primary production of bodies of water). - Izdatel'stvo Akademii Nauk SSR, Minsk. 329 pp. (In Russian).
Vollenweider, R. A. and A. A. Samaan. 1958. A note on the use of C^{14} for measuring carbon assimilation in periphyton. Unpublished manuscript. 4 pp.
Westlake, D. F. 1961. Aquatic macrophytes and the oxygen balance of running water. - Verh. int. Ver. Limnol. *14* : 499-504.
— 1963. Comparisons of plant productivity. - Biol. Rev. *38* : 385-425.
Wetzel R. G. 1960. Marl encrustation on hydrophytes in several Michigan lakes. Oikos *11* : 223-236.
— 1963. Primary productivity of periphyton. - Nature *197* : 1026-1027.
— 1964a. A comparative study of the primary productivity of higher aquatic plants, periphyton, and phytoplankton in a large, shallow lake. - Int. Rev. ges. Hydrobiol. *49* : 1-61.

— 1964b. Primary productivity of aquatic macrophytes. - Verh. int. Ver. Limnol. *15* : 426-436.
— 1965. Necessity for decontamination of filters in C^{14} measured rates of photosynthesis in fresh waters. Ecology *46* : 540-542.
Whitford, L. A. 1960. The current effect and growth of fresh-water algae. Trans. Amer. microsc. Soc. *79* : 302-309.
— and G. J. Schumacher. 1961. Effect of current on mineral uptake and respiration by a fresh-water algae. - Limnol. Oceanog. *6* : 423-425.
Whitwer, E. E. 1955. Efficiency of finely-divided vs. tape-like aquatic plant leaves. - Ecology *36* : 511-512.
Young, O. W. 1945. A limnological investigation of periphyton in Douglas Lake, Michigan. - Trans. Amer. microsc. Soc. *64* : 1-20.
Zobell, C. E. and D. Q. Anderson. 1936. Observations on the multiplication of bacteria in different volumes of stored sea water and the influence of oxygen tension and solid surfaces. - Biol. Bull. *71* : 324-342.

V

PRIMARY PRODUCTIVITY AND STANDING CROP

RELATIONSHIP BETWEEN STANDING CROP AND PRIMARY PRODUCTIVITY

INGO FINDENEGG

Biologische Station Lunz der Österreich
Akademie der Wissenschaften, Lunz am See, N. Ö., Austria

For bibliographic citation of this paper, see page 10.

Abstract

The effect of production of a given quantity of algae varies from hour to hour. It may be reduced both by increase and decrease of the light intensity to which the species are adapted. Therefore the assimilation rate oscillates according to time of day as well as to weather.

In most cases an inverse correlation was found between the quantity of algal biomass and production per unit of freshweight. Below one square meter of lake surface 1 g of phytoplankton assimilated up to 0,4 g carbon per day when only small quantities of algae were present but only thousandth when crowded masses of cells were present. Another correlation exists between assimilation rate and size of the cells. Nannoplankters generally assimilated much more carbon per unit of biomass than did large diatoms or blue-green algae. Therefore lakes in an advanced state of eutrophication and rich in *Tabellaria* or *Oscillatoria* produce relatively small amounts of organic matter per unit of phytoplankton.

Although much work has been done on planktic primary production in recent years, there is little known about the relationship between standing crop and primary productivity.

The direct count of algae cells and the identification of species is generally considered to be time consuming, laborious, and boring. Therefore preference has been given to methods by which the population density can be calculated from physical techniques, such as measurement of turbidity or of optical density of pigment extracts. We shall, however, never fully understand what is going on in the primary production of a lake until we know what species occur, where and when they appear, and how they interfere with the present planctic community. We also shall never be able to estimate the share of phytoplankton in secondary production if we do not know the nutritional importance of its components. Some small algae may be subject to grazing, whereas bigger species can not be eaten by zooplankton. It must be emphasized that only qualitative counts give a clear idea of the actual biological state of a lake.

Much has been said against the reliability of counting methods. According to my experience, results are reproducible, and correct within plus or minus 10%. Uncertainties sometimes arise from the fact that in fixed material it is not always possible to fully distinguish vital cells from those which are dying or are no longer active in carbon assimilation. Therefore it is advisable to supplement the normal counts of concentrated algae with the inverted microscope by counts of living and not concentrated plankton in shallow one-milliliter Kolkwitz chambers using eventually a water immersion objective. In this way also species that settle incompletely during sedimentation can be recorded with sufficient accuracy.

We shall begin with the vertical distribution of assimilation rates and biomass of the phytoplankton. Fig. 1 shows freshweight and also by means of symbols the qualitative composition of the phytoplankton in different depths of the Mondsee in Austria (14 km^2, 68 m deep, mesotrophic). The assimilation rates are represented by curves. As one can see the weather was sunny on 3 July and prevailingly cloudy on the 6th. On the other hand, the biomass had increased considerably within that time, especially in the 1-3 m layers. Total assimilation (assimilation rate per m^2 of lake surface; see the columns in the right corner) was equal in both series. The lower radiation had been compensated for by the larger standing crop on 6 July. From the curves we learn that optimal assimilation took place in 3 m on 3 July, but in 1-2 m in the second series. Thus light inhibition was stronger on the 3rd. The apparent low assimilation rate at the surface on 6 July is due not only to light inhibition, but also to the deficiency of the phytoplankton that slowly sink down when turbulence is low for some time because of lack of wind.

Fig. 2 relates to the results of 3 series taken in the Swiss lake Walensee which is 25 km^2, 145 m deep and rather oligotrophic. In two of the series the assimilation curves show the typical shape of lakes poor in nutrients: production might be considered almost equally high in the upper horizons of the euphotic zone. This applies to both the middle (16 Sept. 1963) and the right curve (29 May 1964). Correspondingly the vertical distribution of the phytoplankton shows relatively small differences. On the contrary, the series of 27 June 1963 (left part of Fig. 2) has a marked peak of assimilation at the depth of 2 meters due to a maximum of biomass consisting chiefly of *Pandorina morum* and *Uroglena americana*. The appearance of such a large biomass in lakes generally poor in plankton has been observed in several cases during spring (« planktic spring explosion »), and is due to fertilization of the euphotic zone by the spring circulation. The occurrence of these spring explosions of the phytoplankton shows clearly that the absence of a marked optimum of assimilation during almost the whole year is not caused by the high light transmission in this kind of lakes but by the scarcity of nutrients (Findenegg 1964). Otherwise it would not be possible that, with unchanged light transmission, a big maximum is set up at the depth of 2 meters, all at once. Despite theoretical considerations on the penetration of light and photosynthesis this type of primary production is a

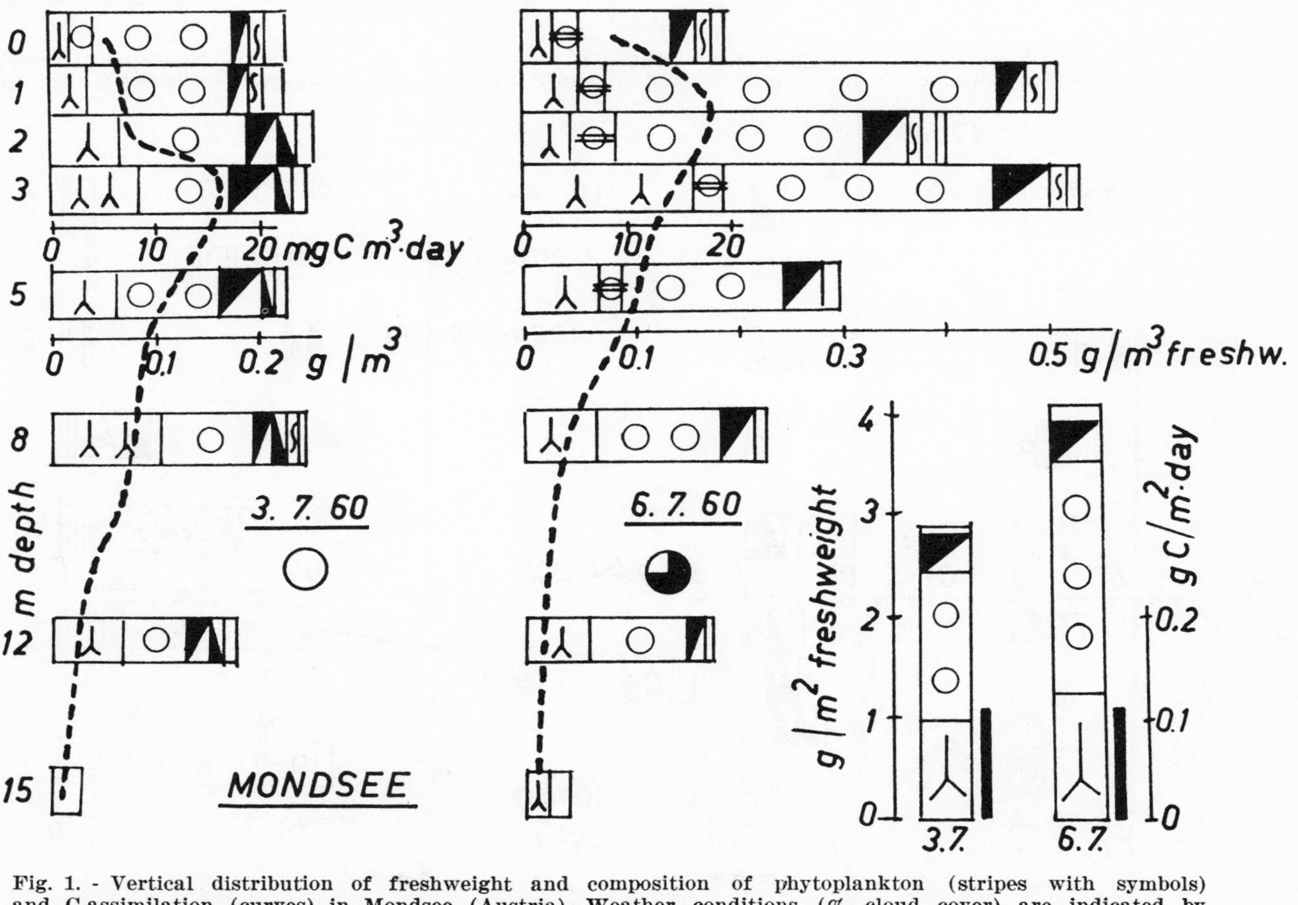

Fig. 1. - Vertical distribution of freshweight and composition of phytoplankton (stripes with symbols) and C-assimilation (curves) in Mondsee (Austria). Weather conditions (% cloud cover) are indicated by circles beneath dates. For key to genera symbols see Fig. 7.

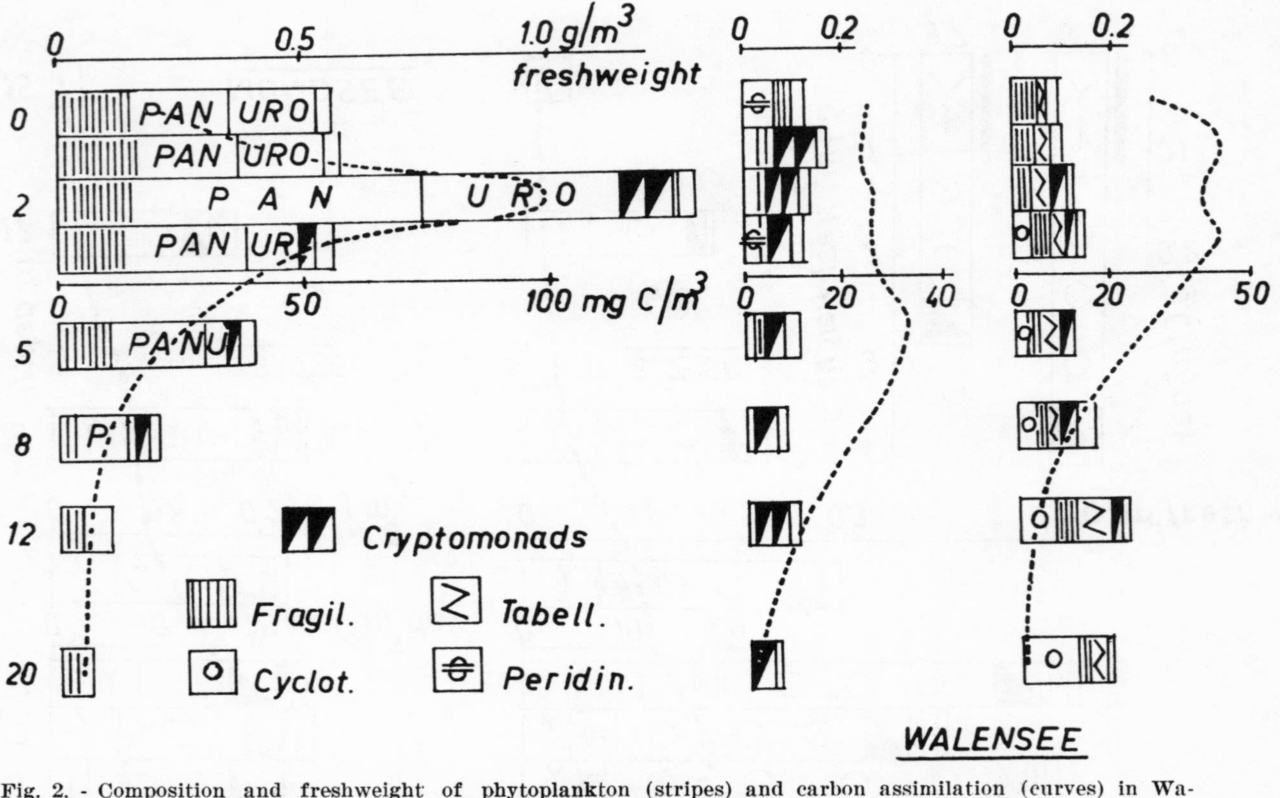

Fig. 2. - Composition and freshweight of phytoplankton (stripes) and carbon assimilation (curves) in Walensee (Switzerland). Dates from left to right are: 27.6., 16.9. and 29.5. 1964.

limnological reality. Moreover, it is of importance for secondary production whether an intensive assimilation is going on in the uppermost layers only, or a low production is extensively distributed within a deep watermass.

From Fig. 2 we still can learn another conspicuous fact: there is a remarkable difference in the relationship between biomass and the amount of carbon assimilated within the 3 series. In the left graph (June 1963), when a large algal mass was present the relative assimilation per unit of freshweight is much lower than in the following series. The quotient carbon / freshweight (the « Aktivitätskoeffizient » of Nauwerck 1963) is 0,08 in the optimal cubimeter of the first set, but 0.3 and 0.28 in the second and third series, when the water was thinly populated. This is only one of the large number of examples of the rule that relative production decreases with increasing standing crop.

Another example is given in Fig. 3, which represents the values found in the upper basin (Obersee) and in the lower part (Untersee) of Lake Constance in June and September 1963 and June 1964. In the Untersee (upper part of Fig. 3), the highest relative and also absolute assimilation quotient was reached in June 1964 when the biomass was smallest. The activity coefficient was 1.2 in the « best » depth then. But it was only 0.2 with the largest standing crop in September 1963 and 0.57 with a medium biomass in June 1963. Also in the upper basin (« Obersee ») relative production was higher in June 1963 when only small amounts of phytoplankton were active than was the case in September. The quotient of assimilated carbon /freshweight was 0.46 and 0.31 respectively.

In comparing this relative production of the Walensee and Lake Constance with that of a Carinthian lake (Fig. 4, Klopeiner See, 1.3 km^2, 46 m deep, primarily oligotrophic but influenced by waste), the small assimilation effect in Klopeiner See is most striking. In all cases the optimal relative assimilation never surpasses essentially 0.1. That is half or less of what I found in the western alpine lakes. The cause of this discrepancy is the fact that quite a different phytoplankton community is present in this lake. *Ceratium hirundinella* and *Peridinium willei, Dinobryon divergens* and *D. sociale, Lyngbya limnetica* and last but not least *Oscillatoria rubescens* are the leading species.

On the contrary, in Walensee and still more in Lake Constance, small forms such as cryptomonads, *Cyclotella* and *Stephanodiscus* form an important part of the phytoplankton. In Klopeiner See these species are only minor constituents. Therefore we may conclude that nannoplankton is more active in assimilation than the larger forms of net-plankton. It is obvious to assume that this higher production rate is due to the more favorable ratio of surface to volume in small cells, which facilitates the uptake of nutrients. Thus nannoplankton species grow and propagate more rapidly, where-

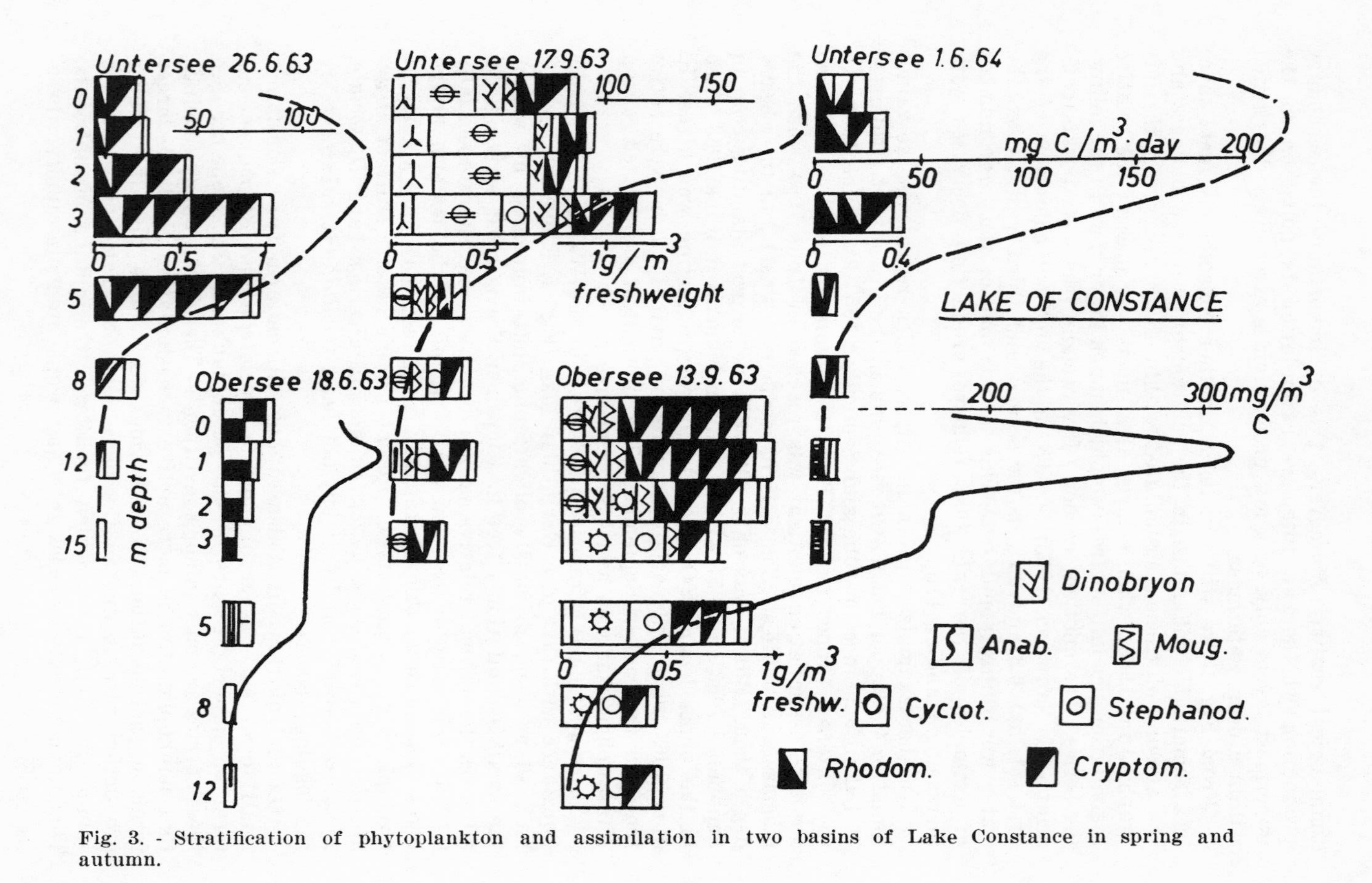

Fig. 3. - Stratification of phytoplankton and assimilation in two basins of Lake Constance in spring and autumn.

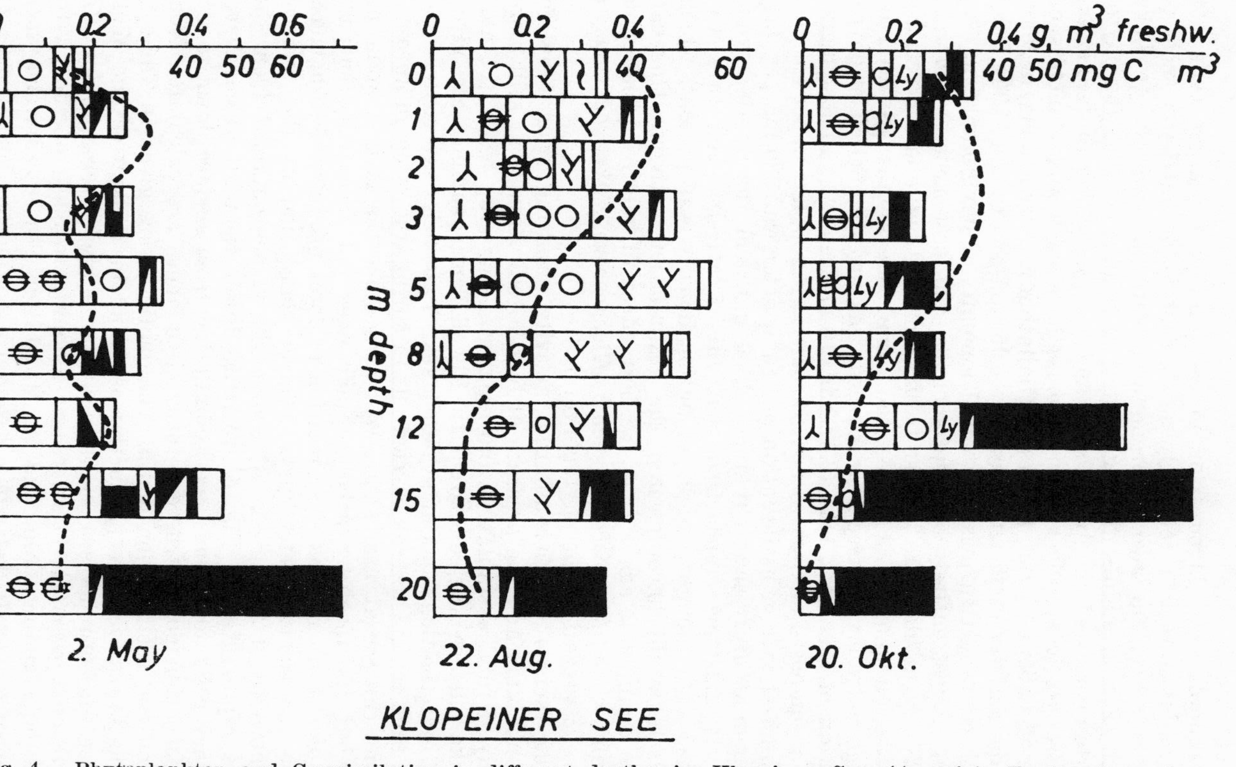

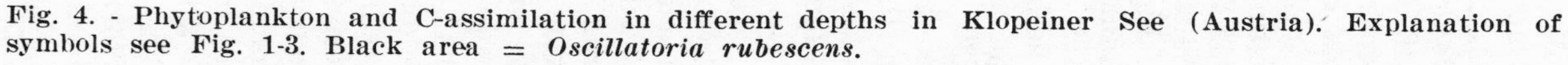

Fig. 4. - Phytoplankton and C-assimilation in different depths in Klopeiner See (Austria). Explanation of symbols see Fig. 1-3. Black area = *Oscillatoria rubescens*.

as big algae generally are rather long-lived and have a low reproduction rate.

Nannoplankton is superior in assimilation to larger cells not only in lakes with a relatively small content of nutrients such as Klopeiner See, but also in pronounced eutrophic waters. In Lake Zürich (Fig. 5, middle and right part) in September 1963 the phytoplankton consisted chiefly of *Oscillatoria rubescens, Peridinium willei* and *Mougeotia* sp. The mean biomass between 0 and 5 meters of depth was about 1 g/m^3, the best assimilation rate was 45 mg/m^3/day, and the activity coefficient was only 0.05 at best. In May 1964 Lake Zürich contained surprisingly small amounts of phytoplankton, consisting almost entirely of cryptomonads. The best assimilation rate was 25 mg/m^3/day, the largest freshweight 230 mg, and the relative assimilation quotient 0.13. At the same time another Swiss lake of high eutrophy, Rotsee near Luzern, was investigated. The phytoplankton of this little lake (0.47 km^2, 16 m deep) was almost monomictic, composed of *Gloeococcus schroeteri* (Fig. 5, left upper corner). The highest rate of assimilation per cubicmeter was 580 mg per day, the amount of freshweight about 1500 mg, and the quotient thus 0.39. It must be emphasized that this relatively good rate was reached even though a high population density was present, which in itself would have lowered the assimilation activity. One can notice again and again that the best effect of relative production is found when *Chlorococcales* such as *Gloeococcus, Ankistrodesmus* or *Chlorella* prevail in the phytoplankton.

The fact of low assimilation activity resulting from crowded phytoplankton and the presence of bigger cells stands out still more clearly if total assimilation and algal standing crop below the unit of lake surface are compared. In Fig. 6 some examples are given from investigations in Carinthian lakes. Two series are represented from each lake. One was carried out in spring, the other in autumn. As one can see at first sight, Wörthersee has the largest standing crop but its absolute carbon uptake does not essentially surpass that of the other lakes. The relation of total production to biomass below the square meter gives a good conception of the relative productivity of the lakes. Of course the values obtained from the series represented in Fig. 6 are much lower than all given above, for they refer to the average assimilation from surface down to the compensation point, whereas those cited above are calculated from the optimal assimilation depth. The quotients are given in Table 1.

The lowest relative production is found in Millstätter and Wörthersee. In both lakes a great many large algae were present. In Wörthersee, *Oscillatoria rubescens* amounts to 80 % of the phytoplankton standing crop, and in Millstätter See big diatoms such as *Synedra acus angustissima, Fragilaria crotonensis,* and to a lesser degree *Ceratium hirundinella* were dominant at that time. The best relative production occurred in Faaker See (18.5.) with a minimal

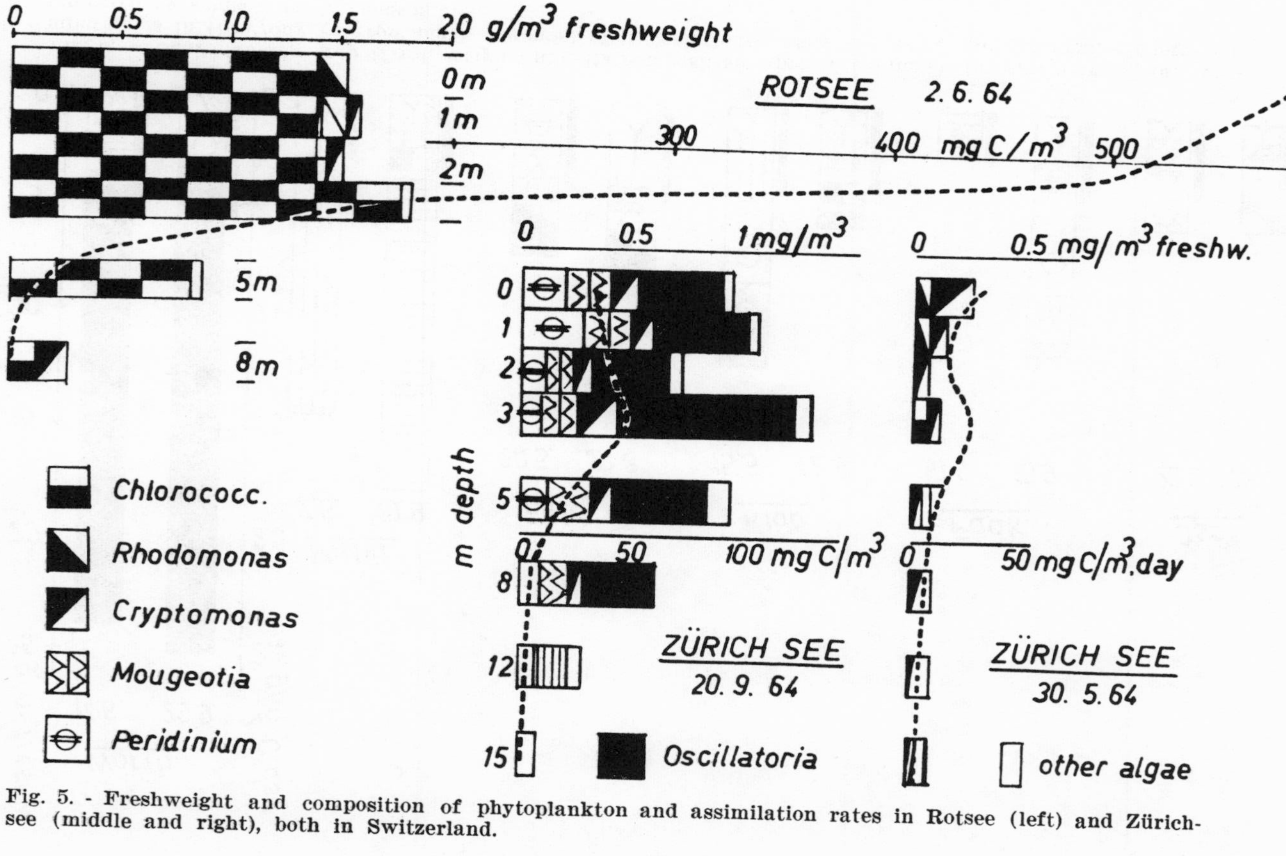

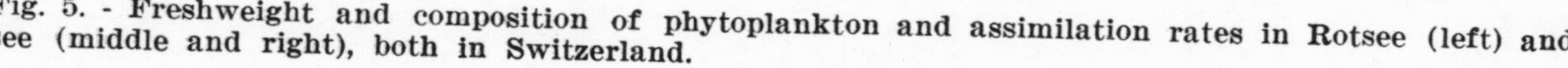
Fig. 5. - Freshweight and composition of phytoplankton and assimilation rates in Rotsee (left) and Zürichsee (middle and right), both in Switzerland.

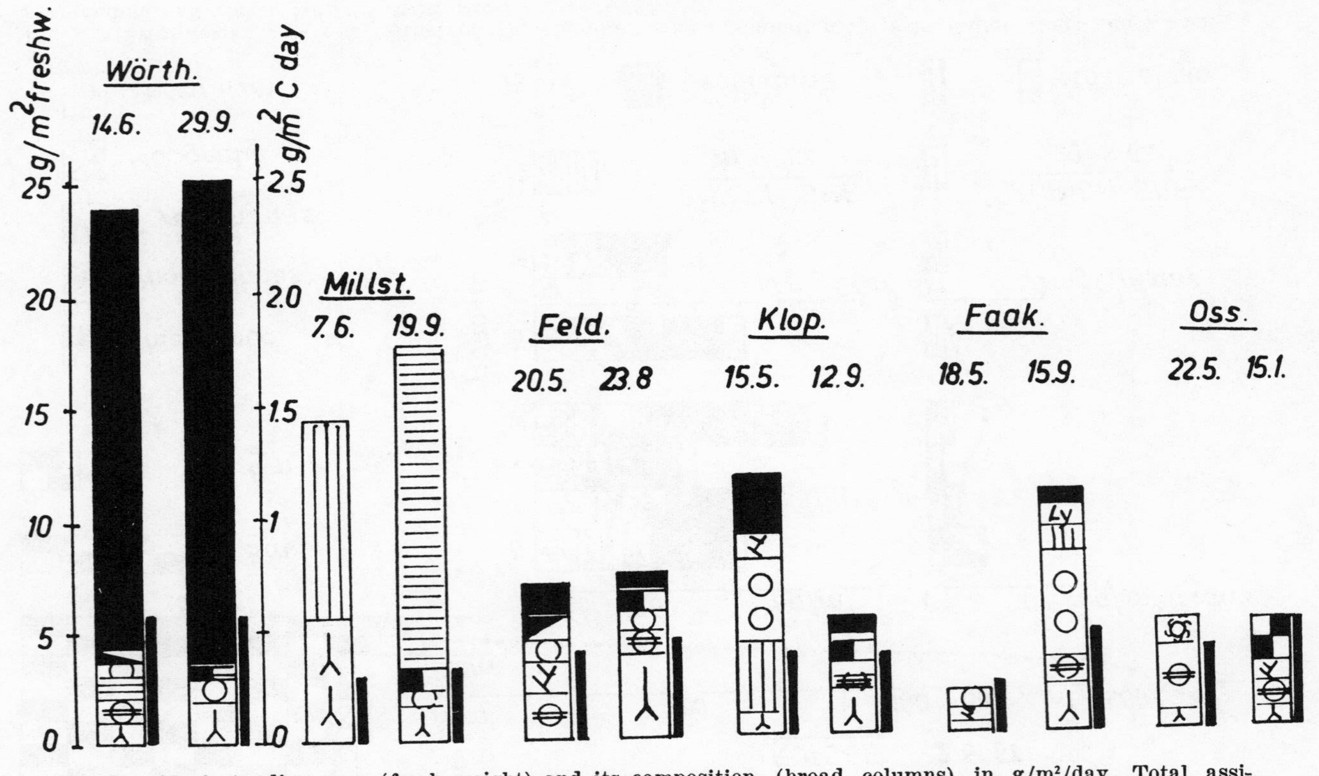

Fig. 6. - Algal standing crop (fresh weight) and its composition (broad columns) in g/m²/day. Total assimilation is in C/m²/day (narrow black columns) of 6 Carinthian lakes in spring and autumn. Genera are indicated by symbols in previous figures.

amount of biomass present, of which 40% consisted of the nannoplanktic species *Cyclotella comensis.* In September the renewal of the rather high standing crop was less favorable, although *Cyclotella* also contributed about 40% to the algal standing crop in this case. Nevertheless, the quotient is more than double as high as in Wörther- and Millstätter Sees with a phytoplankton composed of large cells. Next follows Ossiacher See. With a rather small standing crop present, the activity coefficient is 0.05 to 0.08. Phytoplankton consisted of *Cyclotella* and *Cyclotella* plus *Chlorococcales,* respectively, up to 70% in both series.

Table 1. - Quotients (total assim./biomass) for six Carinthian Lakes.

	Spring	Autumn
Wörthersee (19 km², 84 m deep)	0.025	0.024
Millstätter See (13 km², 140 m deep)	0.021	0.021
Feldsee (0.4 km², 26 m deep)	0.06	0.06
Klopeiner See (1.3 km², 46 m deep)	0.03	0.07
Faaker See (2.4 km², 30 m deep)	0.13	0.05
Ossiacher See (10 km², 46 m deep)	0.05	0.08

In some Swiss lakes things are similar as one can learn from Fig. 7. In Vierwaldstätter See (V) large species such as *Oscillatoria rubescens* and *Tabellaria fenestrata* formed the large standing crop. The activity quotients are given in Table 2.

With respect to phytoplankton mass and its composition, Vierwaldstätter See has the lowest renewal of its standing crop all the

Table 2. - Quotients (mg carbon assim./mg freshweight/m²) for six Swiss lakes.

	Spring 63	Autumn 63	Spring 64
Vierwaldstätter See (114 km², 214 m)	0.014	0.03	0.02
Pfäffikersee (3.2 » 36 »)	0.06	0.09	0.43
Untersee (L. Const.) (64 » 46 »)	0.08	0.09	0.37
Walensee (24 » 151 »)	0.07	0.18	0.09
Zürichsee (89 » 134 »)	—	0.06	0.25
Rotsee (0.5 » 16 »)	—	—	0.16

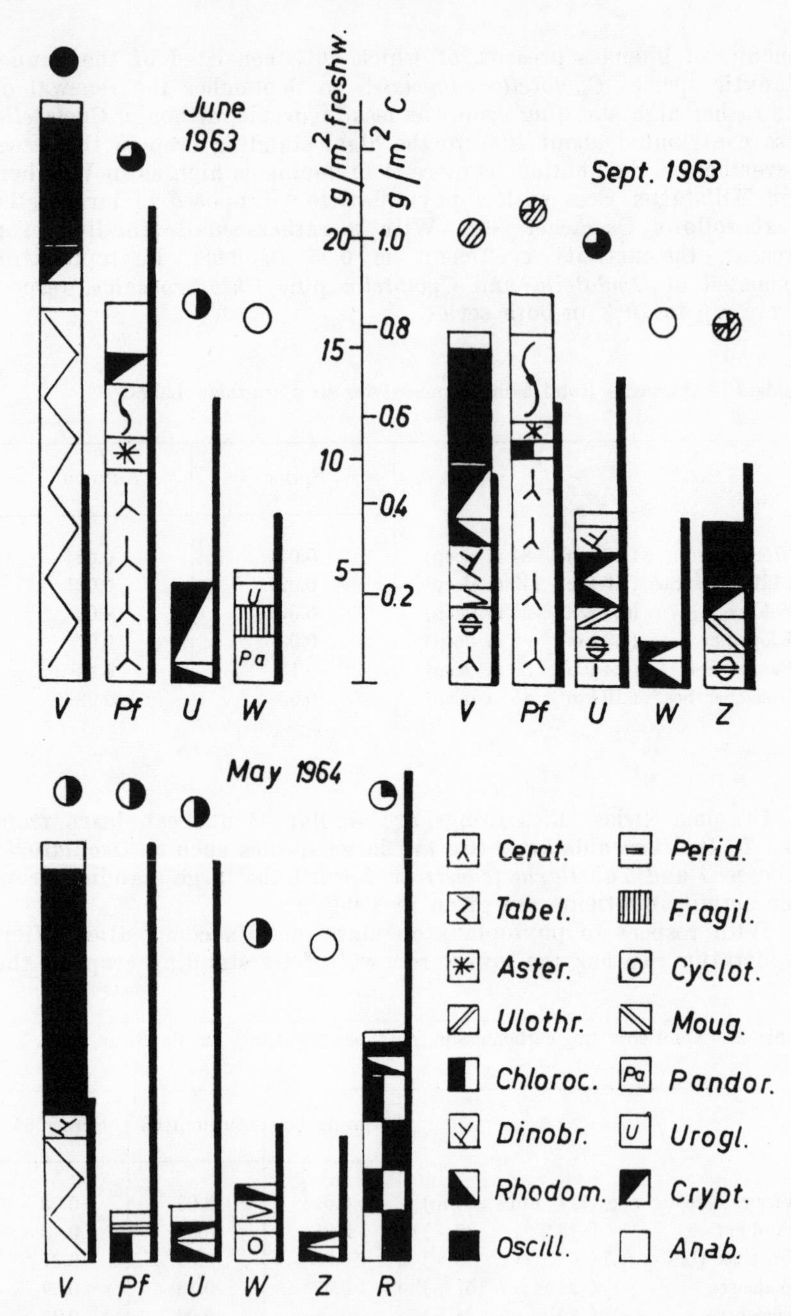

Fig. 7. - Algal standing crop (broad columns) and carbon assimilation (narrow black columns) below square meter in 6 Swiss lakes. (V = Vierwaldstätter, Pf = Pfäffikersee, U = Untersee (L. Const.), W = Walensee, Z = Zürichsee, R = Rotsee near Luzern.

year round. Next to it stands Walensee, which generally has a rather small quantity of phytoplankton indeed but composed of larger algae such as *Fragilaria crotonensis, Pandorina morum,* and *Uroglena americana.* Only in September 1963 a small biomass composed of *Rhodomonas* and *Cryptomonas* produced a good relative assimilation. Also in Pfäffikersee the relationship between production per square meter and freshweight of biomass was unfavorable in the series of 1963 because of the large biomass combined with predominating large species, e.g., *Ceratium, Asterionella,* and *Anabaena.* Yet in May 1964 this lake produced the best activity coefficient observed in all the Swiss lakes under discussion, when a small standing crop existed and *Gloeococcus* was abundant. An analogous result was found in the « Untersee » of Lake Constance and Zürichsee. In Rotsee the only series taken gives an intermediate quotient. The good relative assimilation effect of *Gloeococcus,* which formed about 80% of the phytoplankton, was hampered by the crowded conditions.

Nauwerck (1963) gives values of phytoplanktic volumes or freshweight and of carbon assimilated in Lake Erken in Sweden. The maximum activity coefficient he found was 0.17, the minimum 0.003, the average 0.034. The values of the writer for Carinthian lakes were a maximum of 0.13 in Faaker See and a minimum of 0.021 in Millstätter See. As for the Swiss lakes, the maximum was 0.43 in Pfäffiker See and the minimum value 0.014. This summary means that the renewal of the algal standing crop in the Carinthian lakes seems to be equal to that in Lake Erken, while in certain Swiss lakes it is probably somewhat higher.

From the statement that a close correlation exists between activity of production and magnitude and composition of the phytoplankton, it follows that there must be a certain relationship between production and lake types. In oligotrophic lakes generally small forms such as *Cyclotella* and chrysomonads and also *Rhodomonas* are the chief components of the phytoplankton. Therefore we may conclude that these lakes will renew their algal biomass more rapidly than the eutrophic ones. In spite of this, the biomass remains small because nannoplankton species are short-lived even when they are not eaten up by the zooplankton. Moreover, the scarcity of nutrients limits the community to a number adequate to the decomposition of preceding algal generations. On the other hand eutrophication favors not only the development of large numbers of cells but also the occurrence of large forms, or at least big colonies of crowded cells. This holds true for *Anabaena, Microcystis,* and *Oscillatoria,* as well as for the Pennales among the diatoms such as *Tabellaria,* for *Melosira,* and also for *Ulothrix* and *Mougeotia.* For both reasons the renewal of the phytoplanktic biomass goes on more slowly. But as the species enumerated above grow slowly and are long-lived, a large standing crop exists in those lakes. In addition, large phytoplankton forms are not liable to being grazed.

There is a difference also between the composition of the phytoplankton in small and in large lakes. As a rule small lakes are better wind-sheltered than larger ones because of the wind shelter provided by a superelevated shore. Because of this, turbulence is slight and flotation in water is rendered more difficult. Large diatoms have a relatively high excess weight and an unfavorable specific surface area, so that the rate of sinking is considerable they can not keep floating except when turbulence is effective. Therefore *Fragilaria, Tabellaria, Cymatopleura,* and *Melosira* are missing in the plankton community of small lakes, especially if these are situated in regions poor in wind (Findenegg 1965). Hence, it is understandable that the plankton of small lakes generally is composed either of species possessing their own power of movement or of very small forms which sink very slowly, such as *Cyclotella* or *Stephanodiscus.* If these lakes are subjected to eutrophication, blue-green algae appear instead of *Melosira* or *Tabellaria,* and « water bloom » - forming species become the main constituents of the phytoplankton because of their adaptation to floating by gas vacuoles. As these species (except *Oscillatoria*) generally live in the superficial layers of the lake where light conditions are favorable, this type of lake produces relatively more than larger ones do, if eutrophication is at the same level. Therefore the smallest renewal rate of the phytoplankton biomass generally is found in larger eutrophic lakes characterized by a high percentage of *Tabellaria* or *Oscillatoria.*

In Fig. 8, the phytoplanktic standing crop below one square meter as it was found in the series of 7 Austrian and 6 Swiss lakes at different seasons between 1959-1964 is represented by a semilogarithmic graph. The lowest freshweights were found in Lunzer See and in Lake Constance in many cases. On the contrary, the largest biomass was present in Wörthersee, Zeller See and Vierwaldstätter See.

As an additional note, it is worth mentioning that there is no general correlation between trophic level and algal biomass. Wörthersee and Zeller See (at that time) are polluted by house-hold waste, and so probably is Vierwaldstätter See. According to this eutrophication they contain a large biomass. But although Lake Constance is also polluted to a high degree, its standing crop was not high at all times, but rather medium or low. Ossiacher See too is rather eutrophic, but it is rich in phytoplankton only in summer.

If we compare the amounts of phytoplankton with the quotient of relative assimilation of this biomass per day, graphically represented in Fig. 9, we can say that with regard to the normal sequence a real overturn has taken place. Lake Constance (eutrophic), Walensee and Lunzer See (both oligotrophic) are almost at the top, while Wörthersee, Zeller See and Vierwaldstätter See are at the bottom. The first group is characterized either by small amounts of freshweight, as is the case with Lunzer See, or by larger standing crop with dominating nannoplankton, which holds true for Lake Constance

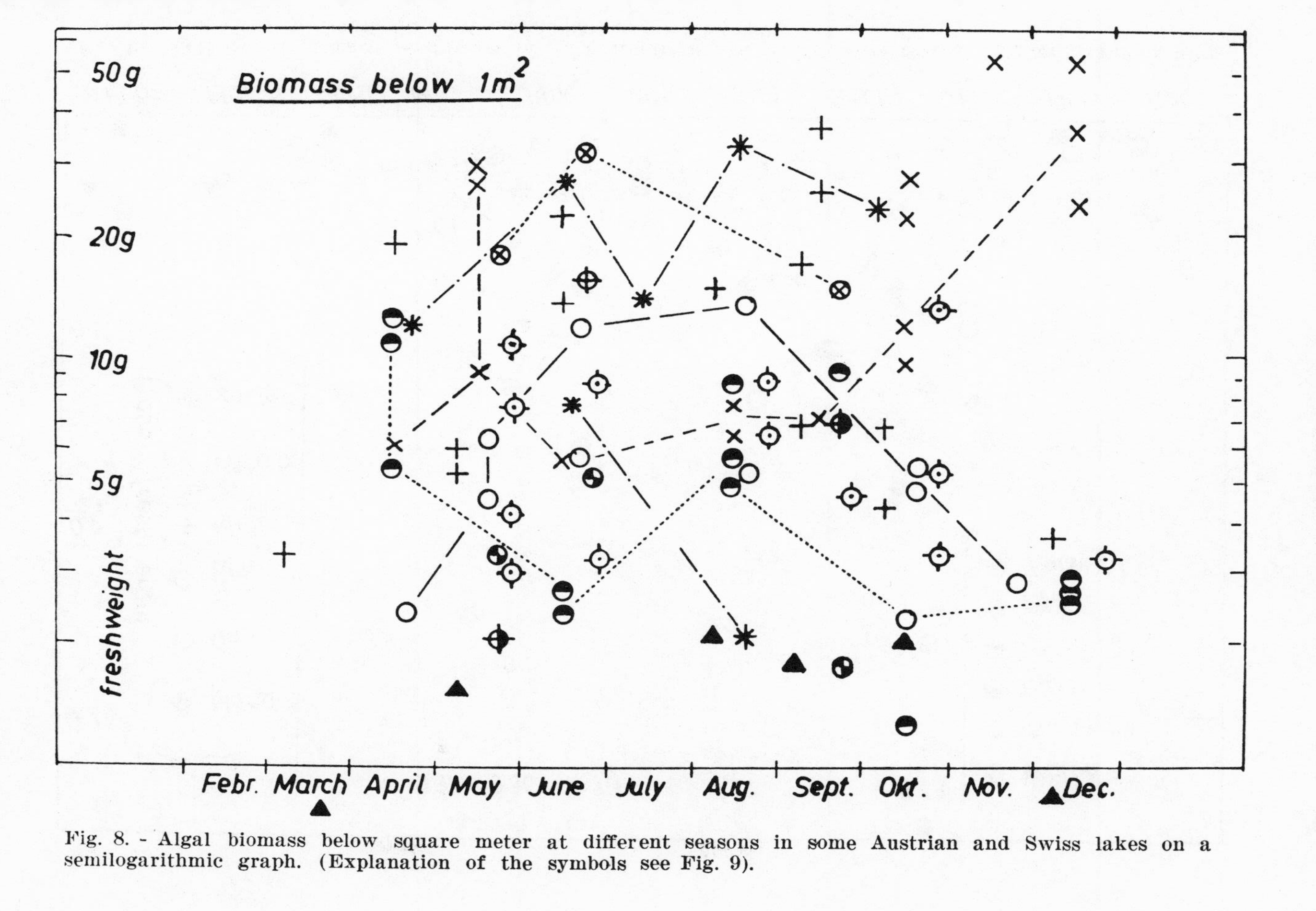

Fig. 8. - Algal biomass below square meter at different seasons in some Austrian and Swiss lakes on a semilogarithmic graph. (Explanation of the symbols see Fig. 9).

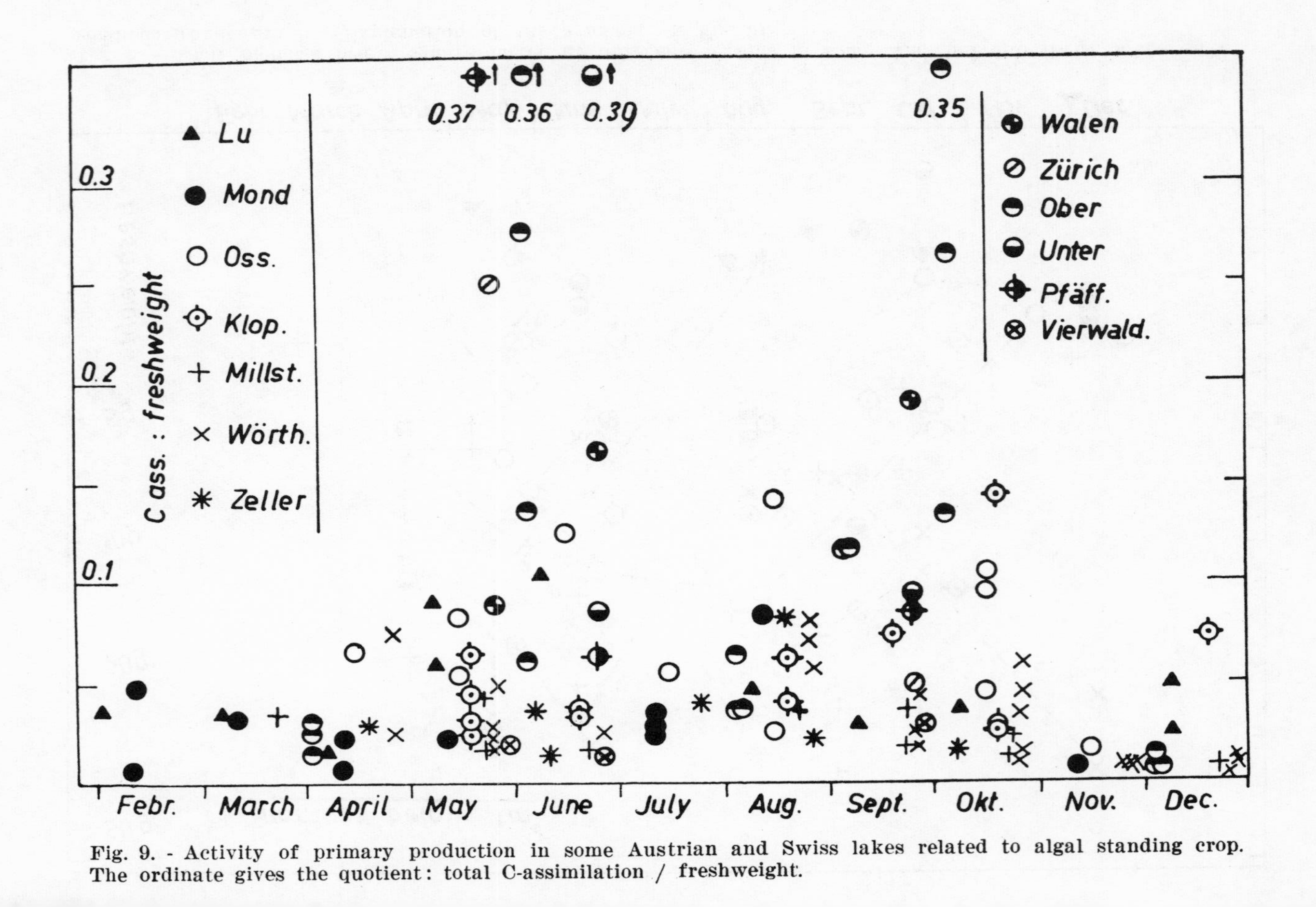

Fig. 9. - Activity of primary production in some Austrian and Swiss lakes related to algal standing crop. The ordinate gives the quotient: total C-assimilation / freshweight.

and to some degree also Walensee. On the contrary, the lakes that show the lowest assimilation activity are pronounced *Oscillatoria*-lakes (Wörthersee), *Tabellaria-Oscillatoria*-lakes (Vierwaldstätter See) or *Tabellaria-Anabaena*-lakes, as was the case with Zeller See before pollution was reduced.

From these examples we learn that the relationship between standing crop and productivity rates varies more or less individually from lake to lake. All we can say in the present state of knowledge is: (1) that nannoplankton is more active in relative assimilation than larger species are, and (2) that increasing population density diminishes relative assimilation rate. The higher activity of small cells can be explained by the favorable ratio of surface to volume, which facilitates the uptake of nutrients. The lowering effect on relative production of crowded masses of phytoplankton may result as well from nutrient competition as from damages caused by concentration of excretes.

REFERENCES

Findenegg, I. 1964. Produktionsbiologische Planktonuntersuchungen an Ostalpenseen. - Intern. Rev. ges. Hydrobiol. *49* : 381-416.

— 1965. Limnologische Unterschiede zwischen den österreichischen und den ostschweizer Alpenseen. Vierteljschr. Natf. Ges Zürich *110* : 289-300.

Nauwerck, A. 1963. Die Beziehungen zwischen Zooplankton und Phytoplankton im See Erken. - Symb. Botan. Upsaliens. *17* (5) : 1-163.

NUMERICAL ASPECTS OF NANNOPLANKTON PRODUCTION IN RESERVOIRS

MIROSLAV STEPANEK
Prague (Czechoslovakia)

For bibliographic citation of this paper, see page 10.

In the course of our survey work on numerous reservoirs in Czechoslovakia, we determined the variations of phytoplankton density, standing crop, photosynthetic productivity, and the depth at which the respective compensation points were found. The relations between these criteria and primary productivity have already been discussed in a number of reviews, and in some of the lectures presented at this symposium. However, the application of the results of other authors to our practical work meets with difficulties of intercalibration and of conversion coefficients. Most of our laboratories are not equipped for determining primary productivity directly. We also suffer from a lack of specialists who could carry on this kind of work. Thus, we are essentially restricted to estimating productivity on the basis of cell counts, which give us some information on population dynamics in phytoplankton. The large amount of such data which has accumulated tells us something about the numerical aspects of phytoplankton development, and about standing crop in various rivers, reservoirs, and ponds. However, it has not been possible to compare these data conclusively on a common basis.

Personally, I have studied the influence of climatic and microclimatic factors on the development of algal populations. The relation between the different portions of the sunlight spectrum and the dynamics of phytonannoplankton populations is of special interest here.

THE DURATION OF SUNSHINE

Close statistical correlation was estimated between the duration of sunshine and the numbers of nannoplankton produced ($r = 0.929$). More than 90% of the reservoir area is illuminated by the sun in the interval from 21 March to 23 October, i.e., practically throughout the vegetation period. The area of water surface illuminated depends on the height of the sun above the horizon. It was estimated that the average duration of sunshine necessary to increase the number of nannoplankton 25,000 cells per ml lengthened from 6.9 hours in 1954 to 10.0 hours in 1957. Equations of exponential curves were fitted to corresponding data, characterizing the above mentioned relations for individual years, and their course was graphed. By analysis of curves using a graphical method it was estimated that the average duration of sunshine necessary for a 50% increase in the numbers of nannoplankton increased during the period from 1954 to 1957 from 2.5 to 4.7 hours.

The relation between the average weekly duration of sunshine and the weekly numerical stages of the nannoplankton was estimated for observed years at a good level of statistical significance ($r = 0.467$, $n = 129$). An equation was fitted to the data:

$$H_{54\text{ to }58} = -5.44 + 2.93 \log Pn.$$

H is the daily average duration of sunshine in hours during the interval between two samplings of phytonannoplankton; Pn is the numerical productivity of nannoplankton expressed by the average standing crop of organisms per ml in a vertical profile of the Sedlice reservoir.

It was estimated that the number of nannoplankton organisms increases only slowly from 100 to 1,000 cells per ml with an increase of the average duration of sunshine in the interval from 0.0 to 3.5 hours. The increasing average duration of sunshine above 3.5 hours (reaching limit values 7 to 10 hours) is correlated with an increase of the average phytonannoplankton count from 1,000 to 100,000 cells per ml.

It was determined that the most significant correlation between average duration of sunshine and standing crop of phytonannoplankton occurred in the water layer from 0 to 2 m. A less significant relationship in the same years was found for a shallower layer near the water surface.

The hypothesis is presented, that the average number of phytonannoplankton organisms per ml of water of the vertical water column of the Sedlice reservoir depends on the duration of sunshine to such an extent that the latter exerts an influence not only upon organism numbers in the euphotic zone, but even in the whole water column. The validity of this hypothesis is supported by statistically significant correlations between the duration of sunshine and the average numbers of phytonannoplankton organisms per 1 ml of the vertical water column in the Sedlice reservoir from 1954 to 1959.

The relation of high significance was estimated between monthly average values of the numbers of nannoplankton and the duration of sunshine for 1954 to 1959. The coefficient of correlation computed for the average duration of sunshine varying from 0 to 8.0 hours was 0.885; the same for the range from 2.0 to 8.0 hours showed $r = 0.929$, and for 2.0 to 7.0 hours $r = 0.969$. It was shown that except in 1956, the numbers of primary producing nannoplankton

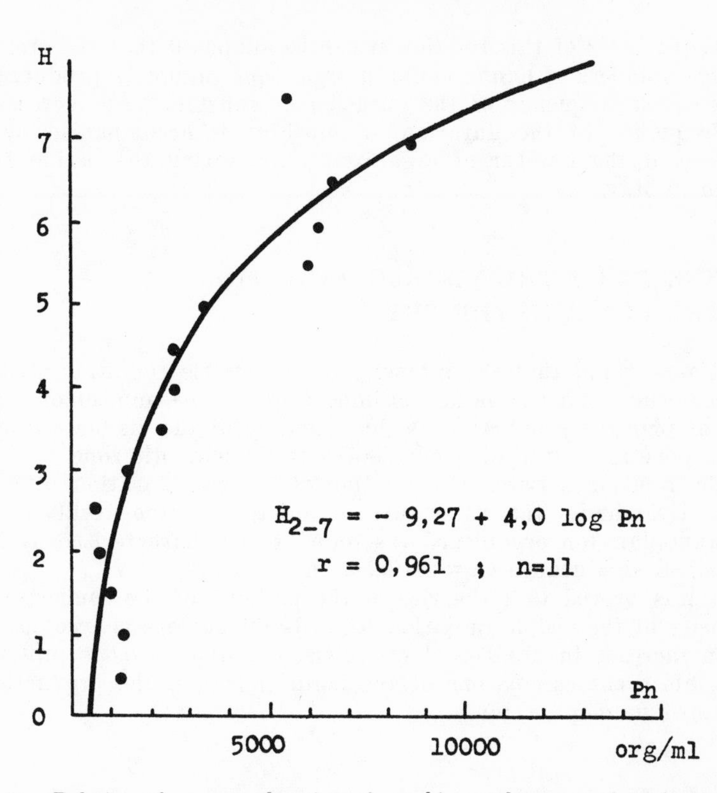

Fig. 1. - Relation of average duration of sunshine and numerical productivity of nannoplankton (monthly averages).

lagged by one unit of the interval of observation (a week) behind the average monthly duration of sunshine. Sliding monthly means were computed for the daily duration of sunshine and the numerical increase of nannoplankton in order to estimate the long term course of both values with the elimination of seasonal influences. It was estimated that the number of organisms showed a continual decrease from 1955 until 1957. The duration of sunshine during this period increased only a little; its sliding means vary about 3.6 hours. On the basis of obtained results it appears that the duration of sunshine had probably no influence upon the long term changes in number of nannoplankton. The relation between the frequency of the duration of sunshine and the increase of nannoplankton was also found. Its course is characterized by a parabola of the second order:

$$aH = 27.19 + 5.31 \text{ Pn} - 0.34 \text{ Pn}^2,$$

where aH is the frequency of the duration of sunshine.

On the basis of this relation it can be supposed that the greatest average numbers of nannoplankton organisms cannot be produced by the greatest frequency of the duration of sunshine. An increase of the frequency of the duration of sunshine is accompanied by an increase of the number of organisms only during the period from March to May.

INTENSITY OF THE VISIBLE RADIATION OF THE SUN AND THE SKY

It was found that the intensity of the visible radiation striking the reservoir surface was not as important for the number of organisms of primary production as the intensity of the visible radiation which penetrates into different depths of the euphotic zone.

The relation between the position of the compensation intensity of the light under the water surface and the average weekly counts of nannoplankton organisms was found to be characterized by high statistical significance ($r = -0.508$, $n = 32$).

It was proved that the rise of the position of the compensation intensity of the visible radiation towards the surface is accompanied by an increase in number of organisms. A hypothesis is presented that this event can be one of the main causes of the productional pulses of nannoplankton.

TOTAL INTENSITY OF THE RADIATION OF THE SUN AND THE SKY

A statistically significant relation was estimated between the total intensity of the radiation of the sun and the sky and the primary production of nannoplankton, characterized by the coefficient of correlation $r = 0.539$ and the number of cases $n = 70$.

It was calculated that the annual sum of the total intensity of the radiation of the sun and the sky amounts to only 55% of the theoretically possible annual sum of the radiation during normal cloudiness. This relation was characterized by a fit of an equation of the exponential form to corresponding data. A hypothesis is stated that a certain limiting value exists of the total intensity of the radiation necessary for the mass development of phytonannoplankton. This was determined to be 300 $cal/cm^2/day$ for the Sedlice reservoir.

It was proved that the relation of the highest significance between the total intensity of the radiation of the sun and the sky and the average number of the nannoplankton was found for the layer from 0 — 3 m ($r = 0.730$, $n = 35$). The relation of the lowest significance was found for the upper layer of water near the surface.

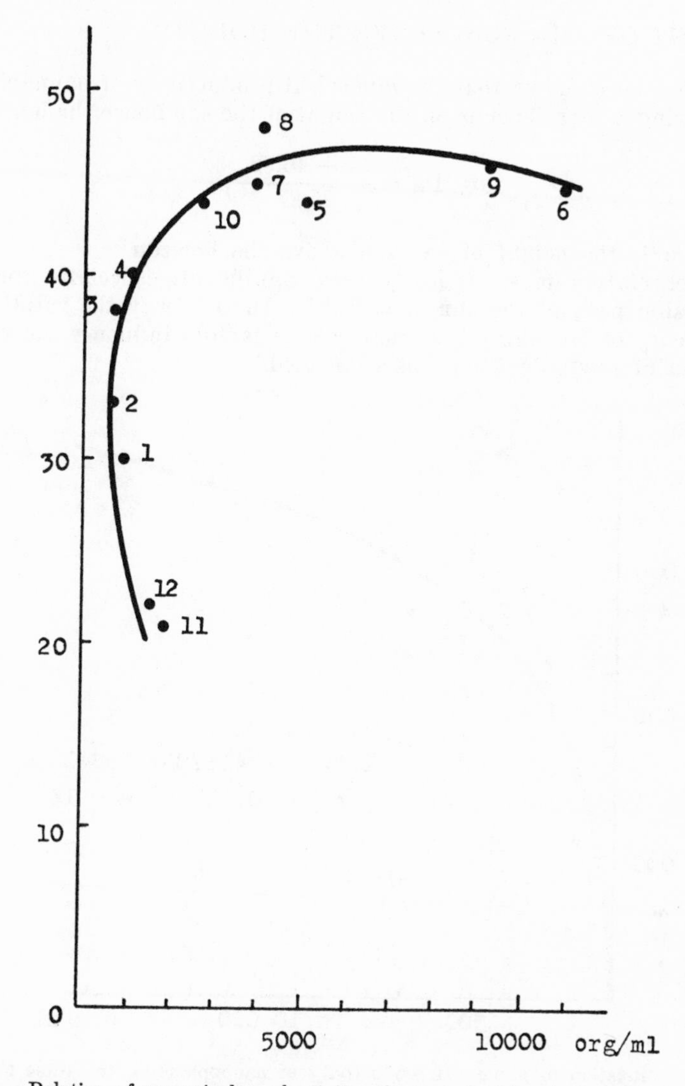

Fig. 2. - Relation of numerical productivity of nannoplankton to frequency of sunshine duration ($\alpha°$) on the Sedlice reservoir.

A hypothesis is proposed that if the caloric content could be registered only for the spectral range 400 to 700 m instead of nearly the whole spectral range, as is the case with records obtained with the Robitsch pyranograph, the statistical significance of the relation between the calories radiated by the sun on the reservoir surface and the numerical productivity of nannoplankton would be higher.

HEIGHT OF THE SUN ABOVE THE HORIZON

It has been shown that the numerical productivity of nannoplankton during a year depends on the height of the sun above the horizon:

$$\log \text{Pn} = \frac{a + 46.92}{25.87},$$

where a is the height of the sun above the horizon.

The relation is statistically very significant, especially for the increasing path of the sun (r = 0.887). In that way the validity of the theory of Maucha (1949) that cosmic factors influence the reproduction of phytoplankton was confirmed.

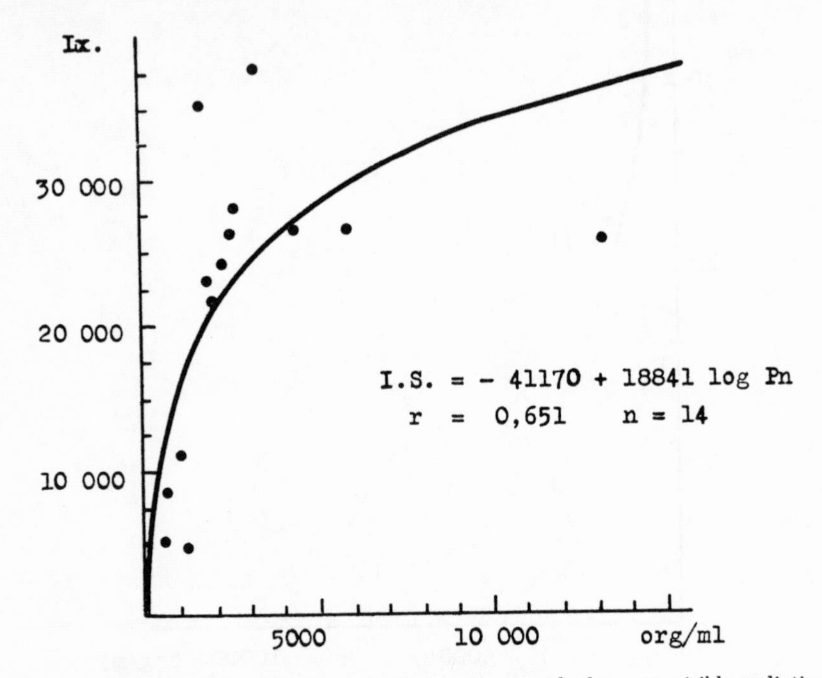

Fig. 3. - Relation of numerical productivity of nannoplankton to visible radiation on the Sedlice reservoir.

The identity of the resulting number of phytonannoplankton organisms was found by separating it from the relation between the height of the sun above the horizon and the duration of sunshine, the intensity of the visible radiation of the sun and the sky, and the total intensity of the radiation of the sun and the sky. So a direct dependence was proved to exist between the numerical increase of nannoplankton on the Sedlice reservoir and the sun.

It was calculated that at the greatest height of the sun above the horizon, the duration of sunshine gives the lowest numbers of organisms, the largest numbers being obtained using data for the total intensity of the radiation of the sun and the sky. This is

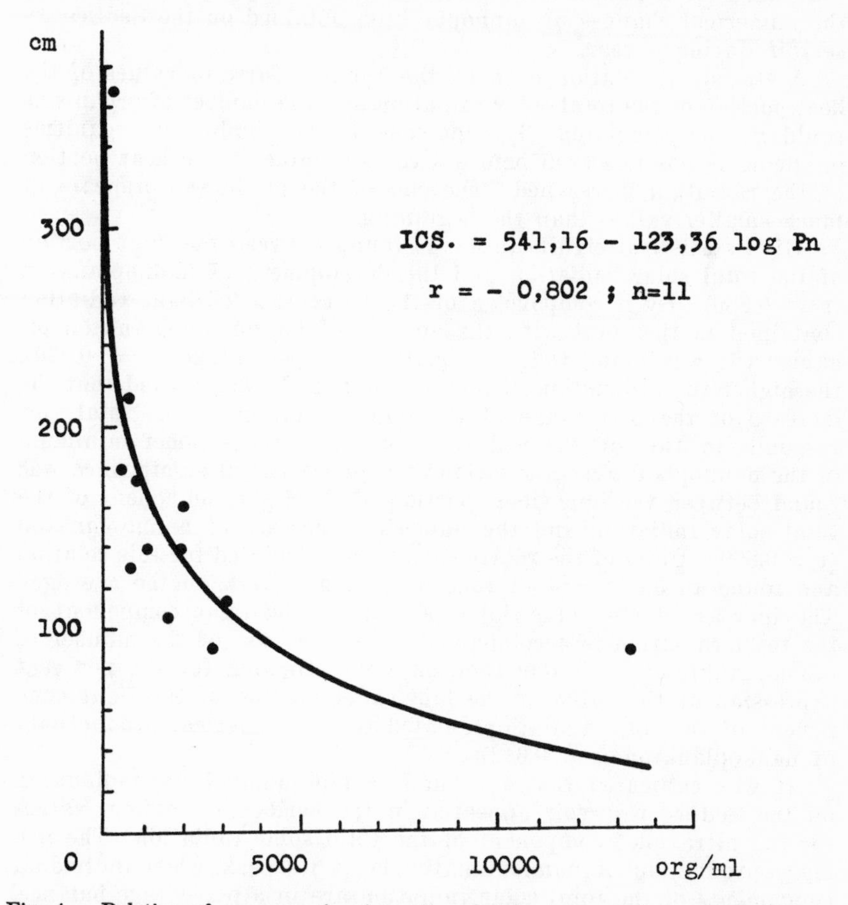

Fig. 4. - Relation of compensation intensity of light to numerical productivity of nannoplankton on the Sedlice reservoir.

probably explained by the fact that the direct rays of the sun can strike the surface of the Sedlice reservoir until the height of the sun above the horizon reaches more than 10°, whereas the total intensity of the radiation of the sun and the sky acts mainly by its second component as soon as astronomical sunrise starts.

COMPONENTS OF THE SOLAR RADIATION

It was possible to think reasonably of the important influence of individual components of the total solar radiation. The data of Dorn were used for a comparison of several years' monthly averages with the numerical changes of nannoplankton obtained on the Sedlice reservoir during a year.

A statistical relation between the annual course of values of the heat portion of the total solar radiation and the number of organisms could not be calculated. The increase of the number of organisms in spring is possible even before a certain value of the heat portion of the radiation is reached. The end of the production appears at much smaller values than the beginning.

High statistical significance was found between the light portion of the total solar radiation and the development of nannoplankton ($r = 0.880$). By a comparison of the percentage of light radiation contained in the total with the number of phytonannoplankton organisms it was found that the smaller this percentage ($r = -0.820$), the higher the production of nannoplankton. It was proved that the increase of the percentage of the light radiation in the total corresponds in July on the Sedlice reservoir to the summer minimum of the nannoplankton. A relation of high statistical significance was found between the blue-violet portion of the light component of the total solar radiation and the numerical increase of nannoplankton ($r = 0.899$). Units of the retardation were calculated for this relation and found to be $L = -1$ (one unit = one week on the average). The increase of the blue-violet portion of the light component of the total radiation is accompanied by an increase of the number of nannoplankton. A similar relation was estimated for the per cent expression of the value of the blue-violet portion of the light component of the total radiation related to the numerical productivity of nannoplankton ($r = 0.642$).

It was estimated that the summer minimum of nannoplankton on the Sedlice reservoir appeared in the period of maximal values for the ultraviolet component of the total solar radiation. The primary production of nannoplankton is at its peak when individual components of the total solar radiation are in a percentage balance. It is pointed out that these relations could possibly be proved finally if true values were available for the components of the total solar radiation recorded on the Sedlice reservoir.

SUNSPOTS

It was estimated that with an increasing relative value of sunspots, the numerical productivity of phytonannoplankton decreases ($r = -0.932$, $n = 35$). For the calculation of this relation, sliding monthly means of both values were used from 1954 to 1959.

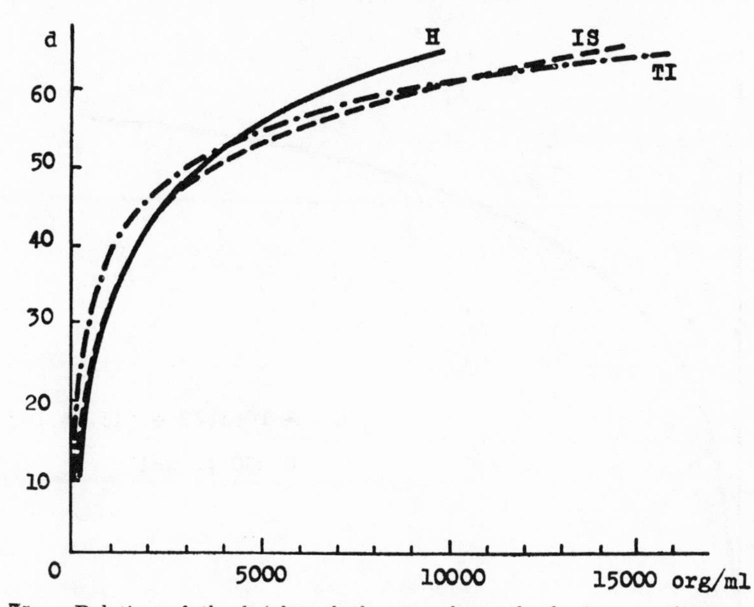

Fig. 5. - Relation of the height of the sun above the horizon to the numerical productivity of nannoplankton on the Sedlice reservoir. (Individual curves were computed using data for the duration of sunshine = H, the visible radiation = IS, and total radiation = TI).

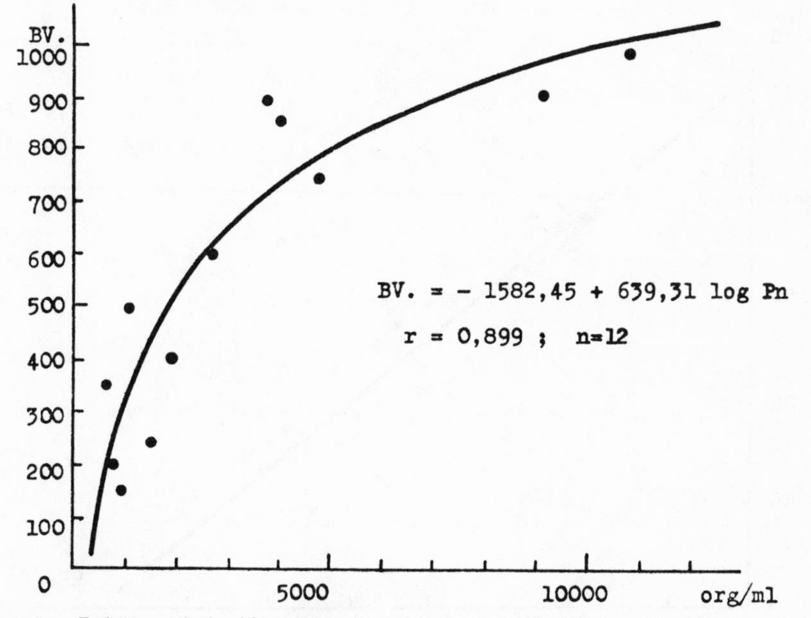

Fig. 6. - Relation of the blue-violet component of total radiation to the numerical productivity of nannoplankton.

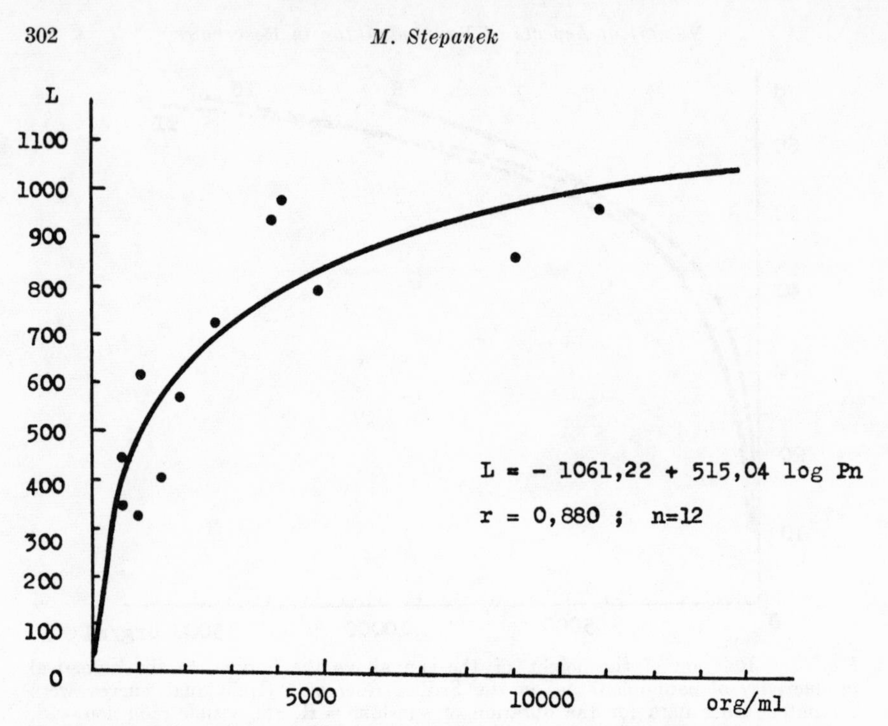

Fig. 7. - Relation of the visible component of total radiation to the numerical productivity of nannoplankton.

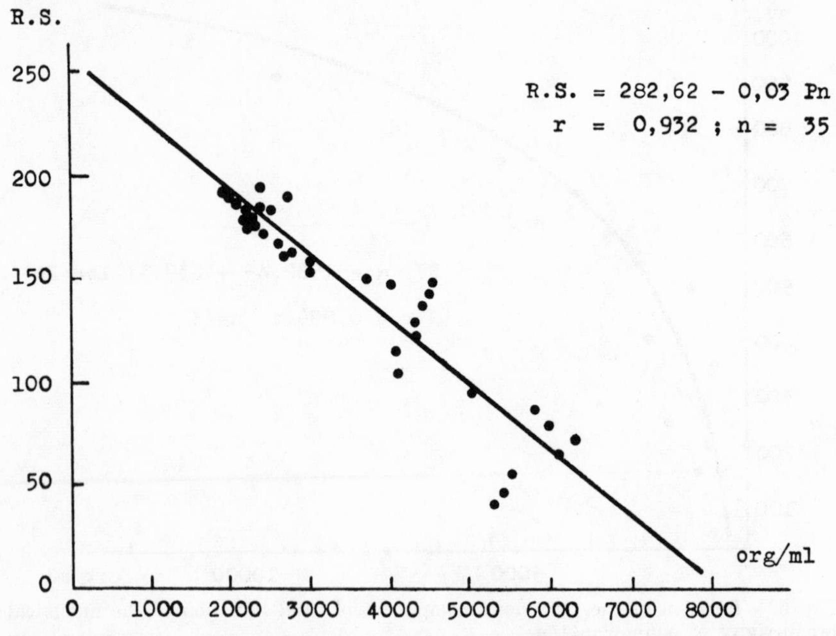

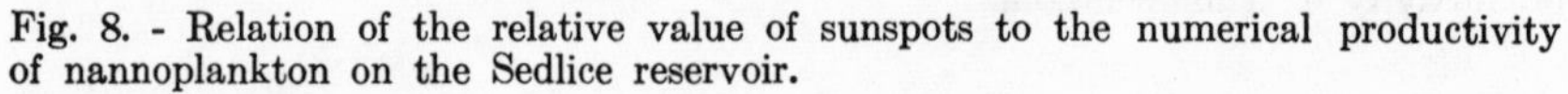

Fig. 8. - Relation of the relative value of sunspots to the numerical productivity of nannoplankton on the Sedlice reservoir.

Fig. 9. - Silon bags.

It was estimated that on the Sedlice reservoir the relations between phytonannoplankton and the solar radiation are generally of exponential nature. We used all our attention to detect every difference in light conditions at various points of the reservoir, which is situated in a deep valley. The reservoir's surface can be exposed to direct sunlight; however, there are some small sidearms of the basin that are mostly shaded in summer. Here phytoplankton density and productivity are quite different from the sunny parts of the lake. These differences were experimentally studied in bags made of silon. The bags are reinforced by solid hoops in order to keep their cylindrical shape in the water. The diameter of a bag is about 2 m, with a length of about 10 m. Four bags have silon bottoms, the remaining four have open lower bases so that natural sediments formed their bottom after their installation. The bags are able to separate from the free water of the reservoir a water column of the approximate volume of 20 m³. The tops of the bags were situated 20 cm above the water surface. The silon material of the experimental bags formed an impermeable membrane for plankton organisms. The stratification of temperature in the silon bags was similar to that of the free

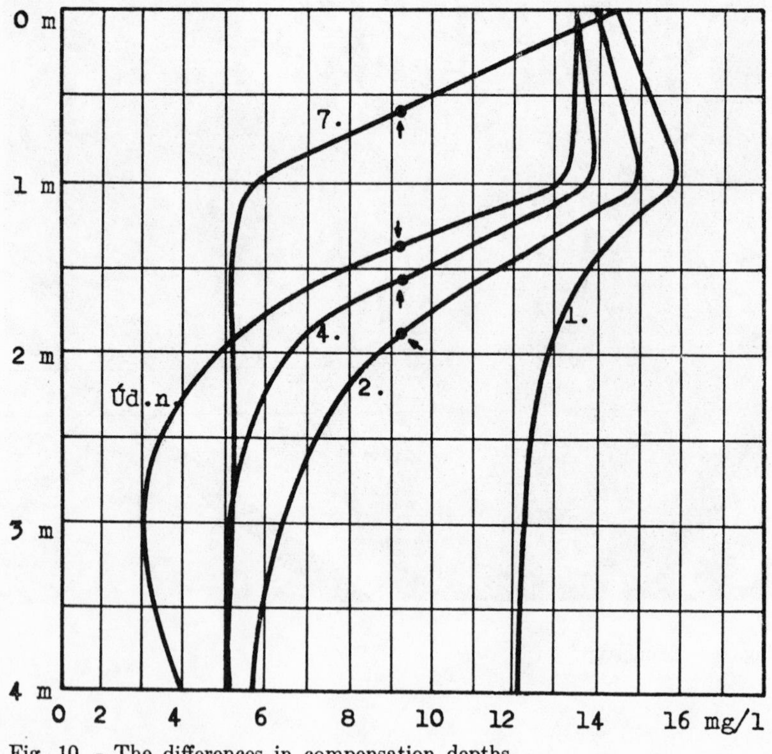

Fig. 10. - The differences in compensation depths.

water of the reservoir. It was estimated that the numerical production of phytonannoplankton in the silon bags, as in the free water, was very closely related to the duration of sunshine. Using the test for homogenity of the sample of coefficients of correlation ($X^2 < X_{0,10}$) and comparing the values of the coefficients of correlation using the z-values, proof was presented of the practically insignificant difference of the relation between the duration of sunshine and the numerical production in individual bags as well as in the reservoir proper. In these bags we determined the compensation depth under different conditions by means of the light and dark bottle technique (Vinberg method). Two different phytonannoplankton biocenoses were used for this experiment. A difference in the position of the compensation points was estimated for individual bags as well as for individual biocenoses. It was demonstrated that individual phytoplankton biocenoses in silon bags are adapted to the hydroclimatic conditions of illumination in which they have developed.

These experiments show how phytoplankton can alter or prepare for the conditions of their own growth, especially with regard to the compensation depth. We also could observe cases in which a biocen-

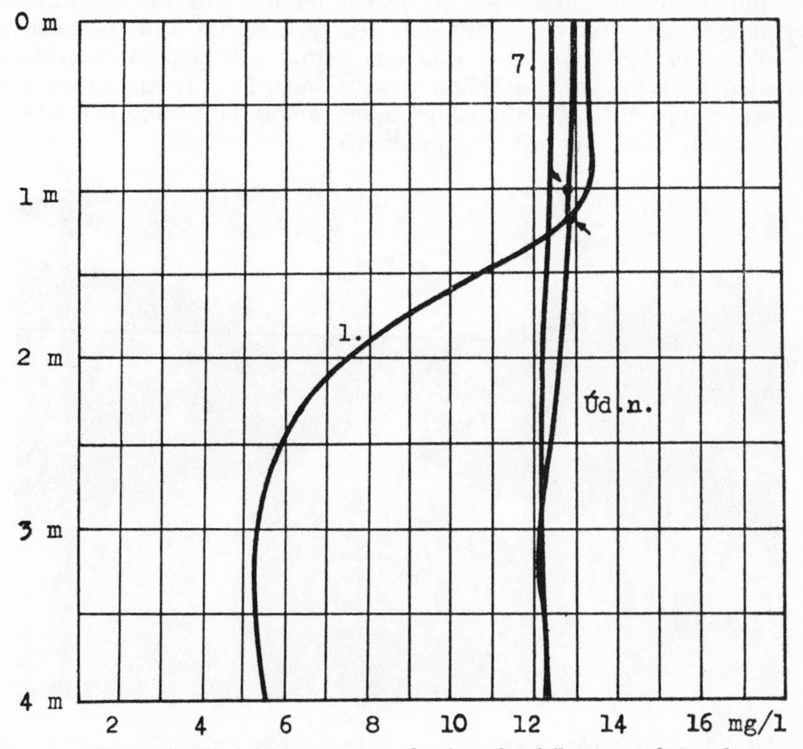

Fig. 11. - Determination of compensation depth under different conditions by means of the light and dark bottle technique.

osis of a certain species composition depressed the growth of other planktonic algae. Water blooms of one species can thus very well push down the productivity of the rest of the community.

The entire system of interacting factors also underlies the influence of other meteorological and hydrological factors. Rain, wind, currents, etc. should be mentioned here. They can effectively mix different water layers of a reservoir and complicate the situation. Our observations show how deeply quantity and quality of sunshine and other meteorological factors interfere with the development and permanence of phytoplankton communities.

Phosphorus and other essential nutrients are rarely limiting in Czechoslovakian waters because most of them are intensely fertilized by sewage. We also observed mutual influences between certain taxonomic groups of bacteria and the biomass of phytonannoplankton.

Our Public Health Service has gathered comparative data on 44 reservoirs and 23 ponds. Unfortunately, we know only the species composition and cell numbers of phytoplankton in relation to bacterial communities, pigment concentrations, and climatic and chemical conditions. We do not at present have the means to apply the tracer technique to our waters and we feel other methods for determining primary productivity are not sufficiently accurate and comparable.

We, therefore, hope that this symposium will help to provide the applied branches of hydrobiology with some help in the evaluation of our data which, on the other hand, might be useful for further theoretical work on primary productivity.

Fig. 12. - The Sedlice reservoir.

REFERENCES

Berg, H. 1957. Solar-Terrestrische Beziehungen in Meteorologie und Biologie. Ak. Verl. G., Leipzig.

Clarke, G. L. 1946. Dynamics of production in a marine area. - Ecol. Monogr. *16* : 321-335.

Cooper, L. H. N. 1955. Hypotheses connecting fluctuations in arctic climate with biological productivity of the English Channel. - Papers Mar. Biol. and Oceanogr., 212-223.

Doan, K. H. 1942. Some meteorological and limnological conditions as factors in the abundance of certain fishes in Lake Erie. - Ecol. Monogr. *12* :293-314.

Edmondson, W. T. 1956. The relation of photosynthesis by phytoplankton to light in lakes. - Ecology *37* : 161-174.

Gessner, F. 1955. Hydrobotanik. Die physiologischen Grundlagen der Pflanzenverbreitung im Wasser. I. Energiehaushalt. - Berlin, VEB Deutscher Verlag der Wissenschaften, 517 pp.

Maucha, R. 1949. Die Photosynthese des Phytoplanktons vom Gesichtspunkte der Quantenlehre. - Hydrobiologia *1* :45-62.

Ohle, W. 1956. Bioactivity, production and energy utilization of lakes. - Limnol. and Oceanog. *1* :139-149.

Rodhe, W. 1948. Environmental requirements of fresh-water plankton algae. Symb. Bot. Upsalien *10* :1-149.

Strickland, J. D. H. 1958. Solar radiation penetrating the ocean. - J. Fish. Res. Bd. Canada *15* :453-493.

Stepanek, M. 1960. Limnological study of the reservoir Sedlice near Zeliv. X. Hydrobioclimatological part. The relation of the sun radiation to the primary production of nannoplankton. - Sci. Pap. Inst. Chem. Technol. *4* :21-140.

— 1961. Limnological study of the reservoir Sedlice. XVII. The development of phytoplankton in silon bags. - Sci. Pap. Inst. Chem. Technol. *5* :275-323.

— 1963. Einige Fragen aus der Problematik der Phytoplanktonpigmente und des Kompensationspunktes im Stausee Sedlice. - Sci Pap. Inst. Chem. Technol. *7* : 387-400.

Zadin, V. I. 1948-56. Zizn presnych vod SSSR. - I-IV. Moskva.

NET PRIMARY PRODUCTIVITY AND PHOTOSYNTHETIC EFFICIENCY IN THE BIOSPHERE

J. R. VALLENTYNE

Division of Biological Sciences
Cornell University, Ithaca, N.Y.

For bibliographic citation of this paper, see page 10.

Abstract

On the basis of recent estimates of net primary productivity in forests and grasslands, Schröder's value for the total net productivity of the land area of the earth must be increased by 30-100%, corresponding to 2.2-3.2 $\times 10^{10}$ metric tons of carbon per year (1.4-2.1 tons C/ha.yr). Using a photosynthetic quotient of 1.25 and evaluating net productivity at 60% of gross, Riley's estimate for the marine environment corresponds to a net productivity of 6×10^{10} tons of carbon per year (1.7 tons C/ha.yr). The comparable value from Steemann Nielsen and Aabye Jensen for the marine environment, after multiplication by 1.82 (1.47 to correct for an error in the estimate of self absorption and back scattering in standards; 1.25 to correct for organic C-14 compounds liberated as extracellular material during experimental determinations), is 2.2-2.8 $\times 10^{10}$ metric tons of carbon per year (0.6-0.8 tons C/ha.yr). In terms of visible radiation incident on the earth's surface the overall efficiency of net primary productivity is approximately 0.4% for land, approximately 0.2% for the oceans, and in the range of 0.2% to 0.3% for the earth as a whole. A detailed account of these calculations will appear in « Geochemistry of the Biosphere », Chapter I in the 6th edition of « Data of Geochemistry » to be published in the professional paper series of the United States Geological Survey.

THEORETICAL ASPECTS OF THE COMPARABILITY OF PRODUCTIVITY DATA

D. F. WESTLAKE

Freshwater Biological Association
River Laboratory, Wareham, England

For bibliographic citation of this paper, see page 10.

Abstract

The International Biological Programme aims to produce comparable biological parameters for the productivity of many different types of community. The actual determinations made will rarely be comparable directly, and consideration must be given to ways of facilitating comparisons. Difficulties in comparisons may arise from differences in terminology, units, criteria, or the methods themselves. A suitable terminology is suggested. It is recommended that biomass should be given as kg or g of dry matter or dry organic matter per m^2, and that productivity should be given as m.t of organic matter per ha per year or g/m^2 day. It should also be possible to convert these values into other criteria, especially into energy as kcal. Some special problems of the relations of the depths of mixing and of compensation to the effective biomass, and the ratio of net to gross productivity of phytoplankton are considered. It is important to establish the principles of methods which allow comparisons, but to vary the methods in detail to suit particular communities and achieve the greatest possible accuracy.

INTRODUCTION

Attempts to compare the primary productivity of different ecosystems encounter considerable difficulties because the nature of different habitats and the corresponding customs and methods adopted by researchers lead to data which are not directly comparable (Westlake 1963). For example, agriculturalists may give the annual crop of beet roots as tons fresh weight per acre, whereas limnologists may give the daily productivity of lake phytoplankton as grams of oxygen evolved per litre per hour. To know the annual net productivity of the field, the production of beet tops, weeds, and lost or grazed material must be added, and to know the annual net productivity of the lake, the production of higher plants must be added and the hourly gross rates converted and integrated into an annual net total. Then both must be converted into some common criterion such as dry weight per unit area, and into some common units such as metric tons per hectare.

An essential feature of the IBP is « The program will set out to obtain internationally comparable observations of the basic biological parameters » (IBP News No. 2, p. 5). All participants in IBP methods symposia must bear this in mind and aim at recommending methods, based on common principles, which either allow direct comparisons or give data which can be easily and accurately converted into comparable forms. They must look beyond the special interests of their own disciplines and endeavor to adapt their recommendations to the common purpose. As the Pallanza Symposium on Primary Productivity in Aquatic Environments is the first of these symposia, we have a special responsibility to establish sound principles and to avoid recommendations of limited relevance.

TERMINOLOGY

Much trouble and misunderstanding could be avoided if a standard terminology were adopted universally, but this is not absolutely essential and cannot be enforced. However, if this freedom is to be enjoyed, it is essential to meticulously define the terms used, or to state that a particular set of definitions has been adopted, and to adhere to these definitions absolutely.

The definitions that I recommend are basically in agreement with the theoretical discussions by Thienemann (1931), Lindeman (1942), Ivlev (1945), Macfadyen (1948), and Odum (1959), modified slightly to suit common usage (Yapp 1958, Westlake 1963).

The *biomass* is the weight of all living material present in a unit area at a given time. In botanical papers the term may be restricted to plant material or to green plant material, but general ecological papers must state whether the biomass refers to the whole biocoenosis or to a particular part of it such as the herbivores, the fish, or the primary producers.

The *primary production* is the weight of new organic material created by photosynthesis, or the energy which this represents. It is the increase observed in the biomass of green plants over a period, plus any losses (e.g., excretion, respiration, damage, death, or grazing). In some very special cases new organic matter may be synthesized by chemosynthetic bacteria using geological sources of energy which has not previously been fixed by biological activity, and this may also be called primary production. Most chemosynthesis uses inorganic energy derived from an earlier biological fixation of solar energy. There is production of new organic material, but in terms of energy flow it is secondary production. If a primary producer also utilizes organic material (e.g., the uptake of organic solutes by algae), this must be deducted from the biomass change.

The *primary productivity* is the production Q divided by the period of time T, i.e., the rate of production. The difference between production and productivity is the same as the difference between distance travelled and velocity. To travel any distance takes time in the same way as production of organic material takes time. In the case of travel the distance is usually the first concern, but in the case of production one usually conducts experiments for a particular time and then finds the production. A given distance, or a given production, corresponds to different periods according to the velocity or the productivity.

The *gross productivity* is the rate of production of new organic matter, or fixation of energy, including that subsequently used by the plant and lost as carbon dioxide and heat; that is, the observed change in biomass plus all losses, including respiration, divided by the time interval.

The *net productivity* is the rate of accumulation of new organic matter, or stored energy; that is, the observed change in biomass plus all losses except respiration, divided by the time interval. The net production is the organic material or energy available for exploitation by secondary producers or consumers.

Productivity can be regarded as the observed capacity of the biocoenosis, or component, to produce organic material or to utilize energy; or as the predictable future capacity if conditions remain constant. It is equal to the potential maximum capacity only if the conditions over the period of observation were optimal.

The *standing crop* (or stock) is the weight of organic material that can be sampled or harvested by normal methods at any one time from a given area. It usually refers to part of a particular component of the biocoenosis, such as the tops of plants, the net phytoplankton, or the catchable fish; which may sometimes be dead at the time of harvesting. If a particular part of the total is primarily selected for theoretical or functional reasons (e.g., the herbivores), biomass may be used; but if the selection is primarily made by the method (e.g., net phytoplankton), standing crop is to be preferred. There will be a stage in scientific investigations when the distinction becomes blurred, but the main use of standing crop is in situations where economic or practical considerations prevent the determination of biomass (e.g., agricultural or piscicultural harvests).

The *crop* is the total weight of organic material removed from a given area over a period in the course of normal harvesting practice. For example, a wheat crop is the annual maximum standing crop of tops (grain, chaff and straw), an alfalfa crop is often the annual total of several standing crops of tops, and a sugar cane crop may be the standing crop of tops after nearly two years. The crop is much less variable than the standing crop, which depends greatly on the time of measurement.

The *yield* is the crop expressed as a rate.

Sometimes dead material is present in samples of living organisms; for example, the heart-wood of trees or the lorica of *Dinobryon*.

As long as this material remains part of the organism, and can be presumed to perform some useful function, it may be included in the biomass. If there is any chance of doubt about which components of the ecosystem are included in the biomass data, or which parts are excluded from the standing crop, it is important to give some qualification (e.g., biomass of phytoplankton, spring maximum standing crop of phytoplankton retained by a 50-70 μ net).

CRITERIA

What should be used as a criterion of biomass or productivity? Some possibilities which have been used are fresh weight, dry weight, volume, organic weight, carbon, and energy.

Primary productivity is essentially the fixation of inorganic carbon into new organic material by means of solar energy. The organic material is used for construction of organisms in the ecosystems and as a source of energy for their activities. The chemical elements involved, if they remain within the ecosystem, can be repeatedly used, but the energy is degraded to heat which cannot be recovered. Rates of biological processes are basically rates of transformation of energy (cf. Odum 1959, Macfadyen 1964). A high productivity may be attained with a small supply of elements if these are recycled very rapidly, but cannot be attained with a small energy supply. On the other hand, direct determinations of energy flow through the components of an ecosystem are difficult, and many problems are more easily conceived in terms of changes of mass.

The majority of the fresh weight in most plants is water. The proportion of water is very variable, particularly when comparing land and water plants, and bears no predictable relation to the organic matter, carbon, or energy content. Hence fresh weights are usually converted to dry weights. However, the carbon and energy contents of the dry weight show considerable variations, mainly due to differences in ash content, which is often about 20 per cent and may reach 90 per cent. Some of the ash minerals may be useful to consumers, but only a very small part of the fixed energy is associated with the ash content. Both carbon and energy show a much more constant relation to the ash-free or organic dry matter. Only organisms with exceptionally high sugar, fat, or protein content show large differences in carbon or energy content, and these differences are fairly predictable (Westlake 1963).

Productivity is often determined in terms of oxygen changes, but the relations between the production of oxygen and carbon, energy, or organic material are not constant, especially when different photosynthetic quotients are involved. Carbon fixation is often determined directly, and with a known or assumed photosynthetic quotient, oxygen results are readily converted to carbon. The formation of organic matter depends on the unique properties of the carbon atom, and its fixation is at the heart of the photosynthetic mechanism. Hence, carbon is often favored as a general criterion of productivity, but many other elements are equally essential to the plant, and usually amount to more than half of the organic weight. There is a tendency for a high carbon content to be associated with a high energy content, but they are not necessarily closely related.

I recommend organic dry weight as the best general criterion. It is intermediate between the fresh and dry weights commonly determined with macrophytes, but free from their variable water and ash contents, and the carbon and oxygen values determined with phytoplankton on the other hand, which are both rather abstract and remote from biomass and food consumption. There is no greater difficulty in making energy comparisons by conversion from dry organic weight than from any other criterion.

The biomass is most commonly determined first as fresh weight for macrophytes and volume for phytoplankton. The mass of plants is a very variable function of volume, so volume is not a good criterion of biomass, and the disadvantages of fresh weight have already been noted. Chlorophyll is sometimes used as a criterion of biomass and can be determined more accurately than other measures of the phytoplankton, but again the chlorophyll content, in relation to other criteria, is very variable. It is probably best to convert all biomass data to dry weight, or, if the data are to be used as a measure of the food available, organic dry weight is preferable.

Ideally, investigations should proceed in an order of priority. First and foremost, the actual observations should be made by the most accurate method, irrespective of the criterion involved (e.g., say biomass by chlorophyll, productivity by C^{14}). Then the necessary conversion factors to obtain organic dry weight should be determined and then, if possible, factors for energy, dry weight, carbon, and fresh weight. To facilitate the study of secondary productivities, it will be valuable to have data on the quality of foods, such as their nitrogen, phosphorus, protein, and fat contents. These factors need be investigated only sufficiently frequently to reveal large changes. It should be remembered that the errors in the use of a single conversion factor will rarely be so large as to make comparisons of converted data less accurate than comparisons of the original data, but the combined errors arising from the use of a sequence of factors may be very large. Therefore, as far as possible the aim should be either a direct conversion from the primary data, or several accurate conversion factors.

If the investigator does not give conversion factors, it is usually possible to make a shrewd guess at them, but is much better if each investigator makes at least some assessment of the factors because he will inevitably have a better knowledge of the nature of the organisms involved.

UNITS

It is easier to convert data from one set of units to another than to convert between different criteria, but it is still very time-consuming. In general, the size of the units should give convenient figures, and indicate either the actual values determined or the extent of the validity of the results. Thus, the units preferred for biomass would be kg/m^2 or g/m^2, or for productivity, g/m^2 day or m.t/ha · yr. The use of the hectare stresses that it is desirable to take numerous samples over large areas, to avoid « edge errors » and small areas of possibly exceptional productivity. The time unit for production rates should normally be related to the actual period considered to avoid misunderstandings. Thus, it is mathematically perfectly valid to express a rate determined on a summer day as per year, but if an annual production were deduced from this it would be quite wrong. On the other hand, the average daily production for the growing season is a useful comparative value, provided it is made clear that it is an average and the length of the growing season is given.

The units of irradiance should be g cal/cm^2 · min or $watts/cm^2$, not lux, because this unit is only valid for white light as detected by the human eye. Kg cal are suitable for energy flow data, but the values obtained are unwieldy when expressed in terms of hectares and years. The standard nutritional unit (10^6 kg cal) may be the answer (Stamp 1958, Macfadyen 1964).

SOME PROBLEMS OF METHODS

Biomass is defined as weight per unit area, and most investigations will determine this directly. However, most phytoplankton determinations are made per unit volume, and the conversion to surface area units requires a depth factor which may be difficult to define. In a productivity study the role of biomass is as a measure of food available and to permit comparisons of the productivity per unit biomass. Lake and ocean ecosystems are best considered as containing at least two sub-divisions; an upper, euphotic, or trophogenic, region where primary production is possible, and a lower region where biological activity is confined to the consumption of imported energy. The division between these two is usually made at the compensation depth, where gross photosynthesis balances the respiration over 24 hours. During the most productive periods this is often approximately the same as the depth to which the surface waters (epilimnion) are mixed. If the compensation depth is below the limits of the epilimnion the effective biomass of primary producers must include the populations above and below the thermocline. If the compensation depth is above the thermocline the average phytoplankton organism may sometimes be near the surface and capable of positive net photosynthesis, and at other times at greater depths where it can only respire, while it is available as a food source in both regions. The effective biomass must include the organisms which at any instant are below the compensation depth but will subsequently be transported above it.

The productivity per unit biomass will be reduced in such a situation. The gross productivity per unit area will not be affected, but the net productivity will be reduced because of the respiring population below the compensation depth. This must be remembered when designing experiments to determine net productivity, and when making gross to net conversions.

Such conversions will often be necessary, although they may be difficult. Oxygen methods usually give gross productivity, whereas C^{14} and biomass accumulation methods give approximately net productivities. Gross productivity is more fundamental, and essential for efficiency studies, but net production is the material available to the consumers of the ecosystem. Determinations of the ratio, net to gross, should be made for 24 hour periods and as annual and growing season averages. It is particularly important to note that the normal C^{14} techniques give only the daytime net productivity, whereas for ecological purposes the 24-hour net productivity (allowing for night time respiration) is essential. Studies of the photosynthetic quotient will also be desirable, or essential when oxygen determinations are made.

IBP programmes should aim at producing annual production data, because these are the most useful for studies of entire ecosystems and for comparative purposes. Much terrestrial information will inevitably be in this form.

CONCLUSION

Further discussion of such problems with some examples and references may be found in Westlake (1963).

This symposium should aim at establishing the principles of methods, because details will always be influenced by individual problems and the nature of specific communities. If these principles are followed, then the results should be readily comparable throughout all PF programmes and indeed over the whole of IBP. Provided comparability is maintained, every encouragement should be given to modifications of the general methods directed at improving the quality of the data obtained.

REFERENCES

Ivlev, V. S. 1945. The biological productivity of waters. - Usp. sovrem. Biol., Moscow *19*: 98-120. (In Russian; seen in translation by W. E. Ricker, typescript, Freshwater Biological Association library).

Lindeman, R. L. 1942. The trophic-dynamic aspect of ecology. - Ecology *23*: 399-418.

Macfadyen, A. 1948. The meaning of productivity in biological systems. - J. Anim. Ecol. *17* : 75-80.

— 1964. Energy flow in ecosystems and its exploitation by grazing. *In* D. J. Crisp, (ed.), Grazing in terrestrial and marine environments. Brit. Ecol. Soc. Symp., *4*: 3-20. Blackwell. Oxford.

Odum, E. P. 1959. Fundamentals of Ecology, 2nd ed. - Saunders Co., Philadelphia, Pennsylvania, 546 p.

Stamp. L. D. 1958. The land use pattern of Britain. *In* W. B. Yapp and D. J. Watson (eds.), The biological productivity of Britain. - Symp. Inst. Biol., 7: 1-10. London.

Thienemann, A. 1931. Der Produktionsbegriff in der Biologie. - Arch. Hydrobiol. (Plankt.) *22*: 616-622.

Westlake, D. F. 1963. Comparisons of plant productivity. - Biol. Rev. *38*: 385-425.

Yapp. W. B. 1958. Introduction. *In* W. B. Yapp and D. J. Watson, (eds.), The Biological Productivity of Britain. - Symp. Inst. Biol. No. 7: ix-xii. London.

THE RELATIONSHIP BETWEEN CHLOROPHYLL AND PHOTOSYNTHETIC CARBON PRODUCTION WITH REFERENCE TO THE MEASUREMENT OF DECOMPOSITION PRODUCTS OF CHLOROPLASTIC PIGMENTS [1]

CHARLES S. YENTSCH

Woods Hole Oceanographic Institution
Woods Hole, Massachusetts

(1) This is contribution number 1686 from the Woods Hole Oceanographic Institution.

For bibliographic citation of this paper, see page 10.

Abstract

Chlorophyll and phaeophytin have been compared to the light-saturated rate of carbon fixation in oligotrophic, euphotic zones of the southern Atlantic Ocean.

The percentage of non-photosynthetic decomposition products (phaeophytin) increases with depth while the capacity for light uptake decreases. The presence of the non-photosynthetic pigments accentuates low carbon fixation per unit pigment values, especially near the base of the euphotic zone, but cannot account for all of the decreased efficiency of pigment in the photosynthetic process.

Photosynthetic-pigment situations characteristic of those found at the base of the euphotic zone are obtained by placing natural populations living at the surface in darkness. When reilluminated, these populations renew chlorophyll synthesis and photosynthetic carbon production. These observations support the idea that features such as the presence of phaeophytin and low capacity for carbon uptake in light indicate a discontinuity in the photosynthetic unit of phytoplankton because of their residence in zones of limited illumination.

INTRODUCTION

Approximately 8 years ago I began to investigate the relationship between the maximum rate of photosynthesis and chlorophyll content of marine phytoplankton. Plankton ecologists for a number of years had recognized the role of chlorophyll in the photosynthetic process, but few direct comparisons were available. Using cultures of marine phytoplankton and natural populations, comparisons were made which led to the development of a mathematical expression predicting the amount of photosynthetic carbon production from measurements of light and chlorophyll content within the euphotic zone (Ryther and Yentsch 1957). The weakest aspect of this formula was the varying amount of photosynthesis per unit chlorophyll at light saturation. This variation was unpredictable, and occurred in both cultures and natural populations. Research on this variation has shown that it involves the two main photosynthetic categories: 1) enzymatic processes and 2) measurements of active pigments.

The effects of enzymatic processes have been discussed by Yentsch (1962). Nutrient limitation affects both photosynthesis and chlorophyll synthesis, although these two processes may appear at times to be out of phase with one another. This paper is focused on the influence of pigment decomposition products on the computation of the maximum rate of photosynthesis per unit pigment in natural phytoplankton populations.

A highly generalized scheme of the photosynthetic unit and of areas sensitive to decomposition is shown in Fig. 1. The four pyroll rings with chelated central magnesium and phyto-carotenoid-lipid fraction represent the fat-soluble chromophoric group which can be separated from the photosynthetic unit by solvents such as ether, acetone, and alcohol. Probably, the principle connection between the protein (chloroplastin) and chlorophyll is by way of the central

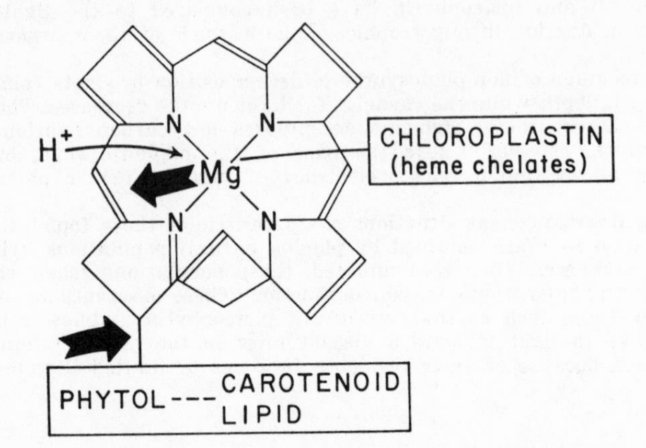

Fig. 1. - Schematic representation of the photosynthetic unit.

magnesium atom; the connection with the lipid fraction is through phytol. The nature of chloroplastin is not completely known. However, a considerable fraction is composed of heme-type proteins and other components of the electron transport system of the photosynthetic unit. Thus, while the fat-soluble portion of the pigment complex is dominated by a magnesium chelate, the water-soluble fraction is dominated by iron chelates. Photosynthesis has never been observed with one of the chelated fractions missing. Modifications during decomposition of the photosynthetic unit occur at the points indicated by the arrows in Fig. 1. These modifications include a loss of magnesium, isolating chlorophyll from chloroplastin, and the separation of phytol from chlorophyll. The resulting pigment decomposition products are phaeophytin, a magnesium-void compound with phytol; phaeophorbide, a magnesium-void compound with no phytol; and chlorophyllide, a magnesium chelate with no phytol.

At the present time, there is no simple means of identifying all of the decomposition products of chlorophyll. However, estimates of the magnesium-void pigments can be made by the method of Yentsch and Menzel (1963). Such measurements have shown that the percentage of phaeophytin increases with depth in the euphotic zone (Yentsch and Menzel 1963; Brown, *et al.,* 1964; Lorenzen 1965; Yentsch 1965).

Other observations have shown that the maximum rate of carbon fixation per unit « chlorophyll » in the euphotic zone decreases markedly with depth (Steemann Nielsen and Hansen 1959). In this paper I explore the possibility that some of the lowered efficiency of chlorophyll in photosynthesis at the base of the euphotic zone is due to the presence of magnesium-void decomposition products of photosynthetic pigments. Experiments utilizing natural populations have been conducted with the purpose of ascertaining the relationships of photosynthesis to chloroplastic pigment obtained under conditions of prolonged darkness followed by full illumination.

METHODS

Sampling

Sea observations were taken aboard the R. V. Atlantis II (cruise No. 14, Oct. 25 - Dec. 10, 1964) at the stations shown in Fig. 2. Water samples were taken at 1, 10, 25, 50, 75, 100, 125, 150, 175, and 200 meters with a two-liter, nontoxic closing bottle. One liter was used for pigment analyses; the remainder was used for C^{14} studies.

Downwelling light was measured using a submarine photometer with a maximum spectral response between 450 and 530 millimicrons. The density of the water mass was calculated from measurements of temperature and salinity following standard oceanographic practices.

Carbon uptake measurements

C^{14} fixation was measured using the technique of Steemann Nielsen (1952). In the case of the open ocean observations, 10μc/ml $Na_2C^{14}O_3$ was added to glass stoppered bottles. In experiments with coastal waters only 5 μc/ml were used. The precision of this technique was about the same as reported by Steemann Nielsen and Aabye Jensen (1957). From the mean of ten samples the deviation is ± 15%. Photosynthetic carbon fixation was measured at 7000 lux from four Sylvania 40 watt « natural » and four « daylight » fluorescent tubes. These illuminated a water bath through which surface sea water was pumped.

Pigment determination

Chlorophyll and phaeophytin (used here in a general sense to include all magnesium-void decomposition products) concentrations were measured following the method of Yentsch and Menzel (1963)

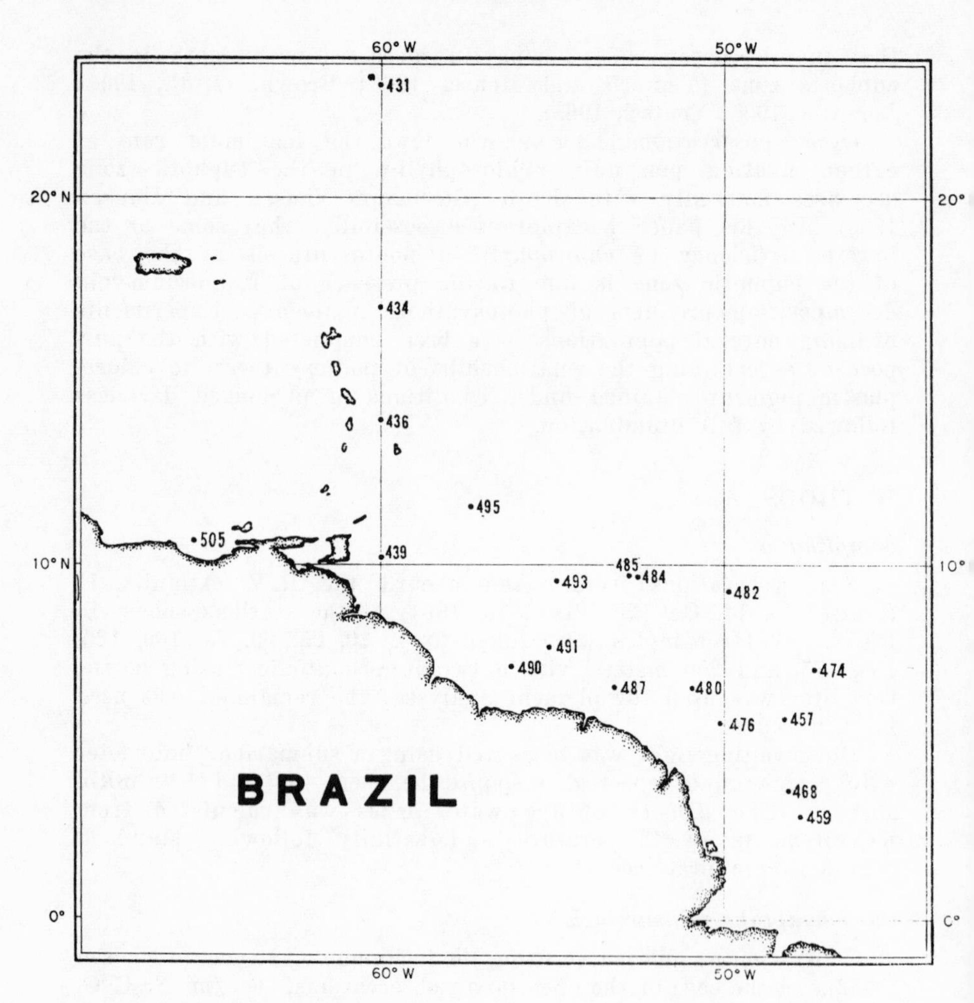

Fig. 2. - Location of stations.

and Yentsch (1965). Gelman filters, type Gm-4, with an 0.8 micron pore diameter were used for filtration. Acidification of acetone extracts was done by adding 0.1 cc of 10% hydrochloric acid to 7 ml of an 85% acetone extract.

The ratio of fluorescence before acidification (Fo) to that after addition of acid (Fa) is 2.3 when all the pigment in the extract is complexed with magnesium (Yentsch 1965), and when the pigment has no magnesium this ratio is 1.0. For the computation of chlorophyll fluorescence, equation number 2 in Yentsch and Menzel (1963) indicates that the relationship between the percentage of pigment in the extract as chlorophyll and the Fo:Fa ratio is linear between the range of 1.0 and 2.3. Dr. James Carpenter of the Chesapeake Bay Institute has shown that this is incorrect and the relationship is a bowed curve, with a maximum deviation of 15% from linearity toward higher chlorophyll percentage, about in the middle of the range of Fo:Fa ration. The corrected equations derived by Dr. Carpenter for the computation of chlorophyll fluorescense (F_{chl}) over the range 1.0-2.3 are:

$$F_{chl} = 1.76 \text{ (Fo-Fa)} \tag{1}$$

$$F_{chl} = \frac{2.3 \text{ (Fa) (Fo/Fa} - 1)}{1.3} \tag{2}$$

Analysis of over 250 samples taken in duplicate from the surface waters around Woods Hole show that 99% of the time, pigment determination is good to $\pm$ 0.3 μg/liter. Although the technique is extremely precise, difficulties arise in calibration which may affect the interpretation of results.

Calibration involves a comparison of the optical density of the red band of chlorophyll to fluorescence emission, which is characterized by the spectral sensitivity of the photodetector and secondary filter in the fluorimeter (Fig. 3). In the Turner fluorimeter the photodetector (931-A) has a quantum efficiency in the region of chlorophyll emission (670 mμ) below 1%, the maximum quantum efficiency being at 360 mμ. Because of the increased sensitivity at shorter wawe lengths, any fluorescence « leaking » through the secondary cut-off filter is accentuated (Fig. 3). Holm-Hansen (personal communication) has observed that changes in the concentration of chlorophyll *c* to chlorophyll *a* alter the relationship between fluorescence and optical density in their fluorimeter, and calibration curves (optical density *vs* fluorescence) show a high degree of scatter, especially at high concentrations. Some of this is reduced by using a secondary filter which cuts off at longer wave lengths (Fig. 3), reducing the fluorescence by chlorophyll *c*. Photomultipliers more sensitive to red light show less scatter; the type 931-A has been observed to vary in red sensitivity by as much as 50 %.

All the above considerations influence the interpretation of the Fo:Fa ratio. In Woods Hole waters the maximum is 2.3, which is about the same as that ratio in cultures of diatoms, dinoflagellates, coccolithophores, and attached brown algae. If for some reason excessive amounts of chlorophyll *c* are present, an underestimate of phaeophytin will occur. Large concentrations of chlorophyll *b* will cause an overestimate. Chromatographic separation of pigments from Woods Hole coastal and offshore waters have shown no or undetectable amounts of chlorophyll *b*.

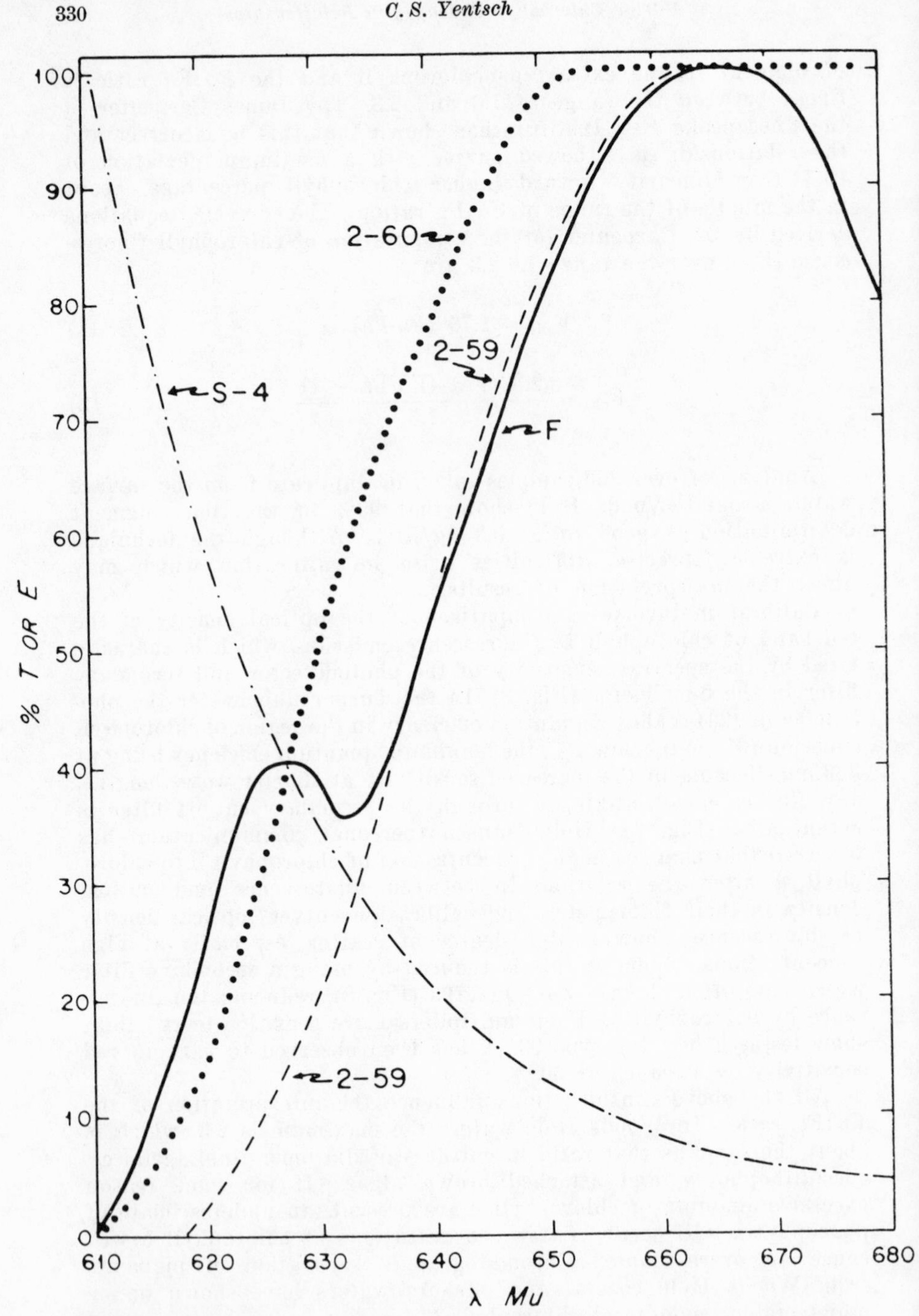

Fig. 3. - Spectral emission of fluorescence from an acetone extract of diatom pigments (F), spectral sensitivity of the 931 A photomultiplier tube (S-4), and spectral transmission of Corning glass filters (2-60) and (2-59).

DEPTH COMPARISON OF THE VERTICAL DISTRIBUTION OF CHLOROPHYLL AND PHAEOPHYTIN AND THE LIGHT SATURATED RATE OF PHOTOSYNTHESIS

Plant pigments

The vertical distribution of plant pigments is compared with the maximum rate of photosynthesis in oligotrophic water masses from various stations in the southern Atlantic Ocean (Fig. 2). Fig. 4 shows that high ratios of Fo:Fa are concentrated near the surface, and low ratios are found near the base of the euphotic zone. The overall range in ratios is 2.3-1.0. When expressed in terms of percentage of total pigment as chlorophyll, this represents a range of 100 to 1 per cent. Therefore, the per cent of pigment near the surface is mostly chlorophyll, while at 200 meters it is mostly phaeophytin. Ratios of 2.0 predominate in the upper 25 meters; below 25 meters, ratios are lower and decrease rapidly with depth between 25 and 100 meters. Below 100 meters, the ratio continues to decrease slowly.

Fo:Fa ratios and the penetration of visible light are shown in Fig. 5. Ratios of 1.8-2.2 or higher occur to the depth where 20% of the surface light is present. Below this depth, ratios rapidly decrease and reach values of 1.4-1.6 at a level where only 1% of the surface light remains. At a level where only 0.1% of the light penetrates, Fo:Fa ratios range between 1.0-1.4. Because water masses are thermally stratified, the vertical distribution of Fo:Fa ratios with depth resembles the density distribution with depth. Therefore, the relationship between Fo:Fa ratios and water density (Fig. 6) is linear. Fo:Fa ratios decrease with increasing density.

The pigment concentration with depth and water density at 8 selected stations is shown in Fig. 7. The total maximum pigment concentration occurs between 60 and 100 meters, and is either at the top of the density discontinuity, or in the discontinuity, and is 4-8 times greater than the concentration at the surface. It is composed of chlorophyll and phaeophytin. With the exception of station 505, the phaeophytin maximum is 20 meters deeper than the chlorophyll maximum. The depth at which phaeophytin and chlorophyll concentrations become equal is below 75 meters at most stations.

Carbon uptake

The C^{14} uptake in uniform light decreases with depth at all stations (Fig. 8). The dark C^{14} uptake shows no significant trend with depth and can be considered constant. Consequently, the

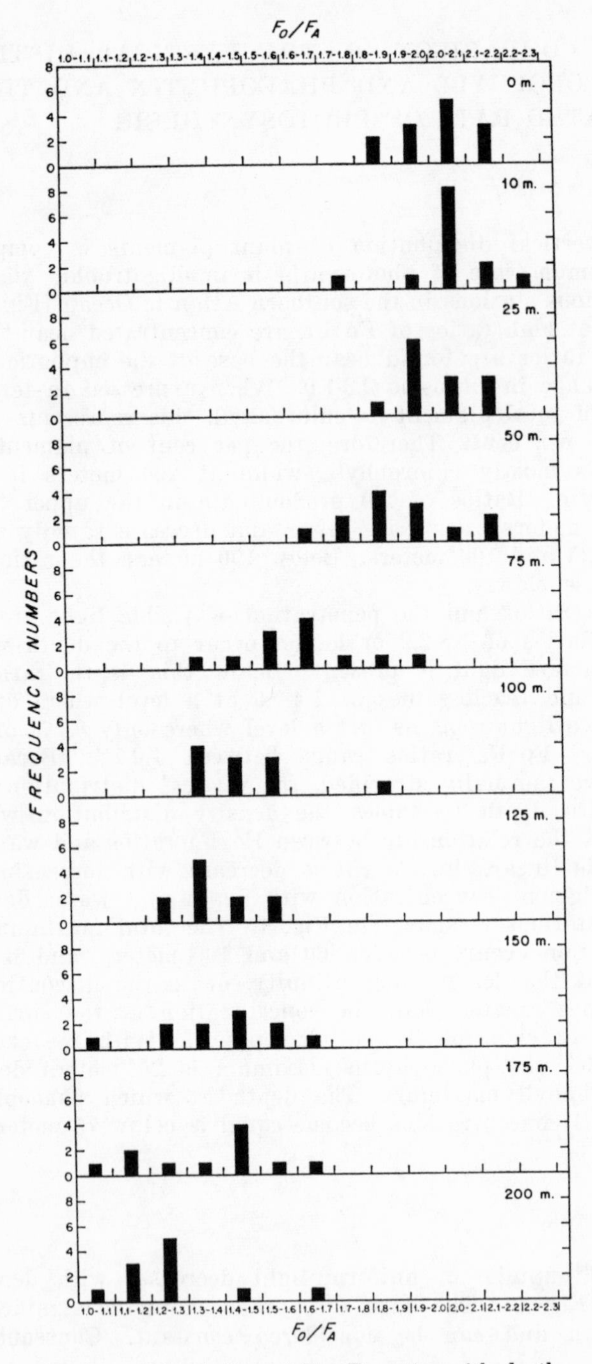

Fig. 4. - Distribution of Fo : Fa ratios with depth.

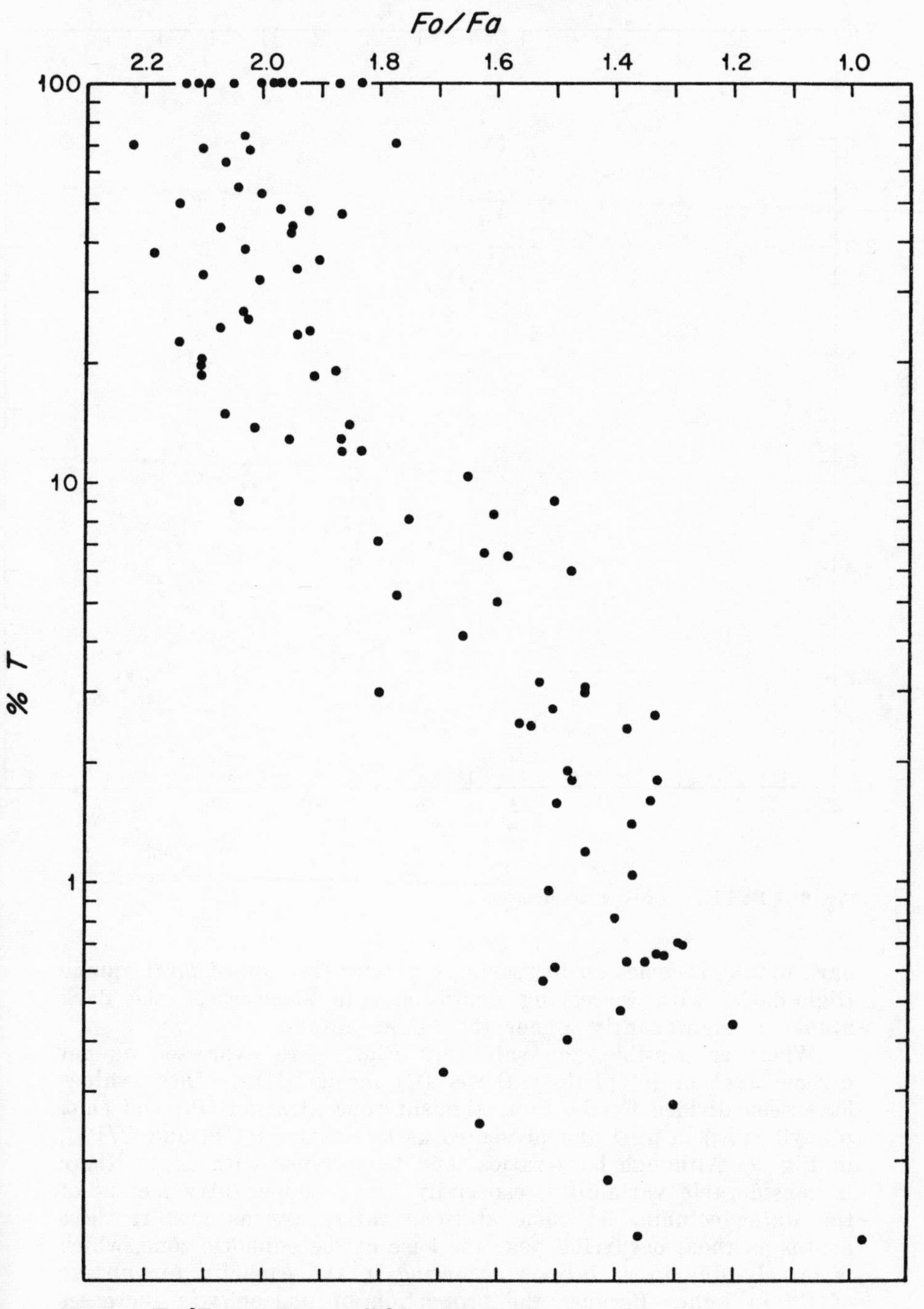

Fig. 5. - Fo : Fa ratios versus light penetration.

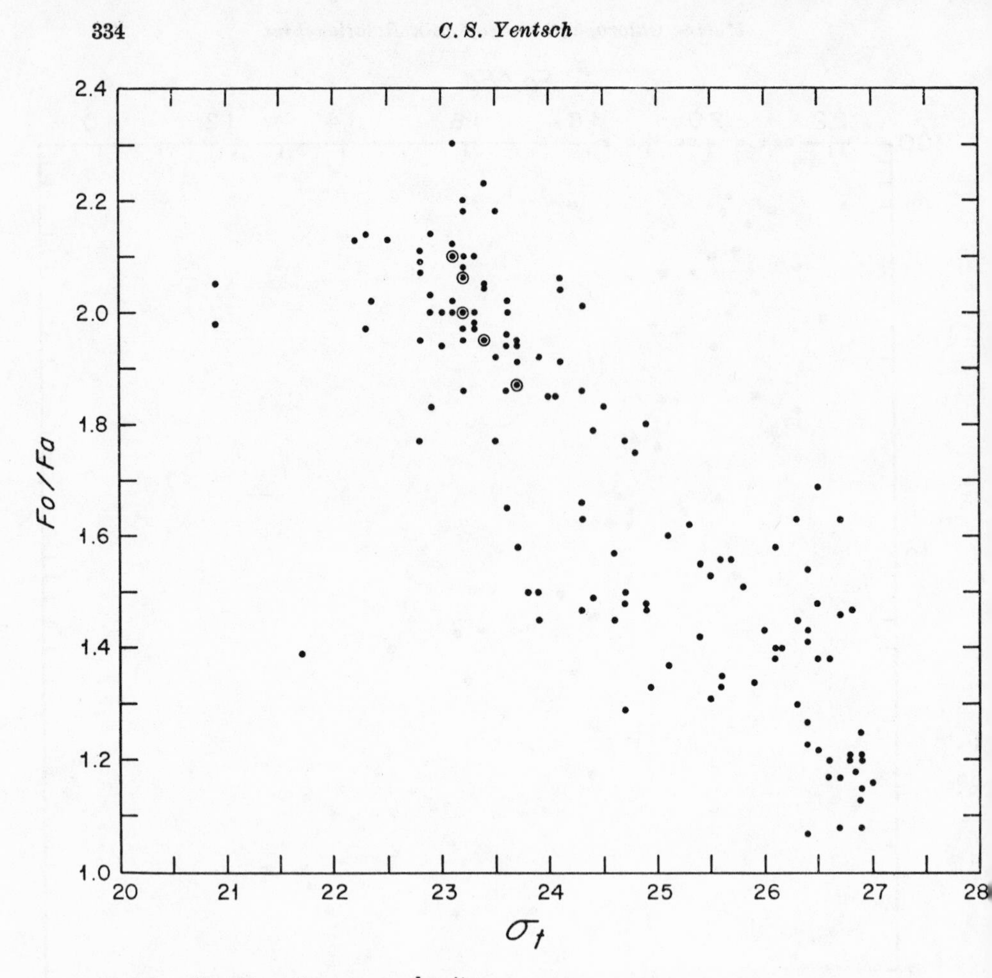

Fig. 6. - Fo : Fa ratios versus density.

dark uptake becomes an increasingly greater fraction of total uptake (light-dark) with increasing depth, and in some cases the dark uptake is significantly higher than light uptake.

Wherever possible, the values in Fig. 8 are expressed as net carbon fixation by photosynthesis (C) in μgC/l/hr. These values have been divided by the total pigment concentration (P_t) and chlorophyll (P_{chl}) in μg/l and presented as two ratios (C/P_t and C/P_{chl}) in Fig. 9. Although both ratios tend to decrease with depth, there is considerable variability, especially in the upper fifty meters of the water column. At some stations ratios are as low at these depths as those occurring near the base of the euphotic zone, which is largely due to variations occurring in the capacity for uptake of C^{14} in light. Because the proportion of phaeophytin increases

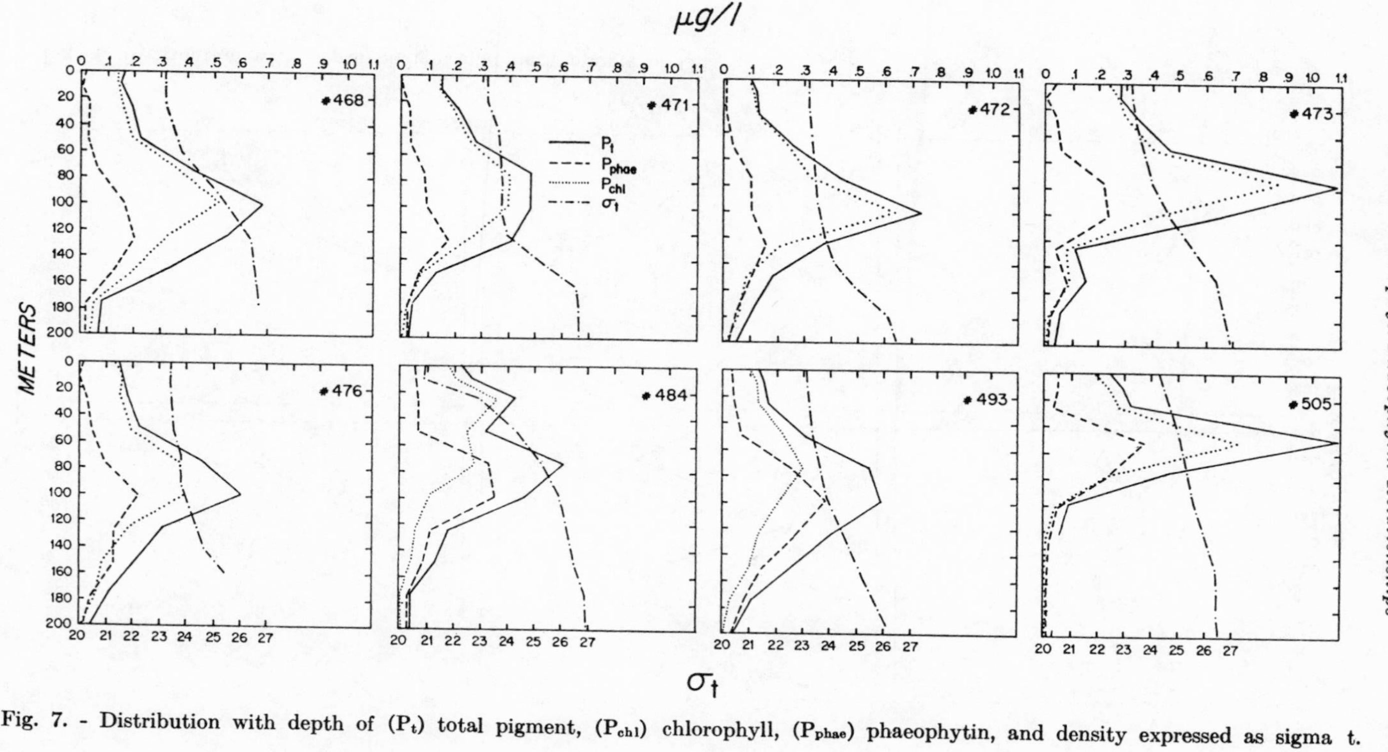

Fig. 7. - Distribution with depth of (P_t) total pigment, (P_{chl}) chlorophyll, (P_{phae}) phaeophytin, and density expressed as sigma t.

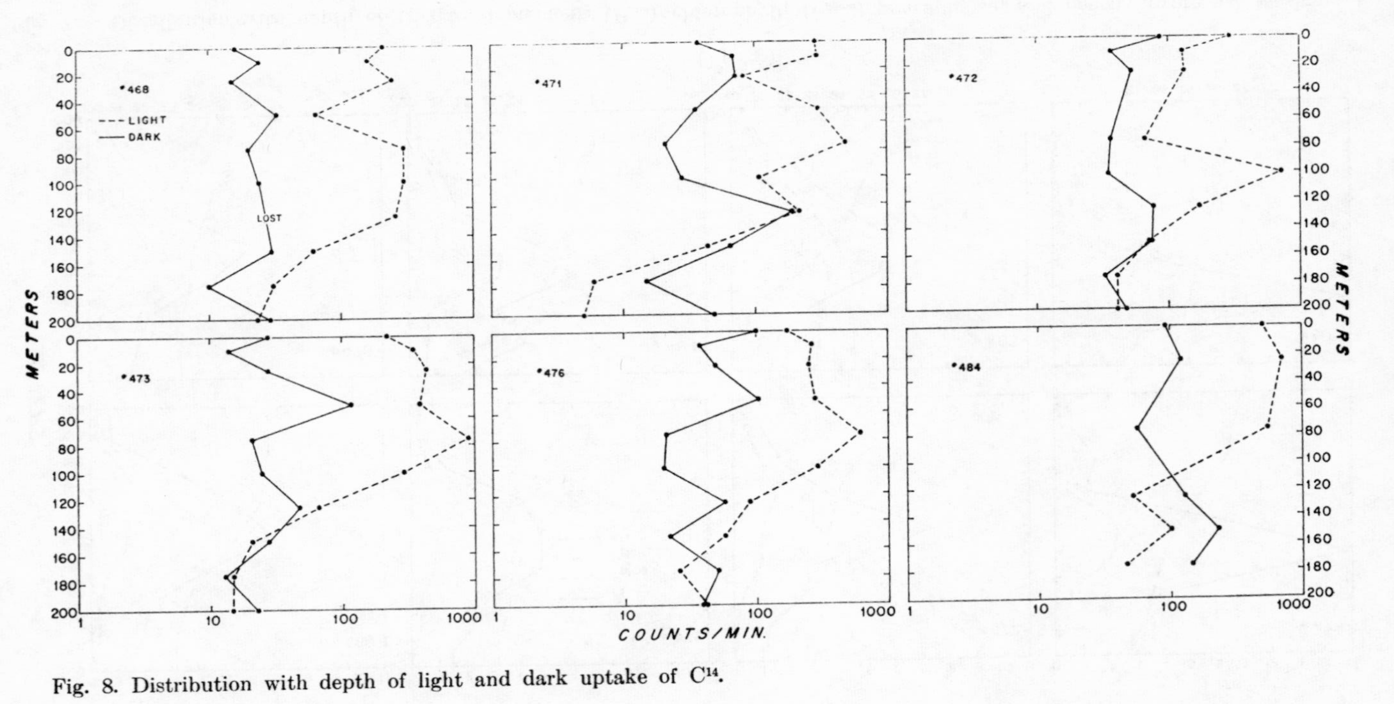

Fig. 8. Distribution with depth of light and dark uptake of C^{14}.

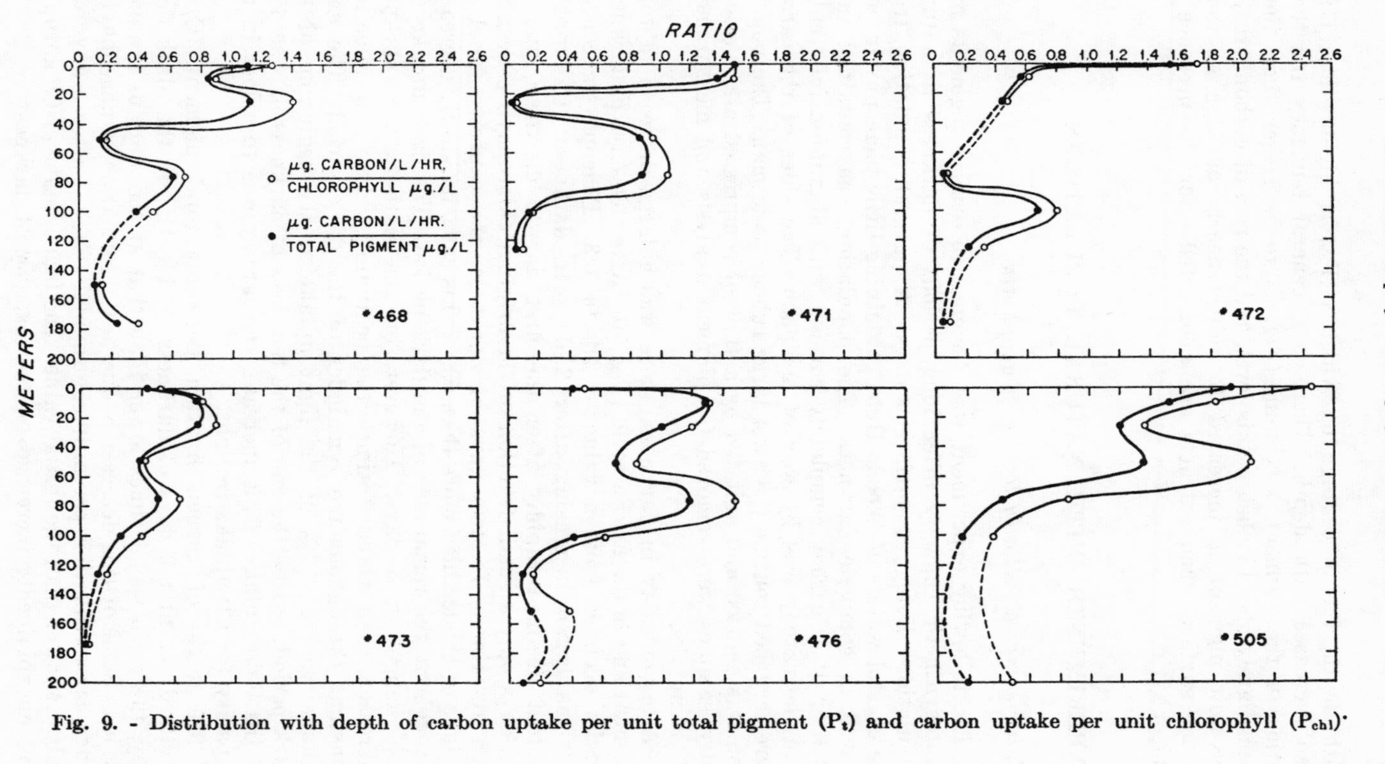

Fig. 9. - Distribution with depth of carbon uptake per unit total pigment (P_t) and carbon uptake per unit chlorophyll (P_{chl}).

with depth, the percentage by which $C:P_{chl}$ is higher than $C:P_t$ also increases with depth. There is a general tendency for these ratios in the pigment maximum (Fig. 7) to be larger than those immediately above or below this level, and the rate of carbon fixation per unit pigment is increased to values nearly as high as those at the surface when extracts are « corrected » for the presence of phaeophytin.

EXPERIMENTS WITH NATURAL POPULATIONS

Water samples initially low in phaeophytin

The following experiment was designed to observe pigments and photosynthetic capacity after long periods of darkness in natural populations living in surface waters. A natural population from the coastal waters of Woods Hole, containing little phaeophytin, was placed in complete darkness. The population was contained in a 12 liter glass carboy completely covered with aluminum foil; entry into the carboy was by way of a siphon. The sides of the carboy were scrubbed routinely with a large rubber policeman. The population was aerated and agitated by a flow of compressed air. Periodically, samples were removed for pigment analysis and measurement of C^{14} uptake.

After one day in darkness, there was a large decrease in total pigment and in the Fo:Fa ratio (Fig. 10). After ten days in darkness, Fo:Fa declined from a value of 2:3 to 1:3. Phaeophytin concentrations began increasing after 2 days in darkness. Chlorophyll content declined rapidly after the first day in darkness, and by ten days only a small fraction of the original chlorophyll concentration remained. In two days of darkness, the capacity for light uptake of C^{14} declined more than 50%; this capacity slowly decreased throughout the remainder of the dark period. The dark uptake did not change for 5 days. However, on the sixth day, it abruptly increased to a value 3 times the beginning value, and remained constant throughout the remainder of the dark period. The dark uptake was only 7% of the light uptake at the beginning of the dark period, but by the end of the ten days of darkness, it was 65% of the light uptake. This increase was largely due to a loss in the capacity for C^{14} uptake in light.

The ratio of carbon fixation per unit total pigment (C/P_t) declined the first 9 days of darkness (Table 1). By the ninth dark day, the ratio was 4 times lower than that at the start of the dark period. Some of this decrease is due to the formation of phaeophytin, since carbon fixation per unit chlorophyll (C/P_{chl}) after 9 days was only 1.4 lower than the initial value. The light uptake of C^{14} abruptly and unexplainedly increased on the last day of darkness.

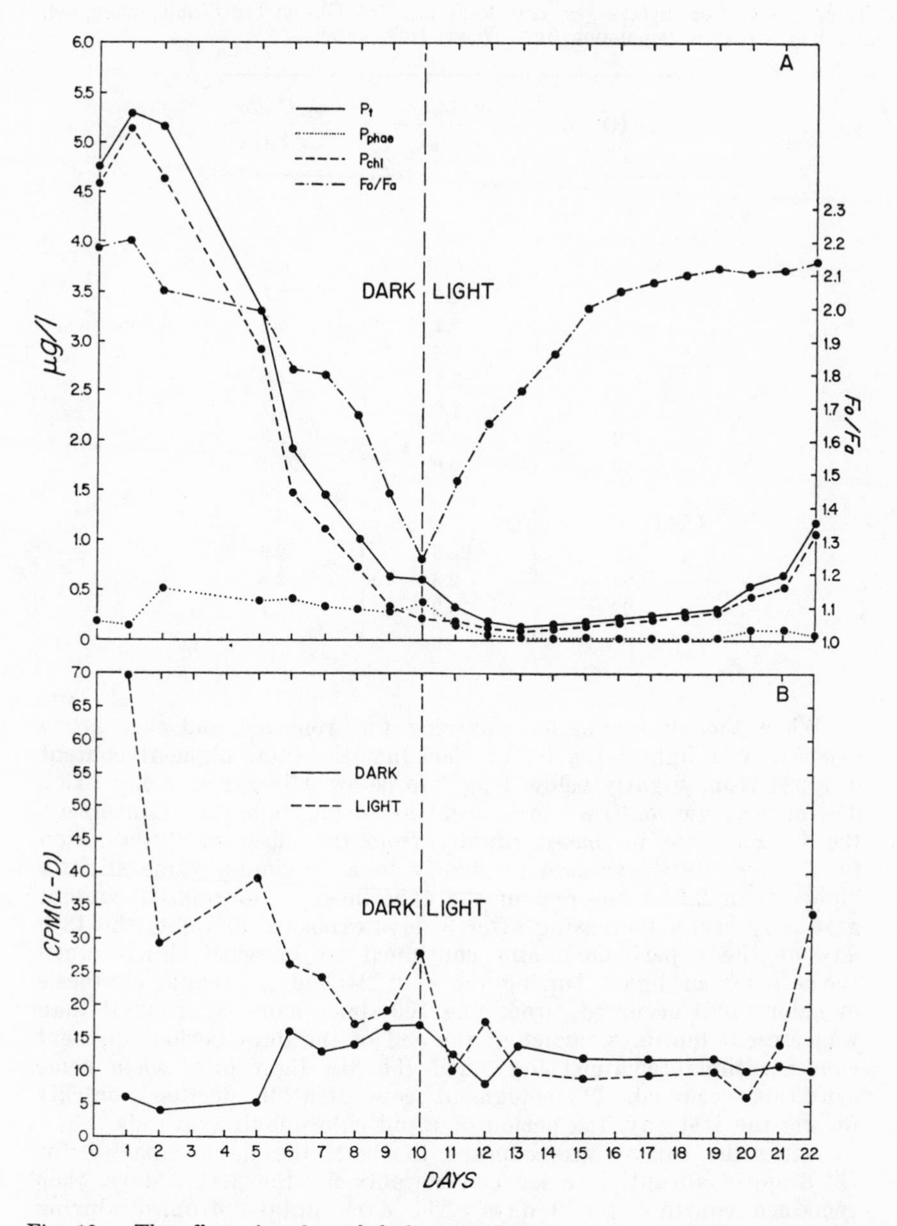

Fig. 10. - The effect of prolonged darkness and then exposure to light on pigments and uptake of C^{14} of a natural phytoplankton population collected in Woods Hole waters, Oct. 28, 1964. Water temperature 15°C.

Table 1. - Carbon uptake per unit total pigment (P_t) and per unit chlorophyll (P_{chl}) in a natural population from Woods Hole waters.

Time (Days)	$\frac{\mu g\ C/l/hr}{\mu g\ P_t/l}$	$\frac{\mu g\ C/l/hr}{\mu g\ P_{chl}/l}$
Dark:		
1	6.05	6.2
2	2.4	2.7
5	5.1	5.8
6	2.6	3.4
7	3.7	4.9
8	1.2	1.7
9	1.5	2.8
10	9.0	25.0
Light:		
12	25.3	36.9
21	2.3	2.8
22	10.2	10.7

When the aluminum foil covering was removed and the carboy exposed to a light intensity of 7000 lux, the total pigment content dropped from slightly below 1 μg/l to below 0.5 μg/l in 1 day. This decline was due entirely to a decrease in phaeophytin. Conversely, the Fo:Fa ratio increased rapidly from the onset of illumination for 5 days, then increased gradually to a maximum value slightly higher than 2.1 at the end of the experiment. Chlorophyll content gradually began increasing after 3 days exposure to light (the 13th day of the experiment), and continued to increase slowly until the 9th day of light. During the next 24 hours, a rapid synthesis of chlorophyll occurred, producing 100 times more chlorophyll than was present ten days earlier at the end of the dark period. Pigment concentrations remained low until the 9th light day, when some synthesis occurred. Phaeopigment concentration declined quickly during the last day, the period of rapid chlorophyll synthesis.

After the carboy was exposed to light, the light capacity for C^{14} dropped abruptly, varied considerably for the first 2 days, then remained constant for 9 days. The dark uptake dropped during the first 2 light days, but not as quickly as the light uptake. On the 3rd day, the dark uptake increased to 30% more than light uptake, and remained at this level until the end of the experiment. On the 10th light day, the light uptake increased rapidly, coinciding with the increasing pigment synthesis. At the end of the experiment, the light capacity for carbon fixation had recovered to half of that observed at the start of the experiment, and the dark uptake was about 30% of the uptake in light.

During the period in which the carboy was exposed to light, net production of carbon occurred only on the 9th, 18th, and 19th days of the experiment (Table 1). Again, for some unexplainable reason, the values on the 9th day were extremely high in relation to all other observations.

Water samples initially high in phaeophytin

The following experiment is the reverse of the previous one. Water samples high in phaeophytin content were exposed to daylight. Because a high percentage of phaeophytin occurs below the euphotic zone, ten-liter water samples were taken from depths of 200 and 500 meters. These samples were placed in transparent 12-liter carboys inside a plastic trough through which surface sea water at 20 ± 2 °C was pumped. The carboys were placed in an unshaded area aboard ship and sampled for 8 days. Mid-day radiation values ranged between 0.8-1.2 calories cm^{-2} min^{-1} during the experiment. The exact amount of illumination these carboys received is not known.

Upon exposure to light, the total pigment content in water samples from both 200 and 500 meters dropped abruptly (Fig. 11). In both cases the drop was due to a loss of phaeophytin and chlorophyll, but the percentage of phaeophytin lost was greater. The Fo:Fa ratio initially was 1.5 or lower at both depths, but by 50 hours it had begun to increase and reached values greater than 2.0 by 100 hours. Chlorophyll synthesis was well under way by 100 hours, and the most rapid synthesis occurred between 140-180 hours. Phaeophytin concentration was constant for a period of 120 hours. Afterwards it declined and became almost unmeasurable.

The trend of C^{14} uptake in the 500 meter sample was consistent with pigment changes. At this depth, dark uptake was about equal to the light uptake for the first 50 hours. At 70 hours, the light uptake began to exceed the dark uptake and stayed at this rate until 120 hours. At 120 hours the light uptake increased rapidly and was approximately 7 times higher than at the beginning of the experiment. After 120 hours, the dark uptake in the 500 meter sample began to increase. It was roughly 50% of the light uptake at the end of the experiment.

In the 200 meter sample, the light uptake was always greater than the dark uptake. Both uptakes were characterized by an initial increase, and then a decrease which was followed by a gradual increase throughout the remainder of the experiment. The dark uptake was also roughly 50 % of the light uptake at the end of the experiment.

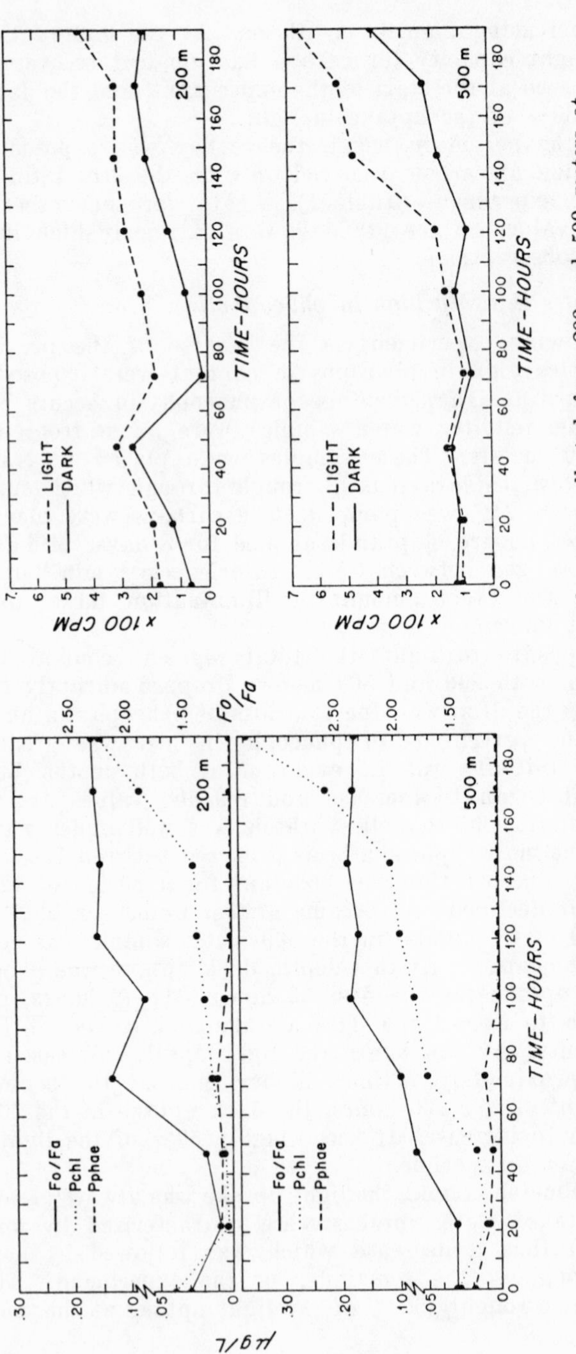

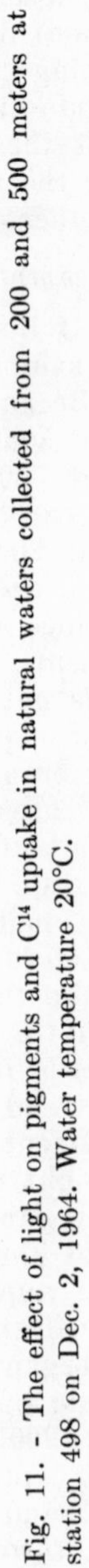
Fig. 11. - The effect of light on pigments and C^{14} uptake in natural waters collected from 200 and 500 meters at station 498 on Dec. 2, 1964. Water temperature 20°C.

At both depths the carbon uptake per unit of total pigment or per unit chlorophyll was extremely variable (Table 2). In the 200 meter samples, there was a tendency for the ratios to be higher between 22 and 100 hours, whether or not a correction was made for the presence of phaeophytin in the extract. In the 500 meter samples, there was no net production in the first 22 hours, and extremely low ratios were encountered during the period of 47-73 hours. By correcting for phaeophytin in the extracts, the photosynthesis per unit chlorophyll values were increased but remained comparatively lower than the ratios that occurred later in the experiment.

Table 2. - Carbon uptake per unit total pigment (P_t) and per unit chlorophyll (P_{chl}) in water collected from 200 and 500 meters at station 498.

Time (Hours)	200 meters		500 meters	
	$\frac{\mu g\ C/l}{\mu g\ P_t/l}$	$\frac{\mu g\ C/l}{\mu g\ P_{chl}/l}$	$\frac{\mu g\ C/l}{\mu g\ P_t/l}$	$\frac{\mu g\ C/l}{\mu g\ P_{chl}/l}$
0	1.1	2.0	—	—
22	4.9	18.8	—	—
47	3.5	8.1	0.5	0.7
73	12.4	13.0	0.3	0.4
100	6.0	7.7	0.4	0.4
122	7.7	7.7	1.25	1.25
146	4.1	4.1	2.45	2.45
171	0.6	0.6	1.0	1.0
194	2.65	2.65	2.7	2.7

DISCUSSION

To thoroughly duplicate conditions identical to those in an open ocean euphotic zone would be a monumental task. Thus the foregoing experiments are limited. Nonetheless, pigment and photosynthetic situations comparable to those in the light-limited portion of the euphotic zone were observed. For example, when a natural population initially low in phaeophytin is placed in darkness, a rapid increase in the percentage of phaeophytin within that population occurs, resulting in a value comparable to those observed at the base of the euphotic zone. Almost simultaneously, there is a loss in the capacity for light uptake of C^{14}. If the dark period is prolonged, this capacity will eventually become equal to or less than the dark uptake. Prolonged darkness also tends to decrease the efficiency of the chlorophyll in the photosynthetic process. Although these changes occur as a function of time, they are identical to changes in photosynthetic-pigment properties which occur as a function of depth in the light-limited portion of the euphotic zone. Thus, with regard to the parameters discussed here, the general picture is one of decreasing continuity in the phytoplankton photosynthetic unit with water depth, which is made apparent by 1) the presence of an increasing portion of the total pigment as nonphotosynthetic decomposition products, and 2) low photosynthetic efficiency. It is suggested that correlations with depth are in reality correlations with decreasing light intensity.

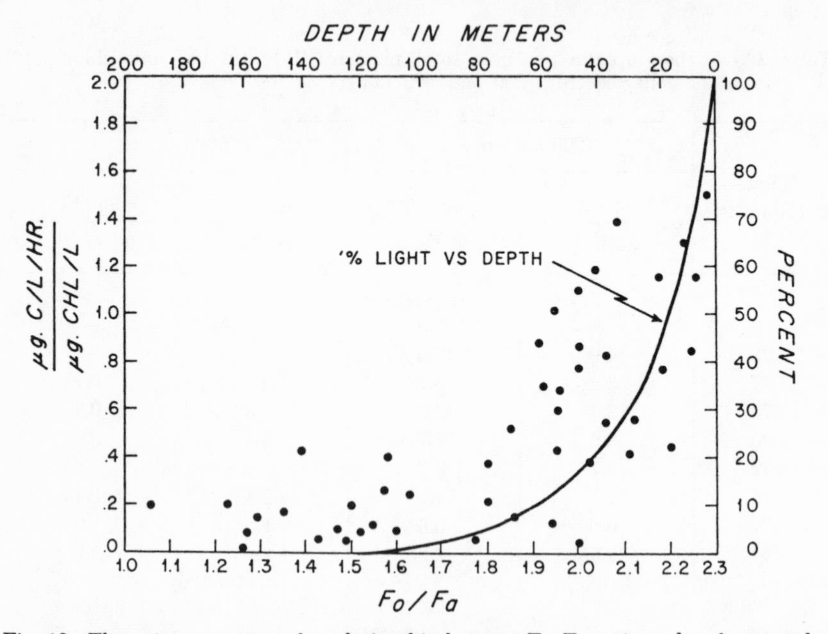

Fig. 12.-The points represent the relationship between Fo:Fa ratio and carbon uptake per unit chlorophyll. Data from all stations are shown in Fig. 1 for all depths. The solid curve represents the relationship between light transmission and depth for the average transparency of these waters ($k = 0.02$).

This idea is supported by examination of the plot of the relationship of carbon uptake per unit chlorophyll to Fo:Fa ratios for all depths at all stations (Fig. 12). Although there is a great deal of scatter, the trend in the data is obviously logarithmic and follows the plot of the average light penetration in these waters. Thus, as the light penetration decreases logarithmically, the capacity for photosynthetic carbon production and the percentage of total pigment as chlorophyll follows this trend.

It is often thought that the formation of the phaeo-type pigment is « the end of the line » as far as decomposition of the photosynthetic unit goes. The « recovery in light » experiments indicate that this is not true. In the most general ecological sense, it would appear that the apparent conversion of chlorophyll to phaeophytin in light is the reverse of the situation which occurs when a population is darkened. Upon exposure to light, phaeophytin rapidly disappears and is followed by an appearance of chlorophyll and then a capacity for the phytoplankton cell to photosynthesize.

Because the percentage of phaeophytin increases with depth while photosynthesis per unit of pigment decreases with depth, one of the goals of this study was to examine these two factors with respect to one another. The results (Fig. 9) show that the presence of nonphotosynthetic phaeophytin, especially near the base of the euphotic zone, skews results towards lower efficiencies of photosynthesis per unit pigment. However, by no means can all of this decrease in efficiency be credited to the presence of phaeophytin. One must conclude that the remainder is due to enzymatic factors concerned with photosynthesis and/or the presence of unmeasured decomposition products. With regard to the latter, it has been observed (Yentsch unpublished) that, at times, when diatom cultures heavily contaminated by bacteria are placed in darkness, chlorophyllide — not phaeophytin or phaeophorbide — is formed. By periodically measuring C^{14} uptake in light, and following the pigment concentration and composition by chromatography, it has been observed that the capacity for carbon uptake in light closely follows chlorophyll *a* content; whereas the total chlorophyll content composed of chlorophyllide *a* and chlorophyll *a*, when compared to C^{14} uptake, yields low carbon uptake per unit pigment values. A number of workers have observed chlorophyllide in natural waters (Patterson and Parsons 1963, Brown *et al.* 1964). Since this pigment may be abundant in extracts from natural populations, a method for simply and accurately estimating its presence in small quantities is needed.

In the region of maximum pigment concentration in the water columns of tropical oceans, previous work has shown that the ratio of total carbon to « chlorophyll » and the maximum rate of C^{14} uptake per unit chlorophyll are lower than at the surface (Steele 1964). According to Steele and Yentsch (1960), the pigment maximum is formed because phytoplankton are nutrient deficient and lose buoyancy. Buoyancy is restored when the cells sink into the region of the density discontinuity and accumulate where nutrient levels are higher. One would expect photosynthetic efficiency to be restored with increased buoyancy. Instead, at most of the stations shown in Fig. 8, there is an indication that the photosynthetic capacity as computed by total pigment is higher than that immediately above or below the maxima, but not as high as at the surface. However, by accounting for the presence of accumulated phaeophytin at this depth, the efficiency of pigment (chlorophyll) in carbon fixation is increased to levels comparable to those at the surface. These observations offer supporting evidence to the theory by showing that some of the plants reaching this level are healthy and capable of efficient photosynthesis.

Acknowledgements

The author acknowledges the able assistance of Mr. Jack Laird and Mrs. Barbara Breivogel in collection, analysis and computation of data. Dr. John Ryther and Mrs. Anne Yentsch read the manuscript and offered suggestions. This work was supported by A.E.C. contract 30-1 (1918) REF. NYO 1918-120 and Grant No. GB 1403 from the National Science Foundation.

REFERENCES

Brown, S. R., S. Ferguson, D. C. Hamilton, and C. R. Meyer. 1964. Chlorophyll *a* derivatives in freshwater lakes. - Proc. 7th Conf. on Great Lakes Research: 140. (Abstr.).

Lorenzen, C. J. 1965. A note on the chlorophyll and phaeophytin content of the chlorophyll maximum. Limnol. Oceanog. *10* : 482-483.

Patterson, J. and T. R. Parsons. 1963. Distribution of chlorophyll *a* and degradation products in various marine materials. - Limnol. Oceanog. *8* : 355-356.

Ryther, J. H. and C. S. Yentsch. 1957. The estimation of phytoplankton production in the ocean from chlorophyll and light data. - Limnol. Oceanog. *2* : 281-286.

Steele, J. H. 1964. A study of production in the Gulf of Mexico. - Jour. Mar. Res. *22* : 211-222.

— and C. S. Yentsch. 1960. The vertical distribution of chlorophyll. - J. Mar. Biol. Ass. U. K. *39* : 217-226.

Steemann Nielsen, E. 1952. The use of radio-active carbon (C^{14}) for measuring organic production in the sea. - J. Cons. perm. int. Explor. Mer *18* : 117.

— and E. Aabye Jensen. 1957. Primary organic production. The autotrophic production of organic matter in the oceans. - Galathea Report *1* : 49.

— and V. Kr. Hansen. 1959. Light adaptation in marine phytoplankton populations and its interrelation with temperature. - Physiol. Plant. *12* : 353-370.

Yentsch, C. S. 1962. Marine plankton, p. 771-797. *In*: Lewin, R. A. (*Ed.*) Physiology & Biochemistry of Algae. - Academic Press Inc. New York.

— 1965. Distribution of chlorophyll and phaeophytin in the open ocean. Deep Sea Res. *12* : 653-666.

— and D. W. Menzel. 1963. A method for the determination of phytoplankton chlorophyll and phaeophytin by fluorescence. - Deep Sea Res. *10* : 221-231.

VI

THEORETICAL PROBLEMS OF PRIMARY PRODUCTIVITY, LIGHT, AND COMMUNITY STRUCTURE

NOTE ON DIFFUSION UPTAKE IN CELLS: SOME MATHEMATICAL FORMULAE

EGBERT KLAAS DUURSMA

Present address: I.A.E.A.
Musée Océanographique, Monaco

For bibliographic citation of this paper, see page 10.

Abstract

Two physical models, concerning uptake of substances by spherical particles are presented. The derived formulae give relations between the total net uptake as a function of time; the principal factors are the diffusion coefficients, which can be calculated from a graphical representation of the experimental results. The models may be applied for living cell diffusion uptake.

In uptake of substances by living cells, ions or complete molecules may react with the cell wall, the cell membrane or the cell constituents in such a way that they change their composition.

However, for the transport of unchanged ions or molecules into cells the process of diffusion is responsible; and even the same process occurs when the mobility or degree of freedom is less than that outside the cell, or when the substances react with the medium by reversible processes.

In many cases there may be a need for a good description of these diffusion processes, and then it is not always simple to calculate the principal factors, which are the diffusion coefficients, i.e., for practical purposes one may be interested in the total uptake of the cells and the determination of the accumulation of the diffusing matter over certain time intervals. For a better understanding of these uptake processes under different conditions, an estimation of the principal factors (the diffusion coefficients) and their dependency on the conditions may help.

For mathematical derivation of the diffusion coefficients from total uptake results, handbooks do not usually supply the answers, and Duursma and Hoede (1966) constructed several physical models, which under special boundary conditions give solutions of the basic diffusion differential equations. These equations can be applied to a number of diffusion and uptake processes like adsorption and absorption of radio-nuclides and other substances on and in particulate matter of the sea and the diffusion in sediments.

Two of these models (Hoede 1964) can be used to study the uptake by diffusion of living cells; the cell in the first model is regarded as a homogeneous sphere and in the other model as a two layered sphere in which the uptake takes place through an outer layer (cell wall and cell membrane) into an inner core (cell content). In Fig. 1 both models are presented together with the calculated formulae.

A (1)
$$M(t) \simeq \frac{4}{3}\pi R^3 N\left[\frac{1}{1+\frac{\rho R}{3}} - \frac{6}{\pi^2}\sum_{n=1}^{\sim}\frac{\left(1-\frac{5\rho R}{n^2\pi^2}\right)}{n^2}\times \exp.\left\{-\frac{Dt\left(n\pi+\frac{\rho R}{n\pi}\right)^2}{R^2}\right\}\right]$$

(2)
$$\log_{10}\left[\frac{V_s}{V} - \frac{M(t)}{C_o V_p (1+k)}\right] \simeq A - \left[\frac{D}{R^2}\left(\pi+\frac{PR}{\pi}\right)^2 \log_{10} e\right] t \quad (n = 1 \text{ only})$$

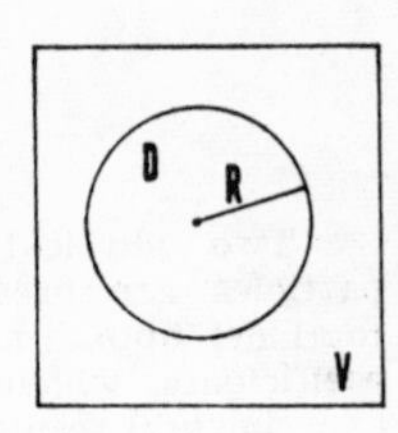

B (1)
$$M(t) \simeq \frac{4}{3}\pi b^3 - \frac{8c^3}{\pi}\left(1+\frac{\delta^2 a}{c}\right)\exp.\left(\frac{-D_2\pi^2 t}{c^2}\right)$$

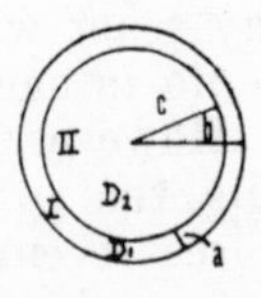

(2)
$$\log_{10}\left(V_p - \frac{M(t)}{C_o(1+k)}\right) \simeq \log_{10} N\left[\frac{8c^3}{\pi}\left(\frac{1+\delta^2 a}{c}\right)\right] - \frac{D_2\pi^2 t}{c^2}\log_{10} e$$

Fig. 1.

A: Sorption by homogeneous spherical particles from a limited source. (1) Calculated formula for N particles (Hoede 1964b); (2) simplified formula for $n = 1$ only. Explanation of the characters: D = diffusion coefficient in the particles; R = radius; V_p = volume of particles; V_s = volume of solution; $V = V_p + V_s$; C_o = concentration of the solution at $t = 0$; $M(t)$ = total uptake in all particles; N = number of particles; n = 1, 2, 3, 4....; t = time (at $t = 0$ the concentration in the particles is 0); A = constant; e = number of Euler (= 2.7183); $(1 + K)$ = distribution coefficient of the concentration inside and outside the particles at $t \rightarrow \sim$; $\rho R = 3\, V_p : V_s$.

B: Sorption in a two layered particle from a constant source. (1) Calculated formula for one particle (Hoede 1964a) valuable for $t \rangle\rangle (c^2 : D_2\pi^2)$; (2) simplified formula for N particles. Explanation of the characters: D_1 = diffusion coefficient in the outer layer (I); D_2 = diffusion coefficient in the inner core (II); b = radius total particle; c = radius inner core; $a = b - c$; $\delta^2 = D_2 : D_1$; t = time (at $t = 0$, C_I and $C_{II} = 0$) $M(t)$ = total uptake; V_p = volume particles; N = number of particles; $(1 + K)$ = distribution coefficient of the concentration inside and outside the particles at $t \rightarrow \sim$ ($C_I = C_{II}$); C_o = constant concentration in the solution. Formula (B) valuable for $\delta < 1$ and $(a : c) < 1$.

The chemical and biological reactions of the diffusing substances with the organic living material in the layers themselves may be considered as being identical when ions diffuse in the interstitial water of sediments and sorption occurs simultaneously on the sediment particles by reversible processes. The simplest example of this reaction occurs when the concentration of the bound substances is proportional to the concentration of the diffusing substances.

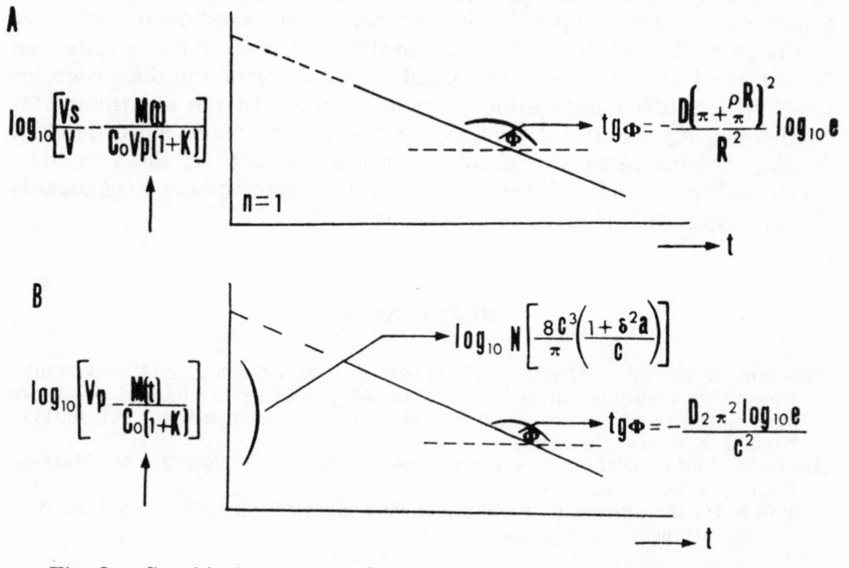

Fig. 2. - Graphical representation of the simplified formulae of Fig. 1.

Then

$$K = \frac{C_{\text{bound}}}{C_{\text{diffusing}}},$$

in which K is supposed to be constant (at constant conditions except concentrations) and for simplicity equal in both layers of model II. The value of K can be determined when uptake and loss of the cells are in equilibrium (net uptake is zero) by

$$\frac{C_{\text{cell}}}{C_{\text{solution}}} = (1 + K) = K'.$$

K′ is the distribution coefficient of the total cell concentration to the concentration of the substances in the solution at equilibrium.

However, physical models and their mathematical calculation only give an answer when boundary conditions of the experiments are exactly identical with those of the models. Progressively more complicated processes or changes in boundary conditions (e.g. not constant solution concentration for model II) are not covered by the theories. Thus, for more complex experiments the boundary conditions have to be made equal to those of the models used.

A great advantage of the above formulae is that for very rapid or very long term processes the diffusion coefficient(s) can be estimated in a reasonable time, especially in model II. If *e.g.* adsorption processes are very rapid, it is impossible in experiments to start filtration within seconds after the addition of the studied substances to the solution. On the other hand for long term uptake processes (weeks) it is often impossible to avoid changes in the environmental conditions, e.g. bacterial growth, so one has to work more quickly.

Fig. 2 illustrates the graphical representation of the formulae, given in Fig. 1, and a direction for finding from uptake experiments the principle diffusion coefficients.

REFERENCES

Duursma, E. K. and C. Hoede. 1966. Theoretical, experimental and field studies concerning molecular diffusion of radioisotopes in sediments and suspended solid particles of the sea. Part A: Theories and mathematical calculations. Neth. J. Sea Res. (in press).

Hoede, C. 1964 a. Diffusion in layered media. Technical Note TN 37. - Mathem. Centre, Amsterdam.

— 1964 b. On the absorption by homogeneous spherical particles. Technical Note TN 38. - Mathem. Centre, Amsterdam.

ECOLOGICAL CORRELATIONS AND THE RELATIONSHIP BETWEEN PRIMARY PRODUCTIVITY AND COMMUNITY STRUCTURE

RAMON MARGALEF

Instituto de Investigaciones Pesqueras
Barcelona, Spain

For bibliographic citation of this paper, see page 10.

Abstract

Any expression aiming to give an indirect estimate of primary production on the basis of present properties of an ecosystem has to include some term reflecting quantitatively the structure of the community (biotic diversity, pigment ratio D_{430}/D_{665}). This is logical, because production has the nature of a derivate relative to time. The expressions of structure, although dimensionless, reflect historic development and thus are related also to time. Accelerations and decelerations in the speed or rate of production are reflected by a decrease or an increase, respectively, of the diversity of the community. Since the pigment ratio D_{430}/D_{665} (or any other analogous ratio) is a good indicator of structural properties of the whole community, it is advisable not to limit analysis of pigment spectra to a single narrow band.

CORRELATIONS AND REGRESSIONS

Information on the properties of the environment and the populations in aquatic ecosystems is growing steadily. A much more rapid increase can be anticipated in the near future, in part perhaps as a consequence of the International Biological Programme. The accumulation of synoptic information, together with the general availability of computer facilities, opens new perspectives and poses new problems in the handling of ecological data.

We assume that we are in possession of a number of measurements, referred to a set of discrete sampling points, and expressing physical properties of the environment or biological properties of the mixed populations. There are many excellent manuals that tell us how to compute both correlations and regressions involving the different variables. To do this may often be useful.

The application of statistical methods requires a sound ecological judgment. Sometimes the data, taken out of their spatio-temporal frame, yield poor statistical correlations. Nevertheless, if such data are plotted, the comparison of cartograms with the distribution of the values of different variables often discloses significant congruence of patterns. In such instances we need a much more powerful tool than the usual statistical methods, perhaps sequential analysis or methods based on information theory. But, for the moment, we shall be satisfied with ordinary multivariate statistical analysis.

Significant correlations may be found among actual or point values or among their derivates. Production has the quality of a derivate of biomass relative to time. Production, thus, is likely to be correlated in a more meaningful way with magnitudes that are themselves also derivates. An example of this is depletion of nutrients. Discreteness of samples and movements of water set great difficulties for a generalized usage of derivates. Here we have another example of the urgent need for methods for rapid and almost continuous sampling.

For the time being we are forced to ignore almost completely the derivates. All too frequently the proper dimensions of the magnitudes that we relate together in the expression of a regression are disregarded. But we must retain the logical requirement that production should be used in relation with something that conveys the notion of time.

In the working of correlations and regressions, the rough data frequently do not approach a normal distribution. In such cases a transformation is required before further statistical analysis. A logarithmic transformation often proves appropiate for parameters referring to populations (chlorophyll content, production, number of cells) and to environmental factors strongly influenced by organisms (nutrient concentrations). Multiplication and diffusion in a non-uniform environment lead commonly to a type of distribution in which density of populations decreases exponentially with increasing distance from a center of maximum density. If samples are taken with a regular spacing or a regular periodicity, chances are that in any series of samples, not the actual densities, but the logarithms of the densities approach normal distribution. Other variables (temperature, salinity) frequently do not require transformation.

The practice of computing regression equations involving different parameters of the same community, as number of cells, chlorophyll, etc., shows how difficult is to remain faithful to the so-called « conversion factors ». For instance, in marine phytoplankton, number of cells cannot be put simply as a function of the chlorophyll *a* amount, but some other parameter, as the pigment ratio D_{430}/D_{665}, must intervene (Herrera and Margalef 1963).

Suppose we want to set the production (P) as a function of several factors, as depth (Z), chlorophyll concentration (C), and the pigment ratio D_{430}/D_{665}. The first impressions of the results of a study still in progress on marine plankton communities are that we may account for a considerable part of the variation of P by writing

$$\log P = a + b \log C - c Z - d(D_{430}/D_{665})$$

The author has a weakness for the pigment ratio D_{430}/D_{665}. It means the ratio of the absorbancies at the stated wavelengths of an acetonic extract of pigments, roughly a ratio « yellows »/« green », that, in addition to other interesting properties, reflects the effect of depletion of nutrients (see further). It seems that in this particular connection the use of said ratio may have advantages over the inclusion in the regression formula of the concentrations of several nutrients, if it happens that not always the same nutrient acts as a limiting factor.

STRUCTURE

On the other hand, the ratio D_{430}/D_{665} gives a measure of « structure ». Present structure is the result of an historical process. Any structure has properties that are quantitatively measurable, as, for instance, diversity. Biotic diversity is any suitable function which has a minimum when all cells belong to the same species, and a maximum when every individual belongs to a different species. It is practical to measure and express diversity in bits per individual (per cell in the case of phytoplankton), according to the expression

$$D = - \Sigma\, p_i \log_2 p_i$$

where p_i denotes the participation or probability of occurrence of every species in the total number of individuals ($\Sigma p_i = 1$). We leave untouched here the point of whether it is more appropriate to speak of diversity without qualifications, or if it is better, as it seems, to refer always to a spectrum of diversity. In any case, eventual interest in diversity should act as a stimulus to not neglect too much a careful taxonomical study of communities.

A pigment diversity could be computed according to the way plant pigments are distributed among different molecular species. It is assumed that the simple pigment ratio D_{430}/D_{665} gives a rough estimate of such pigment diversity.

The biotic diversity, as bits per individual, or the pigment diversity in the form of a ratio between absorbancies, are simple numbers without dimension. They are well correlated with one another (Table 1) and also with other indices of structure. The last line in Table 1 is interesting, because it suggests that it is possible to use an electronic dimensional particle counter (data were actually obtained with such apparatus) to obtain a useful index of community structure. The correspondance between parameters compared in Table 1 is better than expressed by the correlations, if data are plotted on cartograms.

Table 1. - Statistical correlations between pigment ratio D_{430}/D_{665} and other structural properties of plankton populations. Examples from marine environment. Untransformed data.

Area	Property compared with pigment ratio	Pairs of values	Coefficient of correlation
Western Mediterranean	Biotic diversity of net phytoplankton	54	+0.40
Southern Caribbean	Biotic diversity of net phytoplankton	68	+0.30
Tyrhenean Mediterranean	Per cent of «big» particles in total seston	127	+0.27

Biotic diversity, pigment ratio, particle size ratio, and other indices that could be devised, are expressions of properties concerning organization of the community. As such, they are also expressions of maturity or historical development, and in consequence related to time. For this reason it can be understood why and how they are associated with production, a derivate relative to time.

PRODUCTION AND PIGMENTS

Observations in sea, freshwater, and in laboratory cultures, both published (Margalef 1963, 1964, in press) and unpublished, substantiate the relationship between carbon uptake and diversity. A high inorganic carbon uptake per unit biomass is always associated with a low biotic diversity or with a low pigment ratio D_{430}/D_{665}.

Let us examine more closely an example concerning fresh water communities (Margalef 1964) that seems to be typical; the more extensive material from marine environments fits the same pattern (or sort of expressions at which we arrive), although the coefficients have different numerical values. Here production means inorganic carbon fixation under constant light conditions, and samples are all surface samples. Thus the approach may be conveniently simplified, because light or depth does not enter into the expressions. The correlation matrix between some parameters of interest is reproduced in Table 2.

Table 2. - Correlation matrix between several parameters in 4 different freshwater phytoplankton populations (Margalef 1964). Logarithmic transformations ,

(A) Chlorophyll *a*, mg/m³ (B) D_{330}/D_{665}
(C) Diversity, bits/cell (D) Carbon uptake, mg C/m³/hour
(E) Production per unit biomass, mg C/g C/hour

	(A)	(B)	(C)	(D)	(E)
(A)	1	−0.064	+0.057	+0.964	+0.573
(B)		1	+0.796	−0.319	−0.968
(C)			1	−0.107	−0.723
(D)				1	+0.545
(E)					1

There is a strong positive correlation (+ 0.796) between both structural characters, and both are negatively correlated (—0.968 and — 0.723) with productivity per unit biomass (turnover). The correlations of production may be worthy of closer examination as concerning the often discussed relationship between production and pigments.

Production is positively correlated (+ 0.964) with amount of chlorophyll *a*, as was expected, and negatively correlated with pigment ratio (—0.319) and, more feebly, with biotic diversity.

An appropriate estimate of production was

$$\log P = 1.047 + 0.728 \log C - 0.615 \log (D_{430}/_{665})$$

equal to

$$P = 11.1\ C^{0.728}/(D_{430}/D_{665})^{0.615}$$

Since C is approximately proportional to D_{665}, we could equally well write

$$P = 67.7\ (D_{665})^{1.343}/(D_{430})^{0.615}$$

In Fig. 1 this expression is compared with the often held assumption that production is proportional to the amount of chlorophyll *a*. The study of more extensive material will afford an estimate of the variability of the « constants » involved in our expression, and whether it is worthwhile after all to retain such expression in the proposed form.

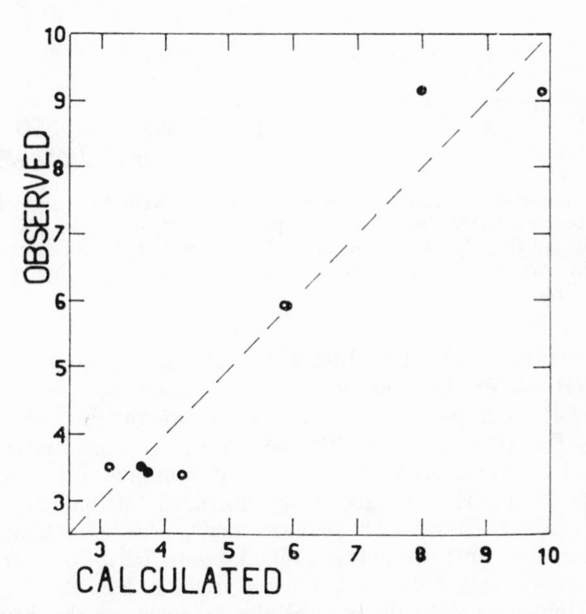

Fig. 1. - Dams in NE Spain (Margalef 1964), surface phytoplankton. Comparison between the observed values of inorganic carbon uptake, in mg C/m³/hour, and the calculated values using the expressions

$P = 11.1\ C^{0.73}/(D_{430}/D_{665})^{0.62}$, black circles

$P = 3.7\ C$, white circles

Note that in the center of the figure a black and a white circle are almost superposed.

As a result of previous work on marine plankton (Margalef 1960) and laboratory cultures (Margalef 1963), the biomass (B) was expressed as a function of the absorbancy of pigment extracts at two wavelengths, F being a convenient factor:

$$B = F\ (D_{430})^3/(D_{665})^2$$

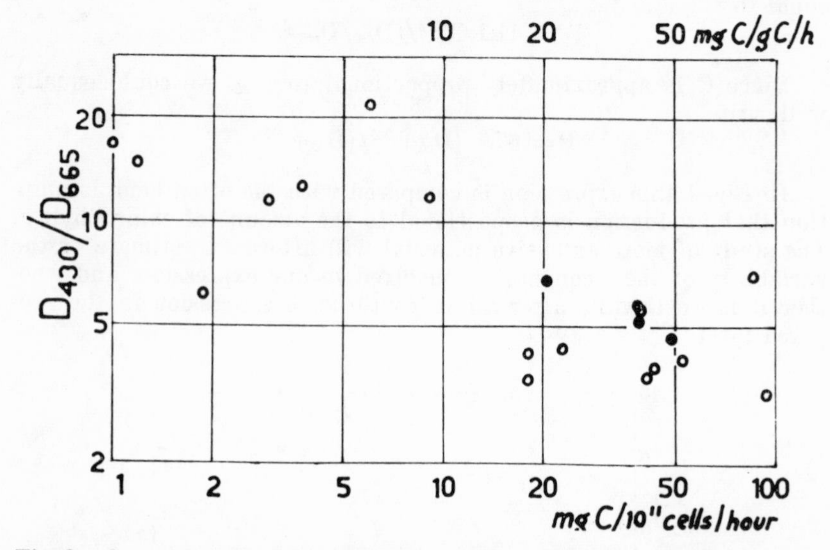

Fig. 2. - Comparison of the pigment ratio D_{430}/D_{665} with the production per unit biomass. Black circles, surface phytoplankton of dams in NE Spain; productivity in mg C/g C/hour; T = 20°C, L = 5000 lux. White circles, laboratory cultures in sea water; productivity in mg C/10^{11} cells/hour; T = 16°C, L = 5000 lux.

By division of the two last expressions, we would expect their ratio (P/B = production per unit biomass) to be approximately proportional to a power (near the cube) of the inverse of pigment ratio D_{430}/D_{665} (Fig. 2). This fits well with the high negative correlation found between production per unit biomass and pigment ratio (Table 2). In a series of laboratory cultures (Margalef, in press; 17 pairs of values, data not transformed), the correlation between production per unit biomass and pigment ratio was found to be — 0.588; these data have been plotted also in Fig. 2.

I am insisting deliberately on the interest of the pigment ratio D_{430}/D_{665}, or of any other descriptive of the general form of the absorption spectrum. Almost everybody agrees that proposed expressions for computation of the concentration of pigments other than chlorophyll *a*, from spectra of acetonic extracts, are unreliable. But we must think twice before limiting analysis to the measure of a narrow band of the spectrum, and expression of pigments to concentration of chlorophyll *a*. In so doing, we may lose irretrievable information about « structure » of the photosynthetizing equipment of phytoplankton. An empirical approach to the problem would be to run complete spectra of a great number of extracts of phytoplankton of different types (as production, species composition, etc.) and to look successively for 1, 2, 3 ... wavelengths permitting the best discrimination between spectra. This could be done by purely statistical methods, and *a posteriori* a biochemical interpretation of the selected points with high indicator value would follow.

STRUCTURE AND FUNCTION

To return to the main subject, we are led to expressions similar to the ones discussed if we introduce biotic diversity in lieu of the pigment ratio. This latter ratio has been preferred here for obvious reasons connected with the extensive use of plant pigment analysis in ecological research. In any case, study of multiple regressions indicates that production is a function, in the mathematical sense, of the structure of the community too, and not only of a simple expression of biomass or total chlorophyll. In a biological sense, production is a genuine function of structure. If the structure of the community remains the same, its function, that is production per unit biomass, remains also constant. Any change of the community structure, in space or time, is linked to a change in the rate of production.

The correlations between inorganic carbon uptake and structural properties of the community can be easily clarified with a dynamic model. Suppose that the supporting capacity of the environment increases: immediately the ratio primary production/total biomass increases. As the cells of the different species multiply relatively free of impediment, the differences in their respective potential rates of increase and in the speed at which rates of increase can be modified, manifest themselves. The consequence is that the representation of the different species (p_i) becomes more and more inequal and the value $-\Sigma\, p_i \log_2 p_i$ drops. The process can continue up to the final dominance of one or a few species, usually of small cells.

As for the pigments, those having a key position or synthesising and decaying more rapidly than others, as chlorophyll *a*, are comparable to the species endowed with a high potential rate of increase. They are simply parts of the system allowing a higher flow of energy through them. Undoubtedly, the rearrangement of the pigment composition is a complex process, as is easily evidenced in thin layer chromatograms; but the bursts of production are always associated with a more rapid increase of the « greens » than of the « yellows », and there is a drop in the pigment ratio D_{430}/D_{665}. Changes in this ratio reflect changes not only in the species composition of the community, but also in the physiological state of unispecific populations. The sensitivity of the pigment ratio to nutrient depletion and supply makes its response useful in the assessment of elements limiting growth (Castellvì 1964).

From the preceding it can be understood why, under similar conditions of light, production per unit biomass is strongly correlated to pigment ratio D_{430}/D_{665}. We may anticipate that this ratio is subjected to a daily rhythm, but further work is necessary to ascertain its generality, importance and meaning. It changes easily under different environmental influences. Although the physiological adaptation of plankton algae to different environmental conditions has been often studied and discussed, insufficient attention has been paid to the eventual reflection of the changes on the pigment composition.

When environmental conditions remain stable and biomass increases, production per unit biomass drops, the population diversifies, dominance disappears, and pigment ratio increases. In short, rate of change of quantitative expressions of structural properties of communities is always linked to accelerations or decelerations in the speed of production.

The author would of course prefer a more elegant and general way of harmonizing these findings in an interpretation of the ecosystem as a cybernetic system, but such a way of reasoning seems not to be acceptable yet to a great number of workers. This is irrelevant here, since the purpose of this paper was to stress only that if we want to make indirect estimates of production, based on environmental factors (light, nutrients) and on organisms, then structural properties of phytoplankton populations probably need to be taken into account, in addition to non-structural properties taken separately, such as biomass, number of cells, or amount of chlorophyll *a*.

REFERENCES

Castellví, J. 1964. Un sencillo experimento para demostrar la influencia de la concentración de elementos nutritivos sobre la calidad de los pigmentos de las algas. - Inv. Pesq. *25* : 157-160.

Herrera, J. and R. Margalef, 1963. Hidrografía y fitoplancton de la costa comprendida entre Castellón y la desembocadura del Ebro, de julio de 1960 a junio de 1961. - Inv. Pesq. *24* : 33-112.

Margalef, R. 1960. Valeur indicatrice de la composition des pigments du phytoplancton sur la productivité, composition taxonomique et propriétés dynamiques des populations. - Rapp. Proc. Verb. C.I.E.S.M.M. *15* (2) : 277-281.

— 1961. Correlations entre certains caractères synthétiques des populations de phytoplancton. - Hydrobiologia *18* : 155-164.

— 1963. Modelos simplificados del ambiente marino para el estudio de la sucesión y distribución del fitoplancton y del valor indicador de sus pigmentos. Inv. Pesq. *23* : 11-52.

— 1964. Correspondence between the classic types of lakes and the structural and dynamic properties of their populations. - Verh. int. Ver. Limnol. *15* : 169-175.

— (in press), Laboratory analogues of estuarine plankton systems. - Conference on Estuaries, held at Jekyll Island Georgia, March 31 April 4, 1964.

STANDARD CORRELATIONS BETWEEN PELAGIC PHOTOSYNTHESIS AND LIGHT

WILHELM RODHE

Institute of Limnology

University of Uppsala, Uppsala (Sweden)

For bibliographic citation of this paper, see page 10.

Abstract

The response of pelagic photosynthesis to the vertical attenuation of light follows a general pattern, common to various types of lakes as well as to the sea. On the basis of Talling's model for the photosynthetic integral an attempt is made to define quantitative standard correlations. These facilitate the assessment and interpretation of photic and non-photic effects upon the primary production of phytoplankton. With a slight modification they can also be used for approximate estimates of total photosynthesis in cases where only the optical properties of the water and the photosynthetic rate at light saturation are known.

Thirteen years have passed since Steemann Nielsen (1952) introduced his ingenious method for measuring organic production in the sea, equally useful for lakes and other standing freshwaters. Today it is an indispensable tool for marine biologists and limnologists all over the world. As it is easy to use, the bulk of primary production data is now increasing very rapidly.

For the comparison of such data and for their interpretation we still lack, however, some kind of yardstick, *i.e.* a mathematical but simple rule of general application for the quantitative evaluation of similarities and differences in the pelagic photosynthesis of various waters. Without such a rule it appears hardly possible to detect detailed causal connections between planktic CO_2-assimilation and the factors conditioning its size and structure.

For a photosynthetic model the basis must be the intensity and quality of the light, with CO_2-assimilation as the variable. We can, therefore, expect nature to give us its most lucid answers, if we compare waters within a broad range of light transmission and pelagic productivity. From my own material — most of it still unpublished, I am sorry to say — I have picked out 12, mostly well known, lakes, different enough to exhibit Secchi disk transparencies from 1.7 to 13.7 m and assimilation values between 45 and 1600 mg $C/m^2 \cdot$ day (Fig. 1). At this moment it seems unnecessary to comment more upon the lakes than to say that all limnologists, irrespective of ter-

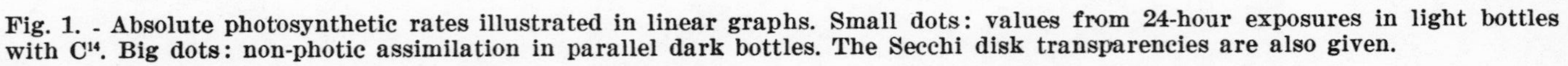

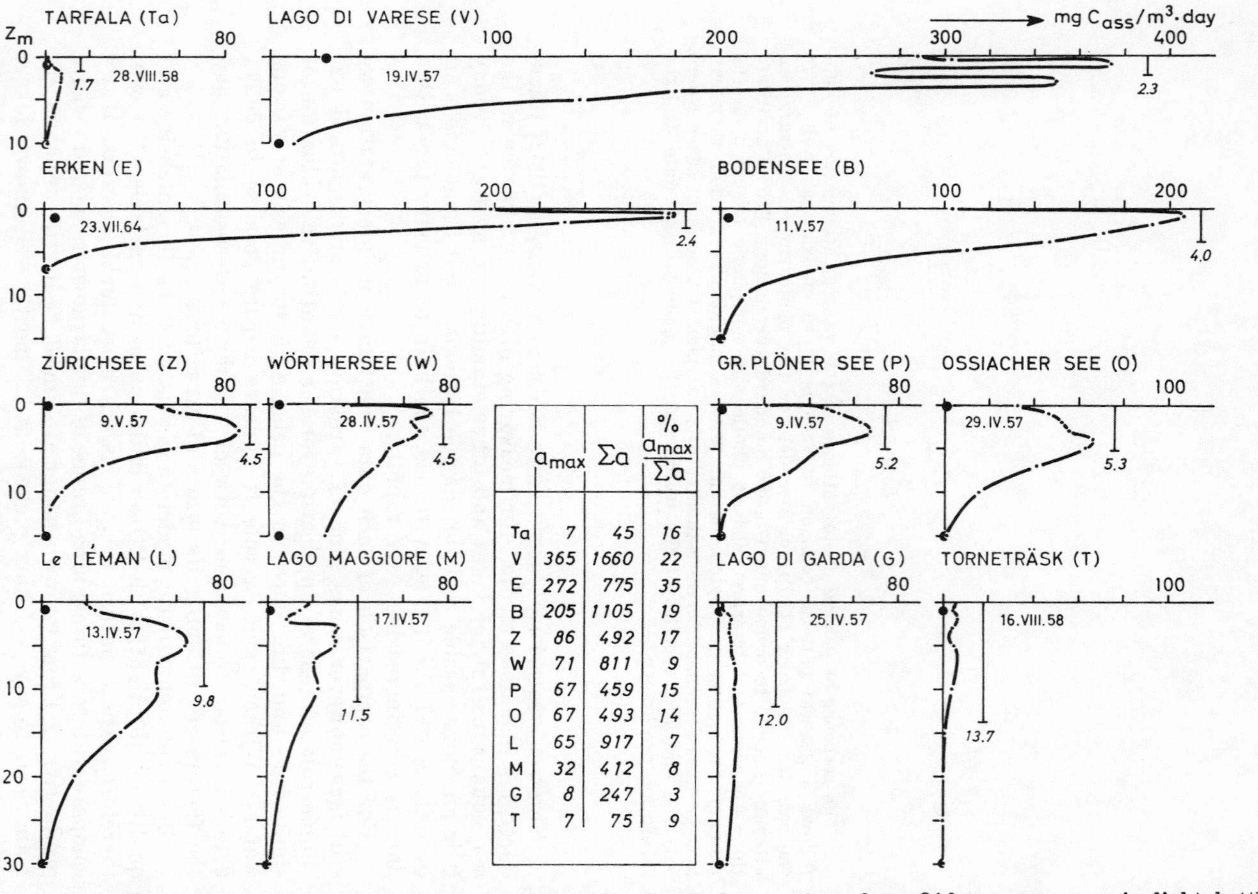

	a_{max}	Σa	% $\frac{a_{max}}{\Sigma a}$
Ta	7	45	16
V	365	1660	22
E	272	775	35
B	205	1105	19
Z	86	492	17
W	71	811	9
P	67	459	15
O	67	493	14
L	65	917	7
M	32	412	8
G	8	247	3
T	7	75	9

Fig. 1. - Absolute photosynthetic rates illustrated in linear graphs. Small dots: values from 24-hour exposures in light bottles with C^{14}. Big dots: non-photic assimilation in parallel dark bottles. The Secchi disk transparencies are also given.

minological confession, would agree to classify the first and the last of them (Lake Tarfala and Torneträsk in Swedish Lappland) as oligotrophic. The very little known Lake Tarfala has been included in this issue because of its high turbidity, due not to phytoplankton as in the highly eutrophic Lago di Varese, but to inorganic silt transported into the lake by a glacier.

The diagrams of Fig. 1 give a good impression of the great diversity of the assimilation curves when they are presented in the conventional form. An index of their different shape is the quotient of maximal assimilation per m^3 (a_{max}) and total assimilation per m^2 (Σ a). As shown earlier (Rodhe 1958), the quotient decreases on the whole with increasing transparency. The high quotient for Lake Tarfala proves that this trend depends primarily on turbidity and not on trophic level, although the separation of these two effects is often difficult or impossible.

In Fig. 2 the optical properties of the lakes are illustrated in semi-logarithmic diagrams by the gradients for the blue (b), green (g), and red (r) light as well as for the entire visible energy (E). The components were measured with Jena-Schott filters BG 5, VG 9, and RG 1, respectively, except for Erken where BG 12 and RG 2 were used. The attenuation of visible energy has been calculated by means of the observed component gradients according to the arithmetic procedure of Åberg (cf. Vollenweider 1960, 1961). All values are in percentages of the intensity ($I'_0 = 100\%$) at about 5 cm beneath the surface.

With regard to the penetration of visible radiation the lakes show great differences: 1% E was found at 4.5 m in Lake Erken but at about 20 m or more in the clearest lakes. Qualitatively, however, 11 of the 12 lakes agree in that green is the most penetrating component (I_{mpc}) of their subaquatic light, whereas in Erken red goes slightly deeper than green.

We can obtain a common denominator for the optical conditions in all the lakes by applying the concept « optical depth » (here called $z_{o.d.}$) suggested by Talling (1957) and using the relative intensities of I_{mpc} for its assessment. Each unit of $z_{o.d.}$ corresponds to a layer, in meters, that causes a halving of I_{mpc}, i.e. $z_{o.d.} = I'_0 \cdot 2^{-n}$, where $n = 0, 1, 2, \ldots.$ From the gradients of I_{mpc} (red in Erken, green in the other lakes) we can easily construct a common scale in which, for instance, the optical depth 3 (z_3) corresponds to an intensity of $1/8 = 12.5\%$ I_{mpc}. On this scale, the 10% level is reached at 3.3 and the 1% level at 6.6 units.

Next we transform the assimilation curves of Fig. 1 from absolute to relative units by expressing the assimilation values as percentages of the corresponding value for a_{max}. Then we plot the resulting figures against their optical depths and insert the curve for I_{mpc} into the semi-logarithmic diagram. In that way photosynthesis and light,

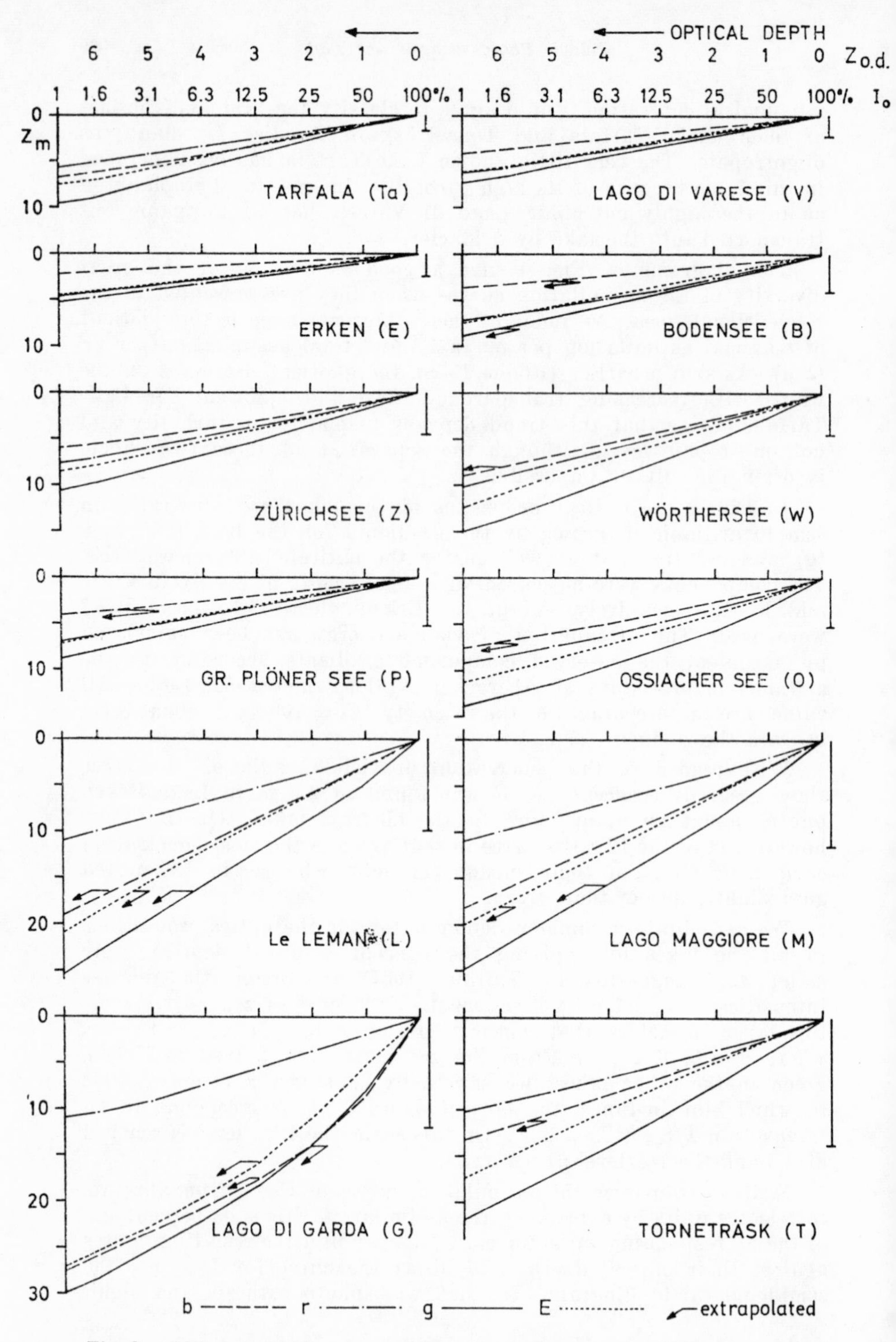

Fig. 2. - Light attenuation in semi-logarithmic graphs. The gradients of total visible energy (E) have been calculated from the observed gradients of the blue, green, and red components of light. The Secchi disk transparencies are indicated.

previously given separately in Figs. 1 and 2, have been correlated into the same graph, Fig. 3.

In this form of presentation, the similarities between the lakes appear more pronounced than the differences. In most cases, namely 9 of the lakes, the slope of the assimilation curves below z_3 entirely or closely agrees with the slope of I_{mpc}. Vollenweider (1960) found the assimilation curves in better agreement with total visible light than with a particular one of its components, but, as stated by Talling (1957, 1965 p. 10), the vertical extent of subaquatic light is due mainly to that part of the spectrum — maybe blue, green or red, depending on the prevailing optical conditions — where the vertical extinction coeffient has its minimum value, *i.e.* I_{mpc}.

Only in Wörthersee, Gr. Plöner See and Lago di Garda, do the assimilation curves clearly deviate from every light gradient. An examination of the phytoplankton distribution (quantitative counts by A. Nauwerck) shows that just these three lakes differ from the others in having a considerable increase and change of the algal population in the lower part of the photosynthetic layer. Within the upper layer, however, several of the other lakes also have erratic curves which may be due to heterogenous stratification of plankton algae.

Fig. 4 is a combined plot of the curves from Fig. 3 for the 9 regular lakes. Its evidence is striking: within the upper layer, above about z_3, the course of the curves is rather individual and partly independent of the light gradient, but as a common feature a_{max} is situated in the region between $z_{0.5}$ and z_2, *i.e.* between 70 and 30% of I_{mpc}. Below z_3, on the other hand, most of the curves run more or less parallel to the slope of I_{mpc} down to z_7.

This means that the pelagic photosynthesis in all these lakes, provided they have a uniform vertical distribution of phytoplankton, does respond in the same manner to light. Findenegg (1964) claimed, however, that oligotrophic lakes behave differently from eutrophic lakes in this respect, and he proposed the distinction of two types: type I (eutrophic) represented by the Ossiacher See, and type II (oligotrophic) by the Attersee. But if his data are treated in the way described above (using, like Findenegg himself, E instead of I_{mpc}), no real difference between his types can be seen (Fig. 5). Such opposite conclusions from the same material stress the importance of the depth scale in studies of the relations between pelagic photosynthesis and light.

Steemann Nielsen performed four *in situ* determinations of photosynthesis in oceans with widely differing productivity, in order to establish a starting-point for an evaluation of the *in vitro* experiments. His figures (Steemann Nielsen and Aabye Jensen 1957, Figs. 9-12) can be redrawn in the manner used here but with blue + green as I_{mpc}. The resulting picture (Fig. 6) is almost identical with that for my 9 lakes (Fig. 4), despite the great differences in

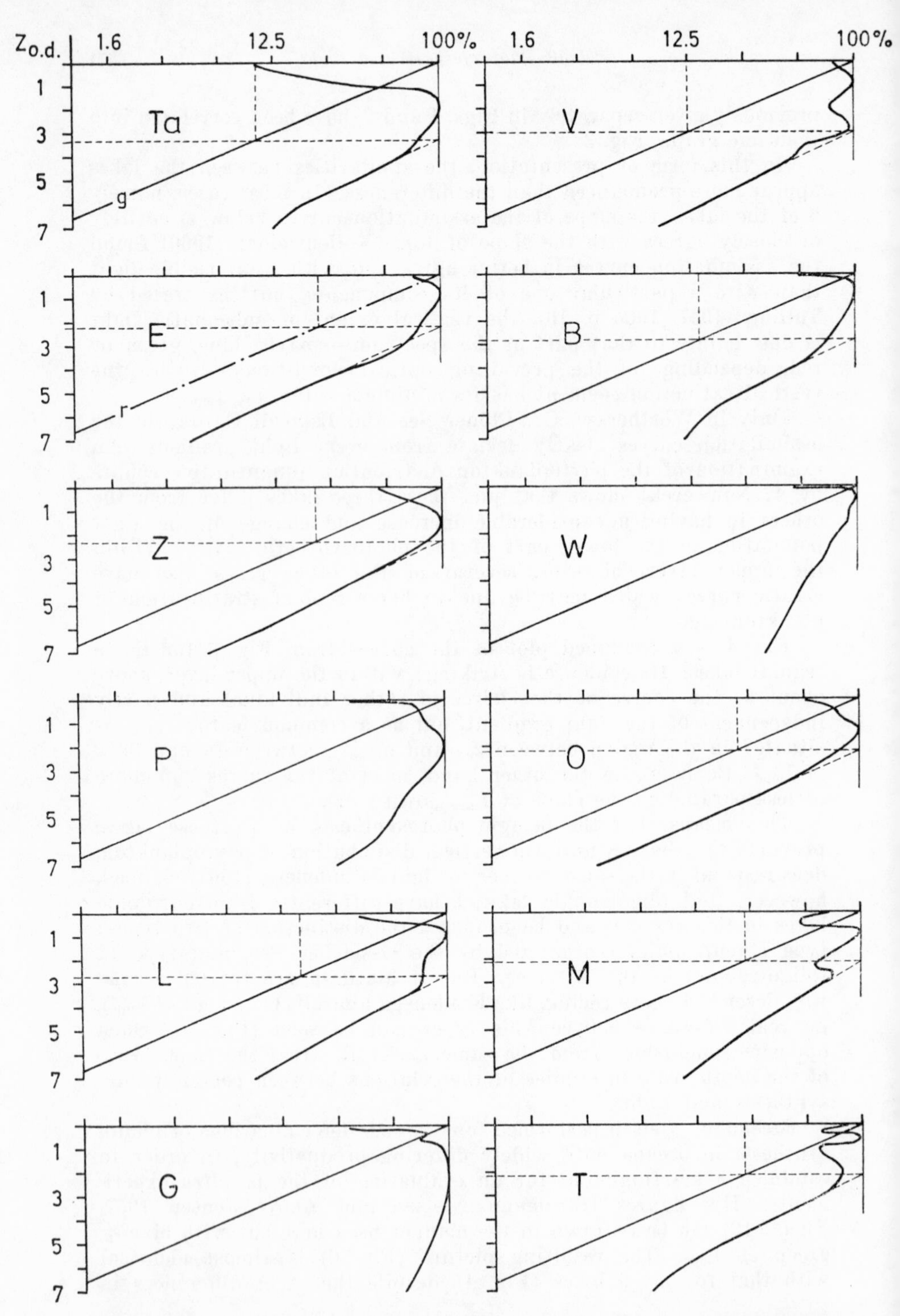

Fig. 3. - Relative photosynthetic rates (from Fig. 1) in semi-logarithmic graphs, plotted against the optical depths of the most penetrating component, I_{mpc} (from Fig. 2). The graphical evaluation of I_K (in units of optical depth and incident I_{mpc}) is indicated.

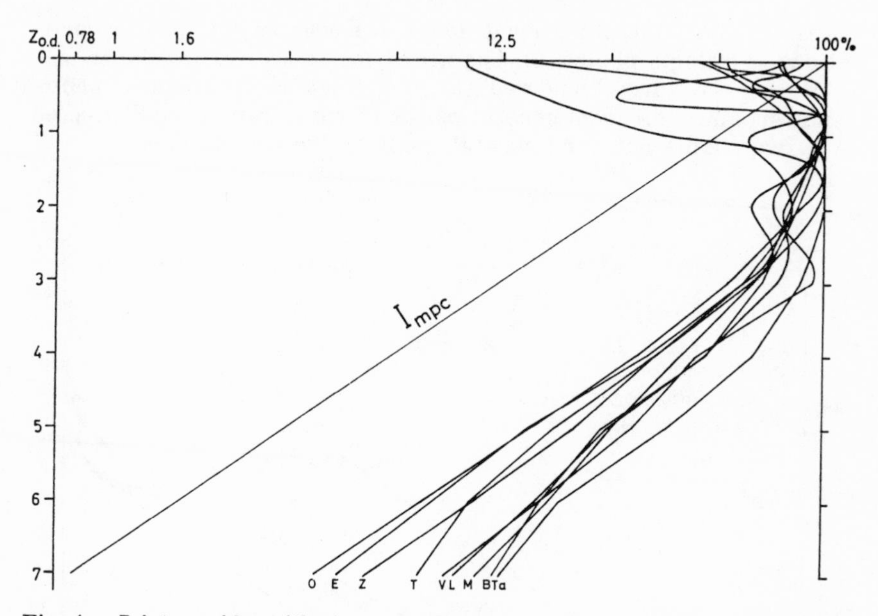

Fig. 4. - Joint semi-logarithmic graph for the «regular» curves of Figure 3.

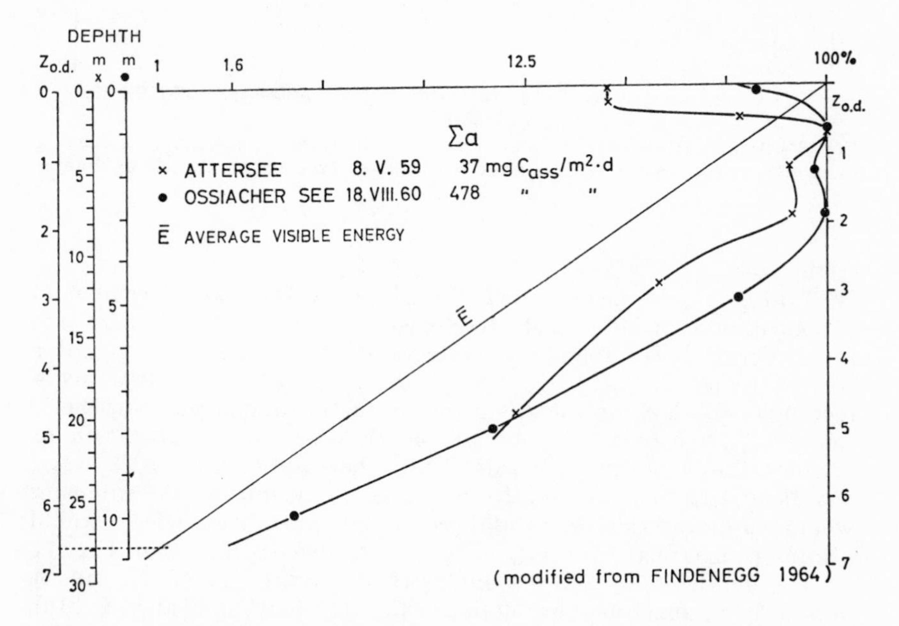

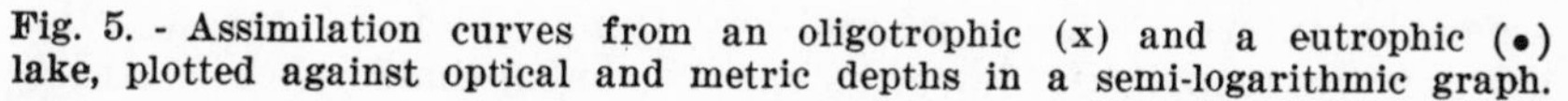

Fig. 5. - Assimilation curves from an oligotrophic (x) and a eutrophic (•) lake, plotted against optical and metric depths in a semi-logarithmic graph.

metric depths and other conditions. Apparently the photic correlations of pelagic photosynthesis are equal in oceans and lakes.

So far we have considered the relationships of planktic photosynthesis and light from general points of view. Now I shall proceed to a few simple but fundamental quantitative correlations.

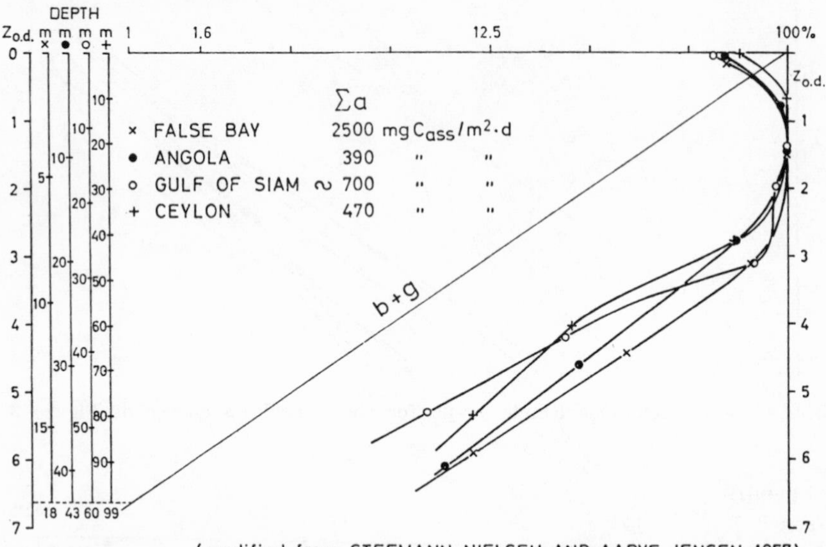

Fig. 6 - Assimilation curves from oceans of different productivity, according to *in situ*-measurements with exposures of sub-surface water at various depths.

Talling (1957) has studied the phytoplankton as a compound photosynthetic system and developed a mathematical model for its structure. He found the integral photosynthesis in a water column to be represented, with regard to its graphical area, by a rectangle, of which one side is given by the maximum photosynthetic rate (a_{max}), the other side by the depth where light saturation of photosynthesis begins. At this depth (here designated $z_{0.5\,I_K}$), I_{mpc} has about half the value of I_K, which is, by definition, the intensity where photosynthesis in the absence of light saturation effects would attain its maximal rate, a_{max} (Fig. 7; cf. also Talling 1957 Fig. 1).

The depth of I_K and its intensity in percentage of I'_0 can be obtained, as suggested by Vollenweider (1960, Fig. 1 and p. 219), from the intersection, in a semi-logarithmic plot, of the a_{max}-line and the extrapolated linear slope of suboptimal photosynthesis

(Fig. 7: D). Applied to our regular assimilation curves (Fig. 3), this procedure yields I_K-values more or less close to 18% and corresponding optical depths around 2.5. Thus our mean value for 0.5 I_K turns out to be 9.1% at $z_{0.5\,I_K} = 3.5$ (Table 1, the first two lines).

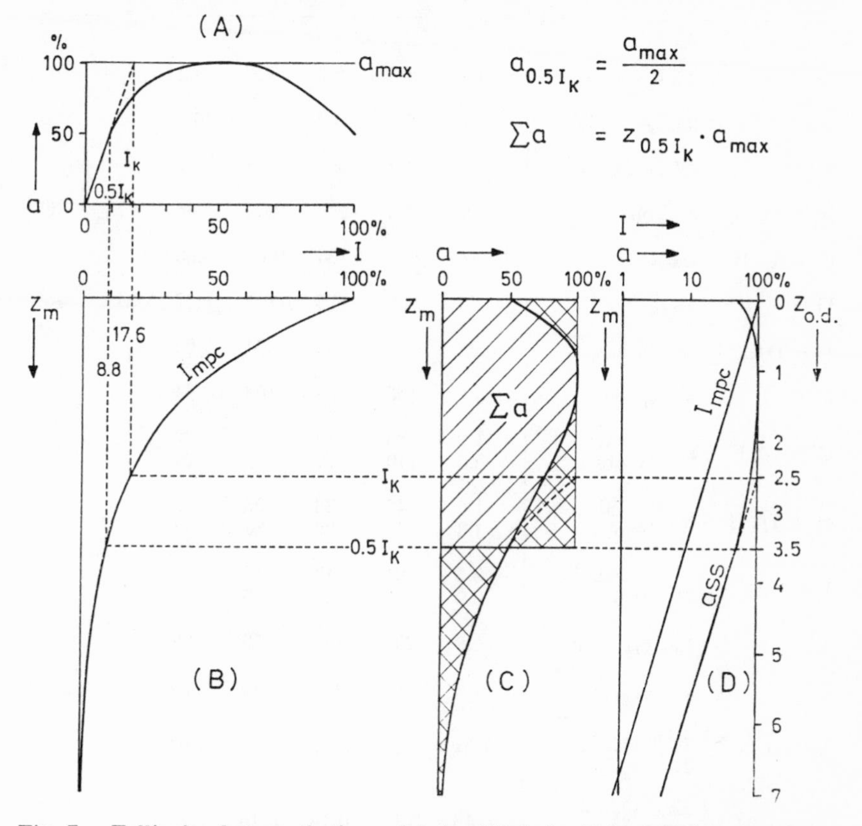

Fig. 7. - Talling's photosynthetic model (modified for light inhibition) and its application for assessing standard correlations (0.5 I_K is here assumed to be at $I_{mpc} = 8.8\%$ I'_0 which corresponds to $z_{o.d.} = 3.5$).

The model presented in Fig. 7 has been made with 0.5 I_K equal to 8.8% which is the exact value ($2^{-3.5} \times 100$) for I_{mpc} at $z_{3.5}$. As an essential feature of the model, the assimilation rate at 0.5 I_K equals 0.5 a_{max}. This holds very nearly for the 9 profiles discussed here: the observed values for their $z_{0.5\,I_K}$ keep rather close to a mean of 0.53 a_{max} (Table 1, item O). We therefore can test the model by means of these assimilation curves.

Our first step is to determine the metric depth of 0.5 I_K, obtainable from the graphs of Fig. 3 and Fig. 2. Multiplication of this

Table 1. - Standard analysis of pelagic photosynthesis, performed on the basis of Talling's model and the observations illustrated in Figs. 1-3.

parameters		Ta	V	E	B	Z	W	P
	$0.5\ I_K : z_{o.d.}$	4.3	4.0	3.2	3.8	3.3		
	$0.5\ I_K : \%\ I_0'$	5.1	6.3	10.9	7.2	10.2		
A	$0.5\ I_K : z_m$	6.4	4.4	2.7	5.2	5.7		
B	a_{max} obs	7.2	365	272	205	86	71	67
C = A · B	Σa_{calc}	46	1606	734	1066	490		
D	Σa_{obs}	45	1660	775	1105	492	811	459
E = D/C	$\Sigma a_{obs}/\Sigma a_{calc}$	0.98	1.03	1.06	1.04	1.00		
F	$a_{0.5\ I_K}$ obs	3.8	165	150	102	44		
G = 1/2 F	$a_{1/4\ I_K}$ calc	1.9	83	75	51	22		
	$a_{1/4\ I_K}$ obs	2.0	110	72	58	24		
H = 1/2 G	$a_{1/8\ I_K}$ calc	1.0	42	38	26	11		
	$a_{1/8\ I_K}$ obs	1.3	59	36	37	14		
I = 1/2 H	$a_{1/16\ I_K}$ calc	0.5	21	19	13	6		
	$a_{1/16\ I_K}$ obs	0.9	31	20	26	7		
K	$\Sigma a_{0.5\ I_K - 7}$ calc	6.9	232	149	156	92		
L	$\Sigma a_{0.5\ I_K - 7}$ obs	7.2	289	145	180	99		
M = L/K	$\frac{\Sigma a_{0.5\ I_K - 7}\ obs}{\Sigma a_{0.5\ I_K - 7}\ calc}$	1.04	1.25	0.97	1.15	1.08		
N = L/D	$\frac{\Sigma a_{0.5\ I_K - 7}\ obs}{\Sigma a_{obs}}$	0.16	0.17	0.19	0.16	0.20		
O = F/B	$\frac{a_{0.5\ I_K}\ obs}{a_{max}\ obs}$	0.53	0.45	0.55	0.50	0.51		
A′	$10\%\ I_0' : z_m$	4.8	3.6	2.8	4.6	5.8	8.0	4.4
C′ = A′ · B	$\Sigma a_{calc'}$	35	1314	762	943	499	568	295
E′ = D/C′	$\Sigma a_{obs}/\Sigma a_{calc'}$	1.29	1.26	1.02	1.17	0.99	1.43	1.56
F′	$a_{0.1\ I_{mpc}}$ obs	5.7	255	140	124	42	44	52
O′ = F′/B	$\frac{a_{0.1\ I_{mpc}}\ obs}{a_{max}\ obs}$	0.79	0.70	0.51	0.60	0.48	0.62	0.78

	O	L	M	G	T	mean	dimensions
	3.2	3.6	3.1		3.1	3.5	optical depth of I_{mpc}
	10.9	8.0	11.7		11.3	9.1	% of incident I_{mpc}
A	7.9	14.5	12.7		10.9		metric depth of I_{mpc}
B	67	65	32	8.1	6.8		mg $C_{ass} \cdot m^{-3} \cdot d^{-1}$
C	529	943	406		74		mg $C_{ass} \cdot m^{-2} \cdot d^{-1}$
D	493	917	412	247	75		mg $C_{ass} \cdot m^{-2} \cdot d^{-1}$
E	0.93	0.97	1.01		1.01	1.00	ratio
F	33	37	19		3.9		mg $C_{ass} \cdot m^{-3} \cdot d^{-1}$
G	17 17	19 21	10 12		2.0 2.1		mg $C_{ass} \cdot m^{-3} \cdot d^{-1}$
H	8 9	10 12	5 8		1.0 1.1		mg $C_{ass} \cdot m^{-3} \cdot d^{-1}$
I	4 4	5 8	3 5		0.5 0.7		mg $C_{ass} \cdot m^{-3} \cdot d^{-1}$
K	76	184	93		17		mg $C_{ass} \cdot m^{-2} \cdot d^{-1}$
L	78	200	118		18		mg $C_{ass} \cdot m^{-2} \cdot d^{-1}$
M	1.03	1.09	1.27		1.06	1.10	ratio
N	0.16	0.22	0.29		0.24	0.20	ratio
O	0.49	0.57	0.56		0.57	0.53	ratio
A′	8.2	13.4	13.4	19.0	11.5		metric depth of I_{mpc}
C′	549	871	429	154	78		mg $C_{ass} \cdot m^{-2} \cdot d^{-1}$
E′	0.90	1.05	0.96	1.60	0.96	1.07	ratio (W, P, G excl. from the mean)
F′	30	43	17	7.8	3.5		mg $C_{ass} \cdot m^{-3} \cdot d^{-1}$
O′	0.45	0.66	0.54	0.96	0.51	0.58	ratio (W, P, G excl. from the mean)

parameter (Table 1, item A) by the corresponding value for a_{max} (item B, from Fig. 1) gives the calculated rate of total photosynthesis, Σa_{calc} (item C), down to $z_{o.d.} = 7$ (= 0.8% I_{mpc}). When we compare the results of our estimates with the observed rates (item D, from Fig. 1), we find quite a satisfactory agreement: the mean of the ratios happens to be exactly unity, but more significant is that the individual deviations from the mean do not exceed more than ± 6-7 % (item E).

Since the range of assimilation covered by our test samples is almost 40-fold (1660/45), the correlation

$$\Sigma a = z_{0.5\, I_K} \cdot a_{max} \quad (1)$$

appears to work generally; the more uniform the phytoplankton, the better the correlation. On the other hand, our three irregular assimilation curves (W, P, and G) are examples of cases where the standard correlation must fail due to the indeterminability of $z_{0.5\, I_K}$.

It will certainly often occur that the depth of 0.5 I_K remains unknown because of insufficient information as to the assimilation gradient. That is usually the case in regional surveys and applied studies, when simplification pays more than highest accuracy and the number of assimilation exposures must be kept to a minimum. Then a more schematic calculation can be justifiable and useful. Not only the restricted material presented here but also the results of Talling and others (cf. Talling 1965) indicate that 0.5 I_K, in general and on an average, is to be found at an optical depth between 3.0 and 3.5. This corresponds to 12.5 and 8.8% of incident I_{mpc}, and we may therefore, without great error, use the metric depth of 10% I_{mpc} for the schematic formula

$$\Sigma a = z_{0.1\, I_{mpc}} \cdot a_{max} \quad (2)$$

In Table 1 (items A′, C′, E′) total photosynthesis has been reestimated according to this simplified correlation. As expected, the calculated values for Σa now differ more from the observed ones than they did when the actual depths of 0.5 I_K were used as parameter, in particular for the most turbid lakes (Ta and V) where $z_{0.5\, I_K}$ is considerably deeper than $z_{0.1\, I_{mpc}}$ In such cases formula (2) must give an underestimation of total photosynthesis. The same occurs if the phytoplankton is distributed as in W, P, and G. To avoid the use of formula (2) in such cases, an additional assimilation exposure at $z_{0.1\, I_{mpc}}$ would be required: if this check reveals a photosynthetic rate much higher than half of the rate obtained at optimal light (at about $z_{0.5\, I_{mpc}}$), the formula is not recommendable (cf. items E′, F′, O′). With these restrictions properly observed, however, our schematic derivation from Talling's theoretical model may prove to serve several practical purposes.

Another application of the model is a comparison of the observed photosynthetic rates below 0.5 I_K with the assimilation gradient due only to light attenuation (Table 1, items F, G, H, I, and Fig. 8). In this region of direct proportionality between light intensity and photosynthesis, the assimilation rates are bound to decrease 50% within every layer corresponding to one unit's interval on the optical depth scale, unless changes in population density or light adaptation do affect the rates. Such aberrations are of minor importance for the gradients illustrated in Fig. 8, but also in these lakes there is a general tendency (most clearly shown in V, B, and M) for the observed rates to exceed the calculated light-dependent gradient. That trend is expected to be enhanced when thermal stratification of the lower euphotic layer favors algal accumulation and adaptation in these strata. Our standard correlation provides a starting point for close and quantitative studies of such relationships.

The areas delimited, at their right, by the dashed and full lines in Fig. 8 represent the calculated and the observed rates of photosynthesis in the layer from $z_{0.5\,I_K}$ down to $z_{o.d.} = 7$ (= 0.8% I_{mpc}). Their values are stated in items K and L of Table 1, and from item M it is seen that the positive effects of « non-light » factors in this layer amount to 10% for the average of all the 9 profiles and to 15-27% for V, B, and M.

For the situation developed in Fig. 7:C (with 0.5 I_K at $z_{o.d.} = 3.5$) it can be derived, with calculus, that the area below $z_{0.5\,I_K}$ enclosed by the assimilation curve covers 18.7% of the assimilation area from surface down to $z_{o.d.} = 7$, i.e. $\Sigma a_{0.5\,I_K-7} = 0.19\ \Sigma a$. Most of the corresponding values from our regular lakes are close to this theoretical value, with a mean of 0.20 Σ a (item N).

Finally, we have to consider the error caused by drawing the lower limit of the photosynthetic integral at $z_{o.d.} = 7$. It can be shown that an additional area equal to $0.088\ \Sigma a_{0.5\,I_K-7} = 0.017\ \Sigma a$ would be obtained, if the assimilation curve (of Fig. 7: C) is extrapolated beyond z_7 to approach zero asymptotically. An error of this magnitude is negligible in the calculation of total photosynthesis.

It has already been stated that the standard correlations for pelagic photosynthesis developed here are based on the model of Talling (1957). They are also in essential agreement with the formulas derived by Steemann Nielsen (1952 p. 132, Steemann Nielsen and Aabye Jensen 1957 p. 61) and by Vollenweider (in Rodhe, Vollenweider and Nauwerck 1958 p. 315). The correlations do not add much to the present theory of pelagic photosynthesis, but it is hoped that they will facilitate and increase its application.

In this aim our standard correlations may serve a twofold purpose: 1) to compare the results of detailed observation with theory

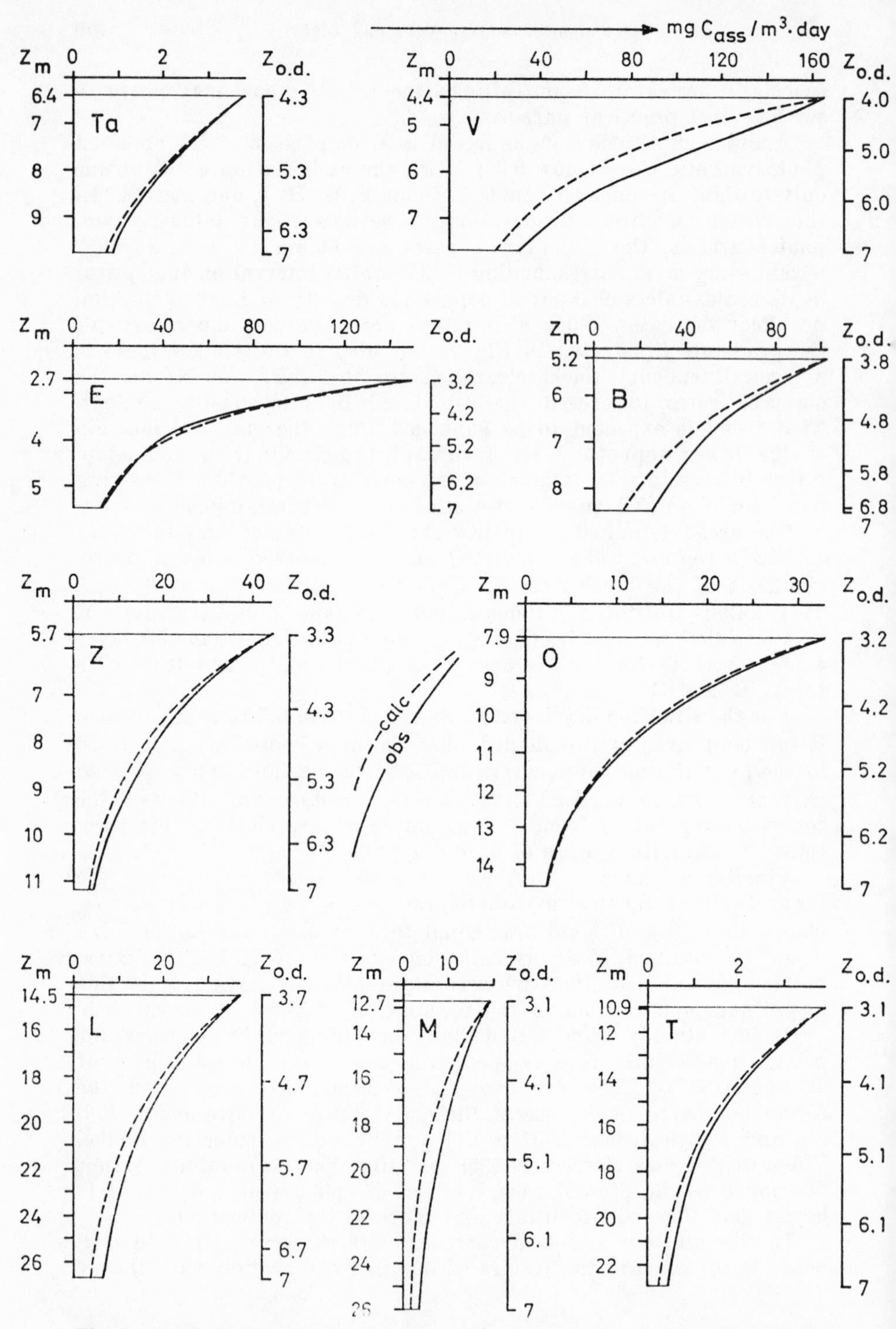

Fig. 8. - Assimilation gradients in the layer of direct proportionality between photosynthetic rate and light intensity ($z_{0.5\,I_K-7}$) . Observed curves from Fig. 1, calculated curves from Table 1.

in order to gain better insight into photosynthetic dynamics of natural waters: deviations are more easily defined and interpreted if referred to a standard; and 2) to enable simple but adequate calculations of pelagic photosynthesis by means of the two basic parameters, optical depth and optimal photosynthesis, from which the entire model can be constructed.

Acknowledgement

My most sincere thanks are expressed to Miss Anna Panders, laboratory assistant at the Institute of Limnology in Uppsala, for skillful help with C^{14}-countings and associated work for more than ten years. Miss Panders has also drawn the figures illustrating this paper, which hardly could have been written without her unselfish assistance.

The preliminary draft presented at Pallanza has been revised after discussions with Dr. J. F. Talling and Dr. R. A. Vollenweider. I thank them cordially for valuable comments and criticism.

REFERENCES

Findenegg, I. 1964. Produktionsbiologische Planktonuntersuchungen an Ostalpenseen. - Int. Revue ges. Hydrobiol. *49*: 381-416.

Rodhe, W. 1958. Primärproduktion und Seetypen. - Verh. Int. Ver. Limnol. *13*: 121-141.

— R. A. Vollenweider, and A. Nauwerck. 1958. The primary production and standing crop of phytoplankton. *In* A. A. Buzzati-Traverso, (ed.) Perspectives in Marine Biology, 299-325. - Univ. of California.

Steemann Nielsen, E. 1952. The use of radio-active carbon (C^{14}) for measuring organic production in the sea. - Jour. Cons. Int. Explor. Mer *18*: 117-140.

— and A. Aabye Jensen. 1957. Primary oceanic production. - Galathea Report. *1*: 49-136.

Talling, J. F. 1957. The phytoplankton population as a compound photosynthetic system. - New Phytologist. *56*: 133-149.

— 1965. The photosynthetic activity of phytoplankton in East African Lakes. - Int. Revue ges. Hydrobiol. *50*: 1-32.

Vollenweider, R. A. 1960. Beiträge zur Kenntnis optischer Eigenschaften der Gewässer und Primärproduktion. - Mem. Ist. Ital. Idrobiol. *12*: 201-244.

— 1961. Photometric studies in inland waters. I. Relations existing in the spectral extinction of light in water. - Ibid. *13*: 87-113.

NOTES ON SOME THEORETICAL PROBLEMS IN PRODUCTION ECOLOGY

J. H. STEELE
Marine Laboratory, Aberdeen

For bibliographic citation of this paper, see page 10.

Abstract

From simple theoretical consideration of plankton dynamics a set of critical factors is deduced. Apart from the usual physical factors, incident radiation, mixed layer depth and nutrients, the results suggest the need for more measurements of light attenuation as a function of pigment concentration, the ratio of maximum photosynthesis to respiration in natural populations and the ratio of filtering rate to respiration rate for herbivores. The most important problem, however, which affects all interpretation of observations, is the need for an independent estimate of the ratio of living plant material to detritus.

INTRODUCTION

The purpose of these notes is to examine some problems raised by simple theoretical pictures of phytoplankton and zooplankton dynamics, particularly in relation to the type of data to be collected in any large sea area. But before beginning a discussion of methods of analysing data collected at sea, such as nutrients, chlorophyll *a* or C^{14} assimilation, there is the problem of the relation of such data to laboratory experimental results. It is hoped that with sufficient information on the response of a wide range of species to variations in conditions as observed in culture, it may be possible to synthesize the response of mixed populations to changes in the natural environment. Apart from the experimental problems of imitating the natural environment, we must also consider the extent to which we can identify the major components of the natural population, a problem which obviously is basic to the interpretation of experimental evidence.

The evaluation of species counts involves two factors — the conversion of the counts to some integrated unit such as volume, or beyond this, carbon content — and the comparison with some chemical estimate such as chlorophyll *a*, carbon, or C^{14} uptake. Far too few such comparisons have been attempted; they suggest, however, that, depending on latitude, a very variable proportion of the photosynthetic population may be found by examination of preserved samples. Thus Hulburt, Ryther & Guillard (1960) for the Bermuda area report that: « ... if the chlorophyll values have any validity, this would indicate that our total cell counts include but a small fraction in terms of biomass as well as numbers of the autotrophic organisms ». Gillbricht (1952), comparing cell counts converted to carbon with chlorophyll estimates, obtained carbon to chlorophyll ratios by weight of 4 : 1 for diatoms and 12 : 1 for dinoflagellates in the North Sea. One must assume that the apparent chlorophyll contents are too high by a factor of about five, suggesting that a large part of the population was not included. On the other hand, in the Arctic Norwegian Sea in June during the diatom outburst, Paasche (1960) found good statistical agreement between cell volume or cell surface area and C^{14} uptake (Berge 1958). Further, the relation of the two fitted fairly well with physiological data on growth rates of phytoplankton although calculated division rates would be about one per day over the water column, which may seem a little high.

Such inconsistencies, as Hulburt *et al.* and others point out, are related to the relative abundance of « nannoplankton » but they can render suspect conclusions based on cell counts and so make it difficult to apply experimental results to natural situations. Also, they make it particularly difficult to try and separate by direct observation the living from the detrital material. Recently, Riley, Wangersky and Hermert (1964) have considered the relation between particulate carbon measured chemically, cell counts expressed as volumes and the area of flakes of detrital material. The apparently small percentage of carbon present as living phytoplankton led the authors to derive a relation between remaining carbon and « detritus » expressed as area of flakes. This relation, however, would imply that the flakes, described as on the average 50 μ at the largest diameter, were 200 μ thick. If the flakes are in fact less than 20 μ thick then a large part of the particulate organic matter must be present in some other form of detritus, or as living cells.

Part of the problem arises from the fact that the identifiable part of the phytoplankton may vary in quantity with the total so that valid regressions can be obtained — such as Gillbricht's or Riley's. This may have the advantage of indicating that the response of this part may be taken as typical of the whole but on the other hand it makes it more difficult to disentangle the interrelations of living and dead organic matter.

Similarly, of course, it means that interpretations of data on the chemical composition of particulate organic matter are at best indirect and generally open to argument — to say the least. This seriously affects the usefulness of such data. It is possible to regard the question purely as one of prediction of relations among the variables measured in nature, particularly those relations between physical and biological parameters. But such « internal » predictions are unsatisfying, unless they can be explained in terms of some model of the phytoplankton population. It can be shown that in the North

Sea and N.E. Atlantic the regression of chlorophyll *a* on particulate carbon can be predicted as a function of phosphate concentration and incident light (Steele and Baird, in press). This can be regarded purely as the ability to predict one biological parameter on the basis of two physical ones, but it is really of interest in terms of possible explanations, either as variations in the ratio of chlorophyll to living plant carbon or as variations in quantity of detritus associated with the plants. It is of importance since ultimately the need is to be able to generalize such predictions.

Thus this problem, the lack of relation between chemical analyses and species counts, affects all discussion of attempts to find patterns in the chemical and bio-chemical data which are the stock in trade of the ecologist. Apart from this, the problem is to relate variables in the physical environment such as light, temperature, nutrients with biological ones, such as rate of production, standing crop, detritus, and to see if one is able to predict the latter from the former.

THEORETICAL CONSIDERATIONS

The data we can expect are, firstly, weather data, beginning with average radiation data for, say, the North Atlantic. Secondly, data on the depth of mixed layer of which Fig. 1 shows some features

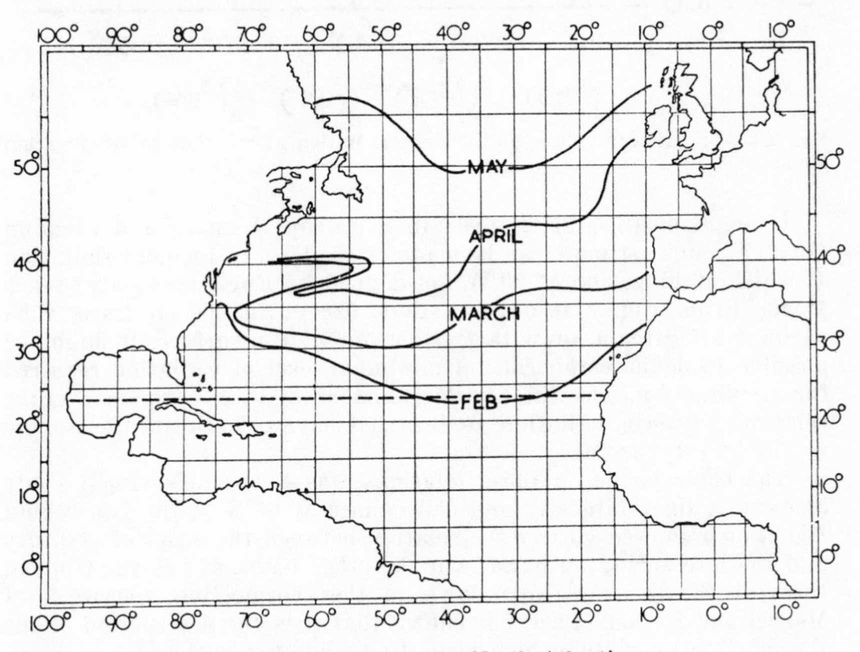

Fig. 1. - The timing of stabilization in the North Atlantic.

taken from monthly averages calculated by Fuglister (1949) from the bathythermographs in Woods Hole collection up to 1949. This illustrates the front of stabilization (> 450 to 150 ft.) that moves up the North Atlantic in spring and must be a major feature in determining the course of the productive cycle. Although the general relations between light, stability and production are « classical » and well accepted, very little has been done to discover what features of the productive cycle could be predicted and exactly what information is needed to make this prediction.

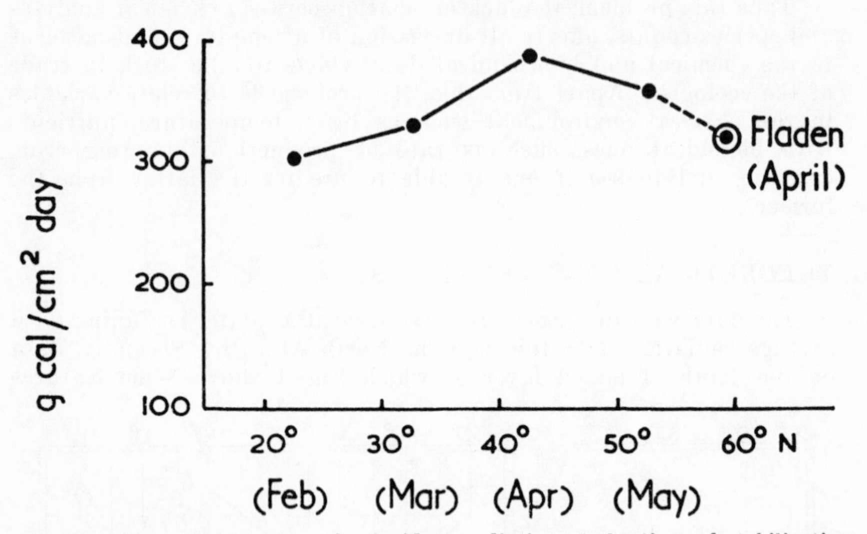

Fig. 2. - Approximate values for incident radiation at the time of stabilization along 30°W.

Using Kimball's tables given in Sverdrup, Johnson and Fleming (1942) a rough estimate can be made of the level of incident radiation at which stabilization at 30°W (as defined by Fuglister's data) takes place. From Fig. 2 it appears that, except in special areas such as the Gulf Stream, or with fresh water at the surface, it might be possible to define empirically a minimum level of radiation required for stabilization. The further development of even a rough average relation between radiation and mixed layer depth in the spring would be very useful.

The basic paper on these relations was Sverdrup's (1953) study of mixing, light intensity and cell counts at 66°N in the Norwegian Sea, which showed clearly the relation between the onset of stability and the late spring outburst. On the other hand, at the sub-tropical limit of the seasonal breakdown in the thermocline, the work of Menzel and Ryther (1960) has shown that it is the destruction of the thermocline in winter which leads to a plankton outburst.

For presentation here Sverdrup's ideas have been adapted in two ways. Firstly, he assumed photosynthesis proportional to light intensity; I have taken a curve for photosynthesis/unit chlorophyll *a* with a maximum p_m (gC/g chlor. day) at a light intensity I_m (g cal/cm^2 day)

$$p = p_m \, I/I_m \exp{(1 - I/I_m)}$$

In the following argument a different shape of curve, such as Talling (1957) has used, would not greatly alter the conclusions. Unlike Sverdrup, p_m rather than « compensation light intensity » has been taken as the factor defining the rate of photosynthesis.

Expressing respiration r in the same units, it can be shown that the critical depth defined by Sverdrup as the depth of the water column within which photosynthesis balances respiration,

$$z_c = \frac{p_m}{r} \cdot \frac{e}{k} \cdot [1 - \exp{(- I_o/I_m)}]$$

where I_o is the incident radiation at the surface and k is the attenuation coefficient.

From this formula, to compare critical depth with mixed layer depth, assuming incident radiation in known, there are three other variables to be estimated. I_m is the lowest radiation at which « saturated » photosynthesis occurs. For the argument here it can be taken as the incident radiation at the surface which will produce saturation just subsurface. For the North Sea there is evidence that this occurs about 90 g cal/cm^2 day (Steele 1962). Rodhe, Vollenweider & Nauwerck (1958) have suggested much lower values for Arctic lakes. k is, effectively, the attenuation of energy utilizable in photosynthesis but this is difficult to relate to actual physical measurements. For the moment, following Sverdrup, it is taken as less than 0.10 for winter phytoplankton densities. The third variable related to phytoplankton physiology is the ratio

$$\frac{p_m}{r} = \frac{\text{max. photosynthesis} \quad (\text{g C/g chlor. } a \text{ day})}{\text{respiration rate} \quad (\text{g C/g chlor. } a \text{ day})}$$

Thus « production » estimates by themselves are insufficient to determine z_c. Using Steemann Nielsen and Hansen's suggestion (1959) that this ratio has a value close to 10 for healthy populations at I_m and taking the 24 hour value to be half this, then

$$z_c \geq 136 \, [1 - \exp{(- I_o/I_m)}]$$

which is shown assuming equality in Fig. 3. The reason for these rather rough calculations is to suggest the kind of result that might be obtained. On this basis, comparing Fig. 2 and 3, the changes in light intensity, about the time stabilization occurs, have little effect on the critical depth, and it is the variation in depth of the mixed layer which is dominant.

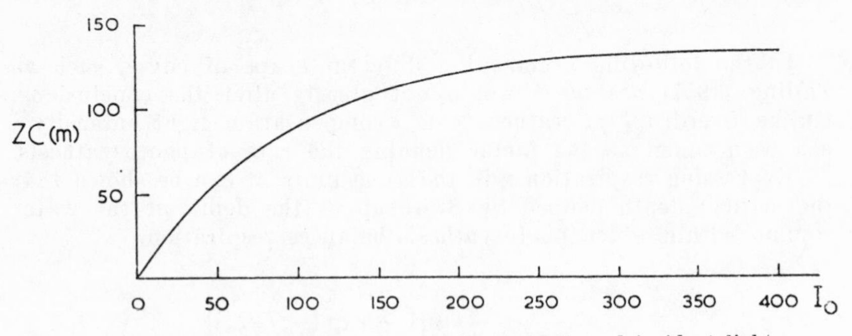

Fig. 3. - A theoretical relation between critical depth and incident light.

Because, in many areas, light may not be limiting, the depth of the mixed layer determines, not whether or not production will occur, but really what level of production can theoretically be achieved. This is best expressed as a relation between mixed layer depth z^* and chlorophyll *a* concentration C which produces maximum production and takes the form:

$$z^* = \frac{p_m}{r} \cdot \frac{e}{k} \cdot [1 - \exp(1 - I_o/I_m)] \left[1 - \frac{C}{k}\,\frac{dk}{dC}\right]$$

Steele & Menzel (1962) have discussed the general consequences of this type of relation. p_m is here taken to be the rate of production of particulate organic carbon within the cell. It is this rate which presumably will determine the rate of growth of the population so that for this purpose we can neglect the problems of extracellular production of organic material.

In this more general form of Sverdrup's relation, the extra factor involves the rate of change of k in terms of the chlorophyll concentration. There have been several attempts to define such a relation (Ichimura & Saijo 1959). Riley & Schurr (1959) in particular give

$$k = .04 + .0088\,C + .054\,C^{2/3}$$

Jerlov's (1951) work on the relation between water types, spectral distribution of attenuation and the absolute values of attenuation coefficients suggests strongly that fairly general relations between pigment concentrations and energy attenuation might exist, at least in the open ocean. It is possible, however, that concentration of particulate matter might also need to be known. The derivation of such a relation with limits set on its accuracy and also on its areas of application would seem one of the main fields of possible development towards applying data on mixed layer depth and incident radiation. For the discussion here, however, Riley's formula is used.

The remaining factor, which determines the chlorophyll concenntration for optimum production, and which is the factor concerned directly with the physiological state of the population, is again p_m/r, the ratio of maximum photosynthesis to respiration over a 24 hour period. Steemann Nielsen and Hansen (1959) have presented evidence which implies that this factor could be considered constant. The experiments consisted of 4 hours incubation in artificial light with respiration estimated from the extrapolation of C^{14} uptake values onto the negative part of the « production » axis. The value for the ratio was approximately 10%. On a 24 hour basis the value could be considered to lie between 5 and 10%.

In Fig. 4 equation (2) is applied to data for December along 30°W. using Fuglister's data and Kimball's tables to derive C^*, the chlorophyll concentration for optimal production in the mixed layer, for varying p_m/r ratios. The general distribution with maximum in the « ridge » at 10°N, and the range of values for $p_m/r = 5$ does not seem unreasonable. However, when the same calculations are made for August (Fig. 5), when light is not limiting, the distribution appears quite at odds with our available knowledge: especially the high values at 30-40°N are utterly different from data for the Sargasso Sea at Bermuda (Menzel & Ryther 1960). In this region chlorophylls in the surface layer are usually less than 0.1 mg/m^3 associated with low nutrient concentrations.

The difference between the estimate of chlorophyll for optimum production and the observed chlorophyll is a measure of the effect of other factors in limiting the population. This divergence will be very severe in areas such as 30-40°N if a constant value for p_m/r is accepted. It would be much smaller if p_m/r varied in relation to the degree of oligotrophy of the environment. There are various possibilities: firstly p_m, the light saturated rate of photosynthesis per unit chlorophyll *a,* could decrease with decreasing nutrient concentration. All the evidence, however, suggests, if anything, an increase from temperate, eutrophic environments to sub-tropical oligotrophic ones. The evidence is not very reliable since, although in temperate areas one usually gets good C^{14}/chlorophyll relations, in lower latitudes these relations are highly variable. This by itself is a very interesting problem. Perhaps the difficulty lies in the phaeophytin content of the « chlorophyll » estimates as Yentsch (1965) has suggested. If p_m does not decrease and if it can be agreed that there is some decrease, say by a factor of 3-5, in the chlorophyll content of phytoplankton, then the other possibility is that respiration in terms of unit chlorophyll increases, since it is reasonable to suppose that respiratory rate is tied to the carbon content of the cell, although Steemann Nielsen and Hansen, in considering p_m/r to be a constant, assume that respiration per unit carbon will decrease. Since this ratio is possibly the main single measure we can obtain of the physiological state of the population, much more relevant than an estimate of « production » on its own, we need to consider (1) whether the « regression » method used by Steemann Nielsen is a valid way of estimating respiration, (2) can it be adapted to a 24 hour basis, and (3) should we try to get a larger number of estimate over a wide range of environments? As

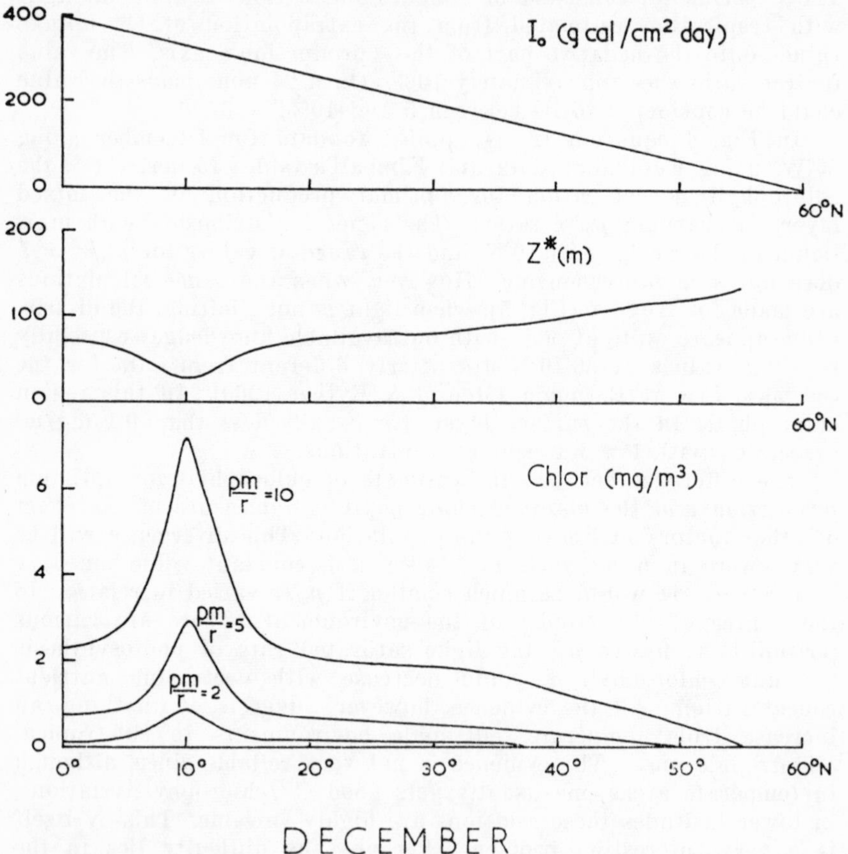

Fig. 4 - Incident light (I_o), mixed layer depth (z^*) and the deduced concentrations of chlorophyll *a* for maximum production at different p_m/r for December.

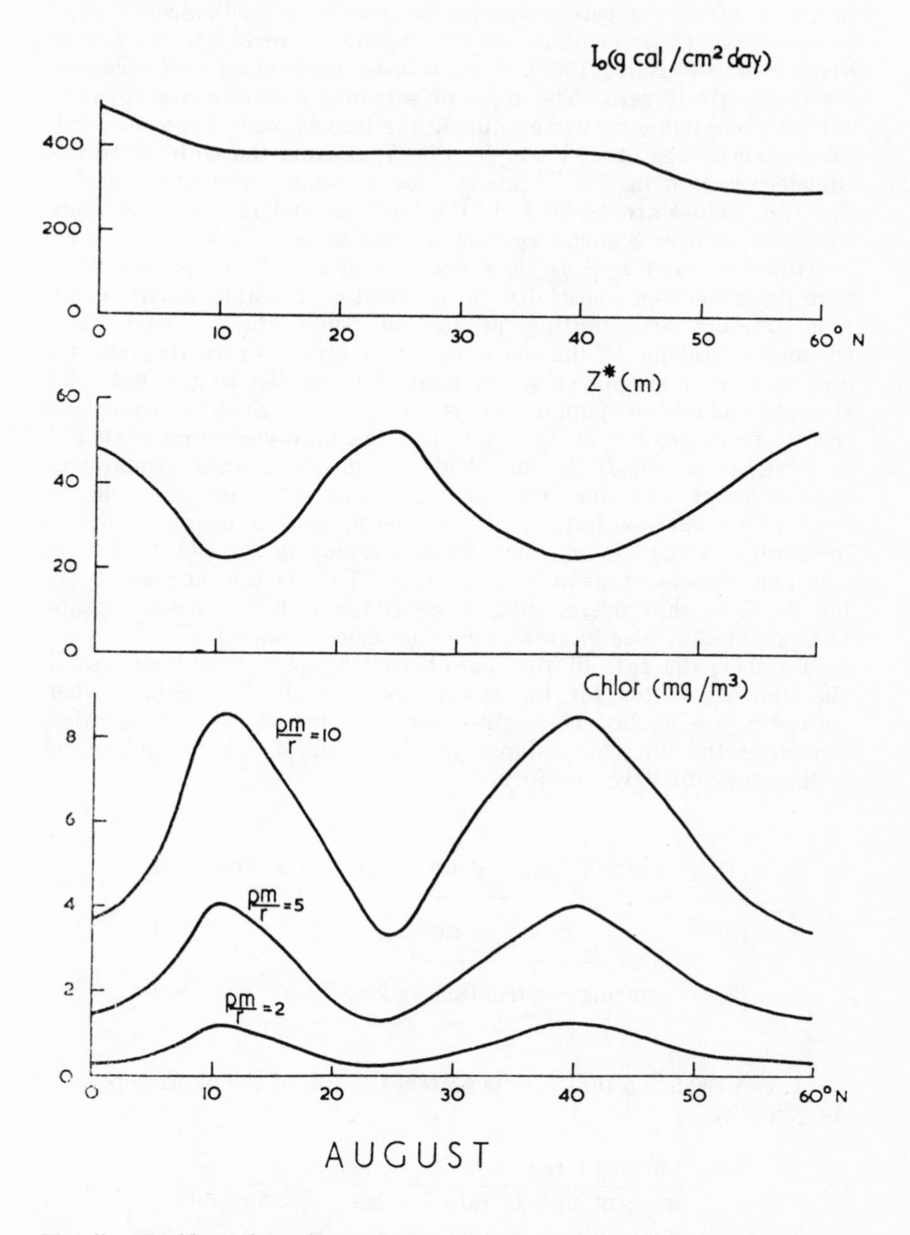

Fig. 5. - Incident light (I_o), mixed layer depth (z^*) and the deduced concentrations of chlorophyll *a* for maximum production at different p_m/r for August.

examples of the possibility of variations in this ratio, Fig. 6 gives values of r/p_m as a percentage for the northern North Sea (Fladen) from *in situ* experiments with the results converted to a 24 hour basis (Steele & Baird 1961). r/p_m is used because on two occasions r was effectively zero. The signs of seasonal variation are apparent with the maximum occurring during the late summer when nutrients were lowest. The other example, Fig. 7, is from the Gulf of Mexico (Steele 1964), using an incubator for a 4-hour incubation period. The r/p_m values are 28, 20 and 21% for 5, 50 and 100 m respectively. The average over 3 such experiments was 26%.

However, even if p_m/r does vary, from Fig. 5 it appears likely that at times and especially in nutrient-poor stable environments other factors are limiting production below the optimum level. Obviously grazing is the main factor likely to exert this control, and on a very simple view one might be tempted to conclude that the phytoplankton population is overgrazed. Such a conclusion results from neglect of the fact that the plant-herbivore system is existing as a steady state. That is, on an instantaneous view, conditions at any one moment may apparently permit a higher level of production but, on a long term, such a change, such an imbalance in the system, may be deleterious to the plant-herbivore relation considered as a steady state. This is not necessarily so, but to show that overgrazing is occurring it is necessary to show that a different *steady state* would be more productive.

Further, the rate of the plant-herbivore cycle is not necessarily the limiting factor for the system as a whole. In regions where nutrients are at low concentrations, the simplest set of equations depicting the nutrient balance in (1) water (2) phytoplankton (3) herbivores will have the form:

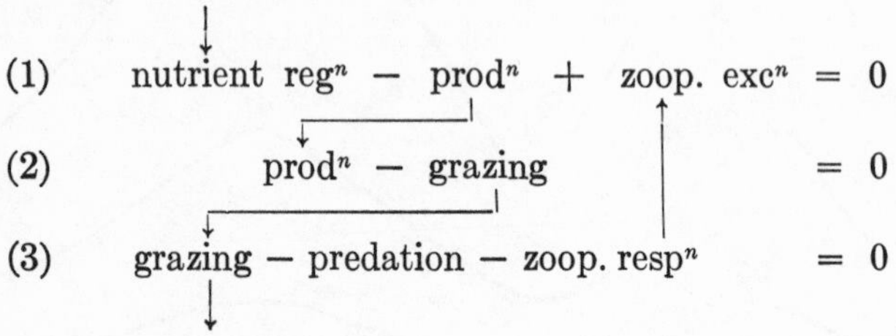

(1) nutrient reg^n − prod^n + zoop. exc^n = 0

(2) prod^n − grazing = 0

(3) grazing − predation − zoop. resp^n = 0

It can be shown that in this system the size of the plant population is given by

$$\frac{\text{nutrient reg}^n\text{ rate}}{\text{nutrient uptake rate}} + \frac{\text{herb. resp}^n\text{ rate}}{\text{herb. filtering rate}}$$

This depends on two components of the system. There is a cycle of nutrients through plant production to herbivorous zooplankton

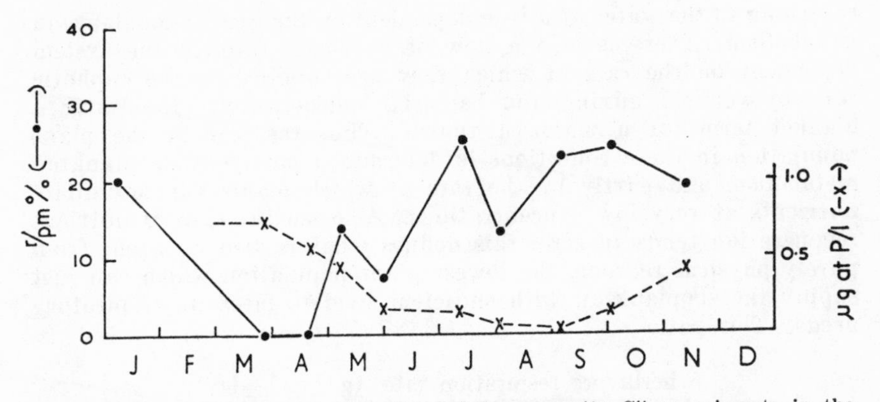

Fig. 6. - Seasonal values of r/p_m deduced from *in situ* C^{14} experiments in the North Sea and compared with phosphate concentrations.

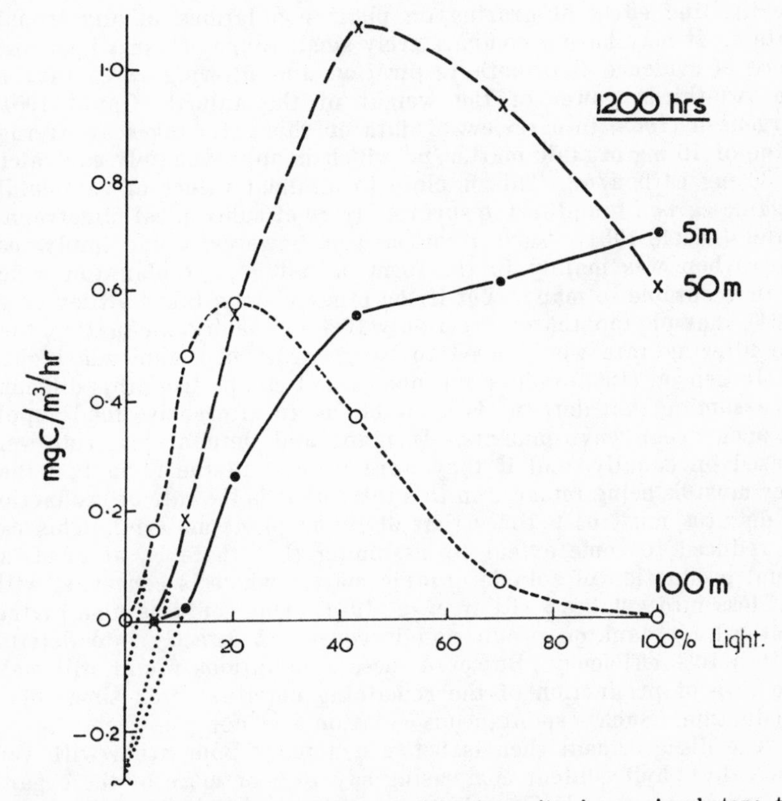

Fig. 7. - C^{14} uptake as a function of light intensity in an incubator; the experiments was started at 1200 hours local time.

returning to the water at a rate dependent on the rate of zooplankton metabolism. There is also a flow of nutrients through the system dependent on the rate at which they are supplied to the euphotic zone by vertical mixing and bacterial regeneration — the latter a blanket term for a major unknown. Thus the size of the plant population in these conditions is determined partly by zooplankton metabolism and partly by the rate at which plants can assimilate nutrients at very low concentrations. Also, as the rate of nutrient regeneration tends to zero, this defines what is also apparent from purely physical reasons, the lowest plant population which can just supply the zooplankton with sufficient food to meet its respiratory needs. This ratio

$$\frac{\text{herbivore respiration rate} \quad (\text{g C/g C} \cdot \text{day})}{\text{herbivore filtering rate} \quad (\text{m}^3/\text{g C} \cdot \text{day})}$$

is probably the main factor that needs to be estimated to discover the limiting effect of grazing on plant populations in oligotrophic waters. It may have a comparatively small range of variations since there is evidence that both respiration and filtering rates vary as the two-thirds power of the weight of the animal (Gauld 1951). Jorgensen (1962) in a review of data on this ratio takes an average value of 70 mg organic matter/m^3 which is approximately equivalent to 30 mg carbon/m^3. This is close to minimal values of particulate organic carbon found in the surface layers of subtropical oligotrophic water (Steele 1964). Such a comparison, however, would imply that the carbon was mainly in the form of living phytoplankton which is unacceptable to many. Yet if the other view is taken (Riley *et al.* 1964), that phytoplankton form only 10% of the organic matter, then the filtering rate would need to be an order of magnitude higher. As Jorgensen (1962) points out, one cannot escape from this dilemma by assuming that detritus is available as an alternative food supply in open ocean environments. If plant and detrital material were grazed on equally, and if they were present in the ratio 1 : 9 then they must be being replaced in this ratio, that is the rate of production of detritus must be 9 times that of living plant material. This can be reduced to some extent by assuming that there is, at most, an equal production of soluble organic matter which is converted without loss into « flakes » (Riley *et al.* 1964). One could assume further that all zooplankton production is converted directly into detritus with a 10% efficiency. But even these assumptions would still make the rate of production of the remaining detritus seven times plant production. Such « spontaneous creation » is not possible.

The disagreement then is between a plant population with very low chlorophyll content comprising say 70% or more of the organic matter, and a population with a « low normal » chlorophyll content comprising say 10-20% of the organic matter with the remaining matter in a refractory form available to zooplankton only very partially or at a very slow rate dependent on bacterial synthesis. This is a difficult question on which, unfortunately, as mentioned at the start, only indirect and inconclusive evidence is available for extreme oligotrophic areas. The only argument in favour of a fairly high proportion of plant material arises from the respiration rate/filtering rate ratio. If the available food were mainly concentrated in 10-20% of the organic matter then the filtering rates would need to be 5-10 times higher which seems unrealistic on the present evidence.

« Detritus » must be important in other respects, for example, just after an outburst, for over-winter zooplankton in high latitudes, and for zooplankton below the euphotic zone; although for the last, the evidence of Vinogradov (1962) of exponential decrease in populations with depth, of increasing size and of carnivorous feeding methods suggests, as he has pointed out, that they are mainly independent of the surprisingly even vertical distribution of deep water detritus.

REFERENCES

Berge, G. 1958. The primary production in the Norwegian Sea in June 1954 measured by an adapted C^{14} technique. - Rapp. P.-v. Réun. Cons. perm. int. Explor. Mer, *144*: 85-91.

Fuglister, F. C. 1949. Average monthly layer depth in the North Atlantic. Woods Hole Oceanogr. Inst. Ref. No. 49-55, 3pp.

Gauld, D. T. 1951. The grazing rate of planktonic copepods. - J. mar. biol. Ass. U. K., *29*: 695-706.

Gillbricht, M. 1952. Untersuchungen zur Produktionsbiologie des Planktons in der Kieler Bucht. - I. Kieler Meeresforsch., *8*: 173-191.

Hulburt, E. M., J. H. Ryther and R. R.L. Guillard. 1960. The phytoplankton of the Sargasso Sea off Bermuda. - J. Cons. prem. int. Explor. Mer, *25*: 115-128.

Ichimura, S. and Y. Saijo. 1959. Chlorophyll content and primary production of the Kuroshio off the southern midcoast of Japan. - Bot. Mag., Tokyo, *72*: 193-202.

Jerlov, N. G. 1951. Optical studies of ocean waters. - Rep. Swed. deep Sea Exped. 1947-1948. III. 1-59.

Jorgensen, C. B. 1962. The food of filter feeding organisms. - Rapp. P.-v. Réun. Cons. perm. int. Explor. Mer, *153*: 99-107.

Menzel, D. W. and J. H. Ryther. 1960. The annual cycle of primary production in the Sargasso Sea off Bermuda. - Deep-Sea Res., *6*: 351-367.

Paasche, E. 1960. On the relation between primary production and standing crop of phytoplankton. - J. Cons. perm. int. Explor. Mer, *26*: 33-48.

Rilev, G. A. and H. M. Schurr. 1959. Oceanography of Long Island Sound. III. Transparency of Long Island Sound waters. - Bull. Bingham oceanogr. Coll., *17*: 66-82.

— P. J. Wangersky and D. V. Hermert. 1964. Organic aggregates in tropical and subtropical water of the North Atlantic Ocean. - Limnol. Oceanogr., *9*: 546-550.

Rodhe, W., R. A. Vollenweider and A. Nauwerck. 1958. The primary production and standing crop of phytoplankton. *In* Perspectives in Marine Biology, ed. Buzzati-Traverso. - Univ. Calif. Press: 299-322.

Steele, J. H. 1962. Environmental control of photosynthesis in the sea. - Limnol. Oceanog. *7*: 137-150.
— 1964. A study of production in the Gulf of Mexico. - J. mar. Res. *22*: 211-222.
— and I. E. Baird. 1961. Relations between primary production chlorophyll and particulate carbon. - Limnol. Oceanog. *6*: 68-78.
— 1965. The chlorophyll *a* content of particulate organic matter in the northern North Sea. - Limnol. Oceanog. (in press).
— and D. W. Menzel. 1962. Conditions for maximum primary production in the mixed layer. - Deep-Sea Res. *9*: 39-49.
Steemann Nielsen, E. and V. Kr. Hansen. 1959. Measurements with the C^{14} technique of the respiration rates in natural populations of phytoplankton. Deep-Sea Res. *5*: 222-233.
Sverdrup, H. U. 1953. On conditions for vernal blooming of phytoplankton. J. Cons. perm. int. Explor. Mer *18*: 287-295.
— M. W. Johnson, and R. H. Fleming. 1942. The oceans; their physics, chemistry and general biology. - New York: Prentice-Hall. 1087 pp.
Talling, J. F. 1957. The phytoplankton population as a compound photosynthetic system. - New Phytol. *56*: 133-149.
Vinogradov, M. E. 1962. The feeding of deep-sea zooplankton. - Rapp. P.-v. Réun. Cons. perm. int. Explor. Mer *153*: 114-120.
Yentsch, C. S. 1965. The relationship between chlorophyll and photosynthetic carbon production with reference to the measurement of decomposition products of chloroplastic pigments. Mem. Ist. Ital. Idrobiol., *18* Suppl.: 323-346.

COMPARATIVE PROBLEMS OF PHYTOPLANKTON PRODUCTION AND PHOTOSYNTHETIC PRODUCTIVITY IN A TROPICAL AND A TEMPERATE LAKE

J. F. TALLING
Freshwater Biological Association
Ambleside, England

For bibliographic citation of this paper, see page 10.

Abstract

Recent information for lakes Victoria (East Africa) and Windermere (England) is used to illustrate some probable effects of climatic and latitudinal differences upon conditions of phytoplankton production and photosynthetic productivity. Average concentrations of phytoplankton (assessed by chlorophyll *a* content, as mg/m^3) in the euphotic zone were often of similar magnitude, although in the temperate lake low values during winter were connected with the seasonal minimum of solar radiation. Seasonal variation in total population density was less pronounced in the equatorial lake, but individual species could vary considerably; a major diatom maximum developed seasonally in both lakes, although under widely differing conditions.

Photosynthetic productivity per unit area ($\Sigma\, nP$) was much higher during the diatom maximum in Lake Victoria than in Windermere, due chiefly to high and probably temperature-dependent rates of photosynthesis at light-saturation per unit of population (P_{max}). Similar rates appeared to be maintained throughout the year and, in conjunction with low light attenuation (k_{min}) and maintained population densities (n), yielded an unusually high estimate for the annual photosynthetic productivity.

Some general measures of production and photosynthetic productivity are discussed, applicable over short or long periods, and useful for climatic comparisons. They include the ratio $n\, P_{max}/k_{min}$, radiation-time integrals, euphotic population ($\Sigma\, n$) - time integrals, and the average photosynthetic activity per unit of euphotic population ($\overline{P}$). The importance of the relative timing of variation in these quantities is emphasized, and analogies are drawn with the results of growth analysis applied to terrestrial vegetation cover. For Lake Victoria, the consequences of a seasonal wind regime probably replace those associated at higher latitudes with incident solar radiation in imposing a fairly regular annual cycle of seasonal variation.

INTRODUCTION

The latitudinal differentiation of climate must obviously exert a major influence upon primary production in freshwaters, as in the sea. Here solar radiation has a critical role, both as the direct supply of energy for photosynthesis and as a more indirect determinant of water temperature, thermal stratification, and associated chemical variation in inland water-bodies. Besides the instantaneous or short-period values of these factors, which are more readily measured and compared with physiological information, the pattern of periodic variation may be often of greater importance. Thus the variation of day-length is a relatively simple, and that of an annually delimited « growing season » a decidedly complex, correlate of latitude.

A systematic and comparative study of limnological productivity in various latitudinal and climatic zones has been suggested as an important aspect of the International Biological Programme. Somewhat similar and much earlier proposals have not been lacking; in particular, a proper execution of the programme suggested by Wesenberg-Lund (1910, pp. 427-429) would have had an incalculable value. Wesenberg-Lund emphasized the importance — and current deficiency — of information from tropical lakes. Even today there is a relative dearth of tropical studies on photosynthetic productivity, and the annual pattern of crop production, to compare with the more numerous accounts for temperate lakes. The following account utilizes some recent information from equatorial Lake Victoria in East Africa, which is described more fully elsewhere (Talling 1965, 1966a), and was obtained through the help and cooperation of the East African Freshwater Fisheries Research Organization at Jinja, Uganda. It is treated comparatively with information from Windermere in northern England (latitude 54°N), to emphasize climatic aspects of the control of primary production in these two lakes.

METHODS

Most of the methods used are listed in other papers which describe work on Windermere (Talling 1960, 1966b) and Lake Victoria (Talling 1965, 1966a). The incident solar radiation was measured using a Moll — Gorczynski type thermopile; underwater light penetration by a selenium rectifier photocell with Schott glass color filters; and water temperature usually by a thermistor-thermometer. Methods of chemical analyses (e. g. for dissolved silica, phosphate and nitrate) are given by Heron (1961) for Windermere, and by Talling (1966 a) and Talling and Talling (1965) for Lake Victoria; essentially similar methods were applied to the two lakes. Nitrate was estimated by the phenol disulphonic acid reaction, which has later been found to yield under-estimates for Windermere (Heron, personal communication). The photosynthetic activity of phytoplankton was measured by oxygen production, using suspended clear and dark bottles and the Winkler oxygen determination. The concentration of phytoplankton cells was obtained by counting with the inverted microscope after iodine-sedimentation. A weighted plastic tube (Lund 1950) was used to obtain integrated water samples over most or all of the euphotic zone in Windermere. The 0-10 m layer was so sampled before 1 June 1964, and the 0-7 m layer after this date.

Some more significant variation existed between the procedures employed in estimating chlorophyll *a*. For Windermere in 1958-59, the diatom cells were filtered out on fine filter paper, from which they were washed, extracted with 90% acetone for approximately 20 hours near 5°C, and chlorophyll *a* calculated from spectrophotometric measurements using the equation of Richards and Thompson (1952). For Windermere in 1964, glass-fibre filters (Whatman GF/C) were used, extracted complete in 90% acetone or in 90% methanol under the same conditions, and chlorophyll *a* estimated from the simplified and revised equations in Talling and Driver (1963). For Lake Victoria during 1960-61 (Talling 1965), fine paper filters were similarly extracted in 90 % methanol, and chlorophyll *a* estimated from the data of Mackinney (1941). In all instances the values obtained depend primarily upon spectophotometric measurements of the optical density of extracts at 665 mμ, and should be closely comparable. An overestimation inherent in the use of the Richards and Thompson equation (Talling and Driver 1963) was largely counteracted by an under-estimate of similar magnitude introduced by losses of diatom cells during transfer (Talling 1966 b). The measurements based on extractions in methanol and acetone have been intercalibrated (Talling and Driver 1963). The former solvent appeared to be about 25% more effective for the summer (but not the spring diatom) phytoplankton of Windermere (cf. Fig. 3), and where available comparisons between the two lakes are based upon methanol extracts.

GENERAL FEATURES OF THE LAKES

The lakes differ in various respects which are largely or completely independent of latitude and climate (see Fig. 1 and Table 1). Although the average and maximum depths are not very dissimilar, differences in area and morphometry are enormous; these are basically due to the relatively recent origin of Windermere by glacial excavation and the more ancient origin of the huge Lake Victoria basin from tectonic movements. In each lake some morphological subdivision of the water-mass exists, with pronounced consequences for the production of phytoplankton. In Lake Victoria a distinction can be made between the offshore lake area and the numerous sheltered bays and gulfs, whereas the elongate form of Windermere is divided into two basins of comparable size by a shallow central region with islands. The present comparison concerns the offshore region of Lake Victoria and the northern basin of Windermere. Locations of the principal sampling stations are shown on Fig. 1.

Some differences in hydrology and chemistry of the lake waters are also relevant to a comparison of phytoplankton production. The replacement of lake by inflow water is rather rapid in Windermere and extremely slow in Lake Victoria. In Windermere North Basin the replacement time (lake volume/mean inflow rate) is of the order of 9 months, and in the latter about 170 years. Indeed, the water income of Lake Victoria is highly peculiar in being derived in large part ($\sim 90\%$) from direct rainfall on the lake surface. Values of the total ionic concentration and (bicarbonate plus carbonate) alkalinity are among the lowest found in the major African lakes (Talling and Talling 1965), but are considerably higher than the corresponding values for Windermere. The comparative poverty of phosphate-phosphorus in Windermere surface water is also notable; the concentrations rarely exceed 4 and during most of the year lie below 2 μg $PO_4 \cdot P$/l (cf. Lund 1950, Heron 1961). Levels in Lake Victoria are generally an order of magnitude higher (Talling 1966 a), and much larger values are commonplace in the East African lakes (Talling and Talling 1965). In offshore surface water of Lake Victoria the concentrations of dissolved silica are relatively high (usually 4-5 mg SiO_2/l) and those of nitrate-nitrogen relatively low (probably usually less than 10-20 μg $NO_3 \cdot N$/l). These two features are probably influenced by climate (cf. Hutchinson 1957, Talling and Talling 1965).

Table 1. - Some characteristics of lakes Victoria and Windermere (North Basin).

Values for the last eight features are from records in 1960-61 for Lake Victoria and in the following periods for Windermere: annual incident radiation, 1964; total ionic concentration and alkalinity, 1955-56; nitrate, phosphate, and silica, 1958-59 (Heron 1961); total iron, July-August 1965; minimum extinction coefficient, 1958-59. Chemical determinations refer to concentrations within the euphotic zones.

	Lake Victoria	Windermere
latitude	½°N-2½°S	54°N
area (km²)	66250	8.05
maximum depth (m)	79	64
mean depth (m)	$\sim$ 40	25
day length (h)	11.9 - 12.2	7.4 - 17.2
incident solar radiation (cal/cm² · year)	154 000	61 750
minimum extinction coefficient, k_{min} (ln units/m)	0.16 - 0.33	0.31 - 0.51
total ionic concentration (meq/l)	1.05 (±0.03)	0.48 - 0.54
alkalinity (meq/l)	0.90 - 0.94	0.15 - 0.19
$NO_3 \cdot N$ (μg/l)	$<$ 5 - 11	70 - 430
$PO_4 \cdot P$ (μg/l)	$<$ 5 - 28	$<$ 0.1 - 3.7
SiO_2 (mg/l)	2.9 - 4.6	0.01 - 1.8
total Fe (μg/l)	$<$ 10	50

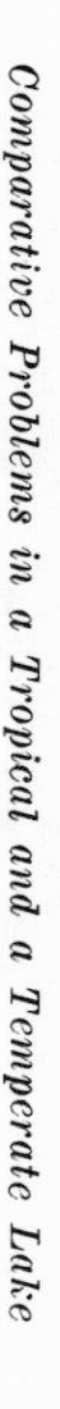

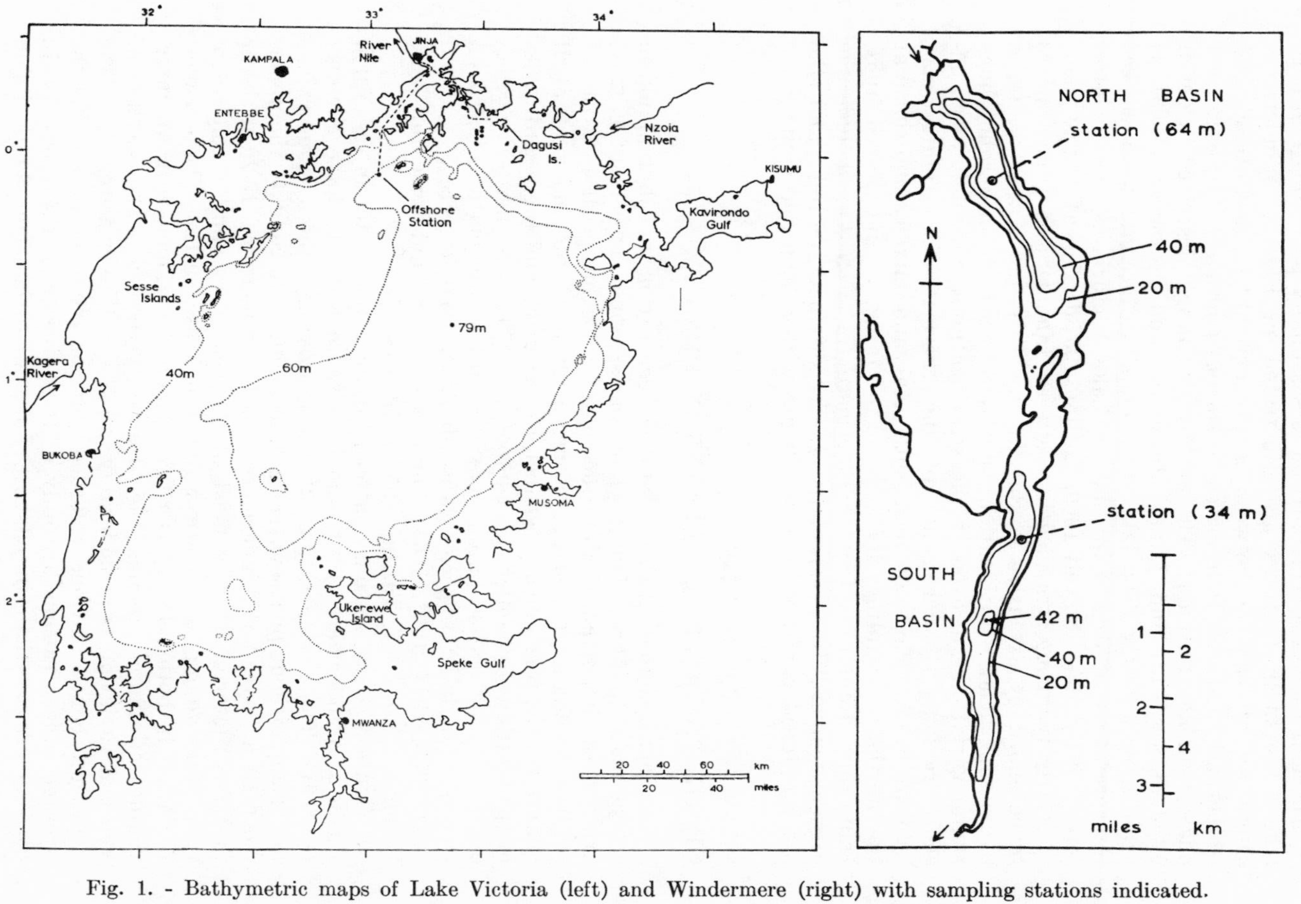

Fig. 1. - Bathymetric maps of Lake Victoria (left) and Windermere (right) with sampling stations indicated.

Thermal (density) stratification is seasonally pronounced in both lakes; the main features are illustrated in Fig. 2. Windermere shows an alternation between a predominantly unstratified and isothermal condition (winter, early spring) and a strongly stratified phase (summer, autumn). The pattern of change, described in detail by Jenkin (1942) and Lund, Mackereth and Mortimer (1963), is typical for a fairly deep lake in a temperate oceanic climate. The interpretation of events in offshore Lake Victoria is more controversial (Fish 1957, Newell 1960, Talling 1957 d, 1966 a), but a phase of relatively strong stratification between January and June appears to be ended by extensive mixing during July and August; later a more superficial stratification slowly reforms. Considerable horizontal variation can exist in the stratification, even in the deep offshore region. Displacements of the thermal discontinuity may lead to brief periods of near-isothermal and mixed conditions at the northern sampling station, as in January and March 1961. Important differences from Windermere (cf. Fig. 4 b) include the higher temperatures, especially in hypolimnetic water, and the much reduced variation of temperature with season and depth.

SEASONAL VARIATION OF THE TOTAL PHYTOPLANKTON DENSITY

Seasonal changes in the total concentration of phytoplankton reflect many factors, among which the seasonal aspects of climate are of great importance. Here the content of chlorophyll *a*, extracted by 90% methanol or 90% acetone, is used as the index of algal quantity. The relationship between its variation and the peculiarities of population dynamics in component species is here neglected, except for the major diatom maxima. Much information, based on counts of individual species, is available for Windermere (e. g. Lund 1949, 1950, 1954, 1961, 1964; Lund *et al.* 1963) and some for Lake Victoria (Fish 1957, Talling 1957 a, 1966 a).

Figure 3 shows seasonal patterns of algal abundance in the North Basin of Windermere (1964) and in offshore water of Lake Victoria (1960-61). The concentrations shown approximate the average values in the euphotic or photosynthetic zones, which are roughly the uppermost 10 m layer in Windermere and the uppermost 15-20 m layer in Lake Victoria. During most of the year very similar concentrations were found in surface water. In gross outline, the seasonal variation in both lakes shows two principal maxima, with an intervening minimum during the June-July period. The extent of seasonal variation is considerably greater in Windermere, as are the maximum concentrations recorded, but for the year as a whole the average concentrations are only a little greater in the North Basin of Windermere than in offshore Lake Victoria. The South Basin is appreciably more productive.

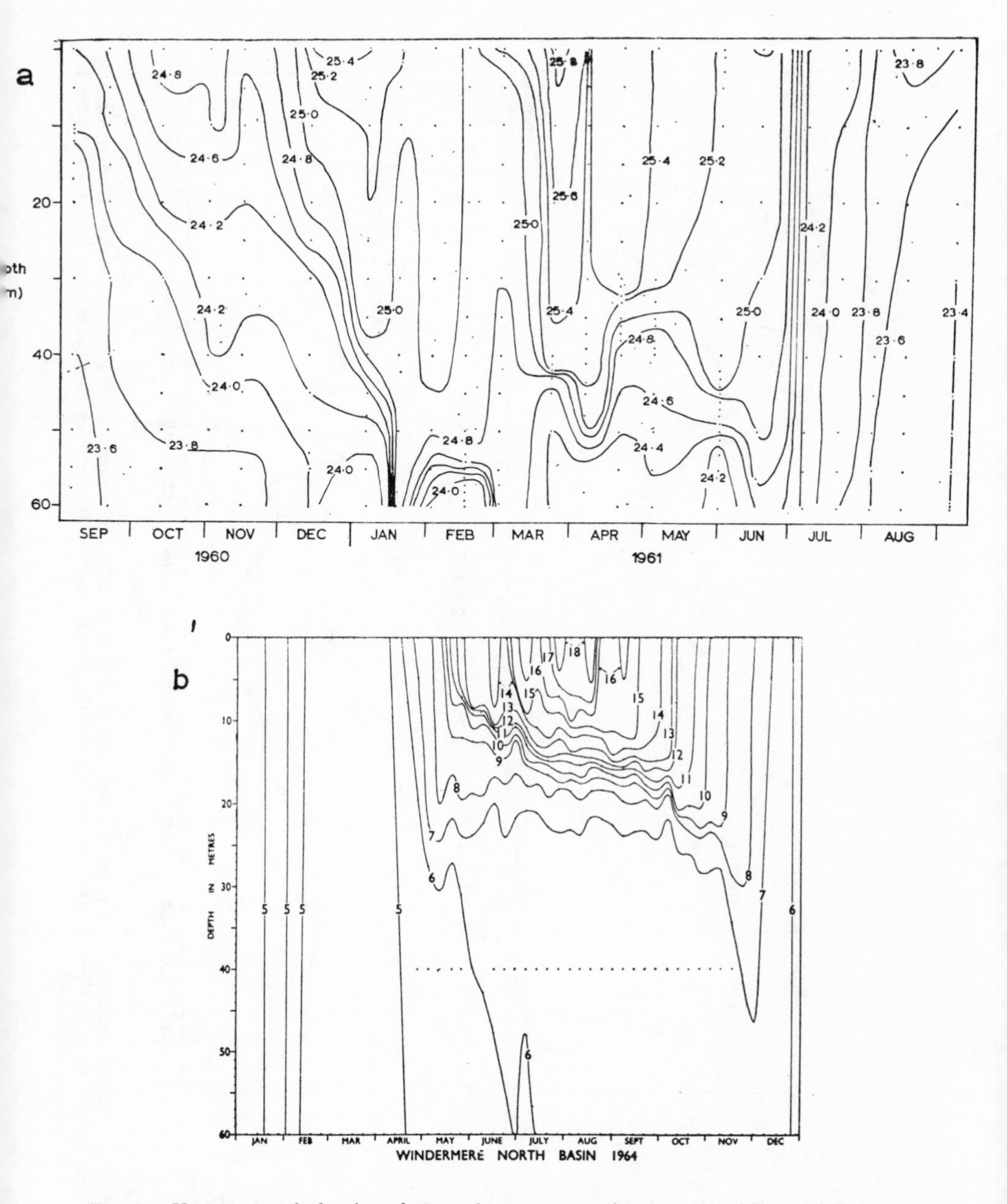

Fig. 2. - Variation with depth and time of temperature (isotherms in °C) in (a) Lake Victoria (1960-1) and (b) Windermere North Basin (1964).

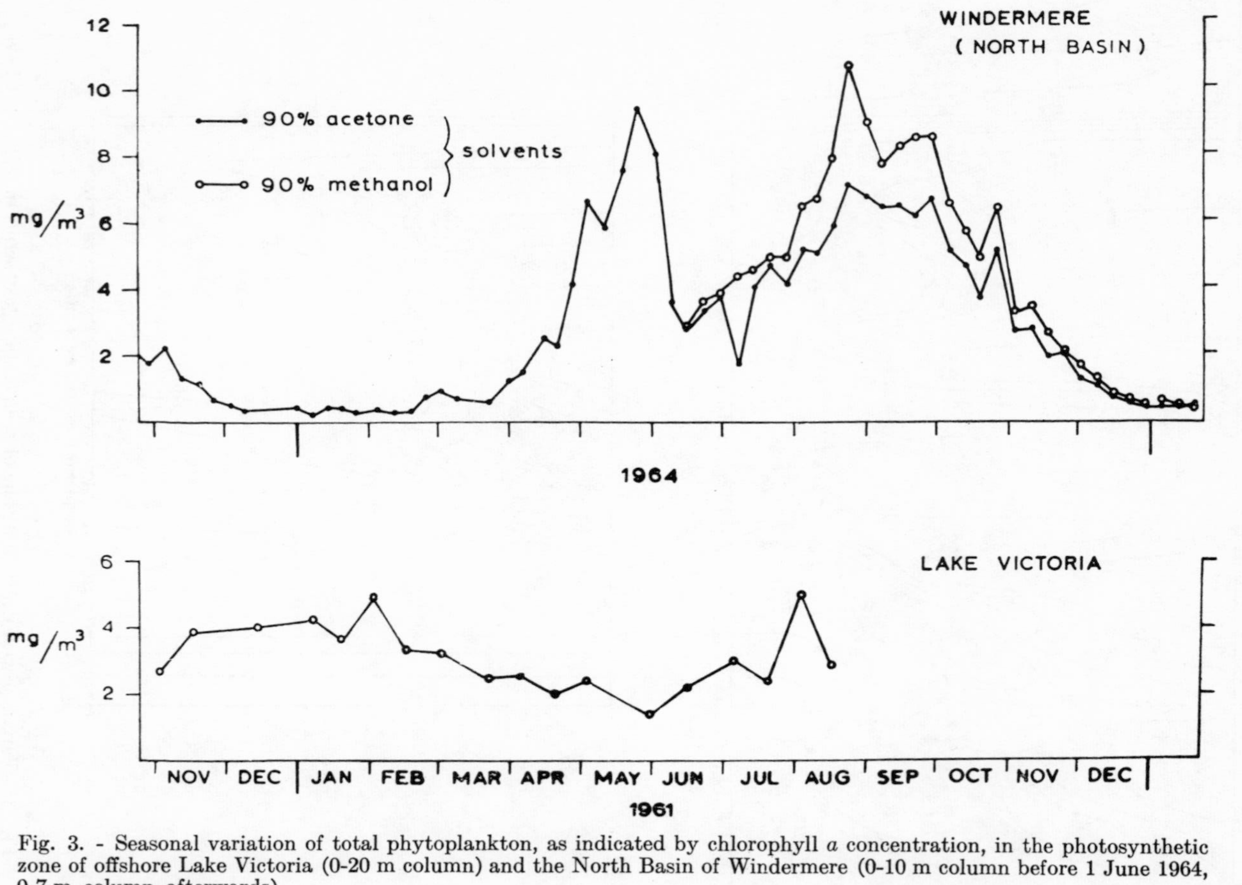

Fig. 3. - Seasonal variation of total phytoplankton, as indicated by chlorophyll *a* concentration, in the photosynthetic zone of offshore Lake Victoria (0-20 m column) and the North Basin of Windermere (0-10 m column before 1 June 1964, 0-7 m column afterwards).

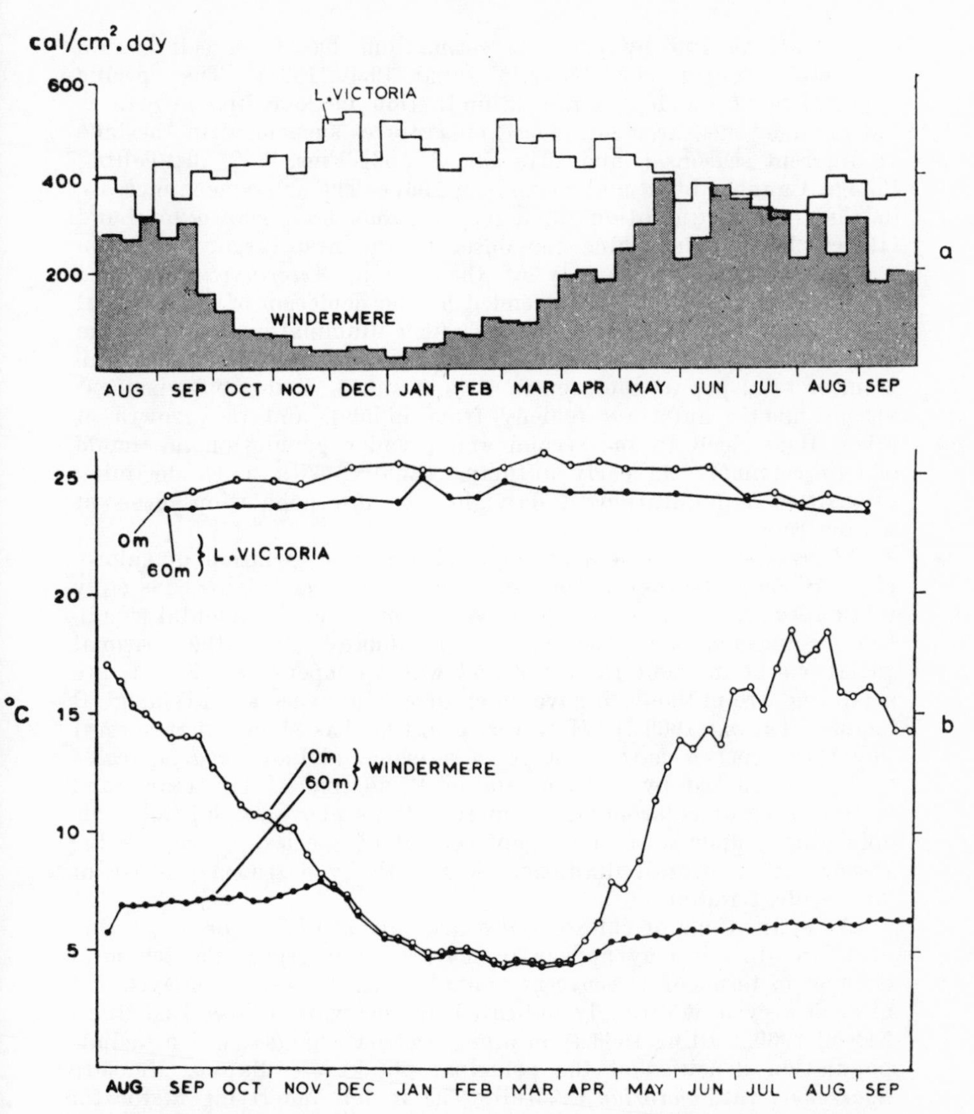

Fig. 4. - Seasonal variation of (a) incident radiation, and (b) of water temperature at depths of 0 and 60 m, for stations on or near Lake Victoria (1960-1) and Windermere North Basin (1963-4).

The winter minimum of daily incident radiation (Fig. 4) can be regarded as the most critical environmental feature for the seasonal pattern in Windermere. The phytoplankton crop is then regularly brought to low levels, as shown by the chlorophyll *a* concentrations and by previous estimations based on cell counts over much longer periods (e. g. Lund 1950, 1964). The specific importance of the low winter illumination in retarding growth is indicated by measurements based on cultures suspended in the lake at different seasons (Lund 1949 Fig. 7, 1964 Figs. 5, 6, 10; Talling 1955 b; Cannon, Lund and Sieminska 1961). The subsequent increase in the incident radiation during spring has been shown by Lund (1949, 1950) to determine the onset of the first large population maximum, composed largely of the diatom *Asterionella formosa* Hass. This growth is usually ended by the depletion of an essential nutrient, silica (Lund 1950, 1964). Further illumination of the diatom cells in the depleted medium induces, via mass mortality, the minimum of total population seen in Figs. 3 and 5. Later replenishment of epilimnetic nutrients (chiefly from inflow), and the growth of other algae, lead to the second and broader population maximum of late summer and early autumn. Finally, with much declining intensities and duration of daylight, the low population levels of winter return.

At first sight, the less pronounced seasonal variation of chlorophyll *a* concentration in offshore Lake Victoria might be simply related to the reduced seasonal variation of environmental conditions connected with the equatorial climate. Thus the seasonal variations of incident radiation and water temperature (Fig. 4) are small and are unlikely to have much direct influence as physiological factors (Talling 1966 a). However, counting has shown that several important species have a range of seasonal abundance comparable to that exhibited by *Asterionella* in Windermere. The year-round maintenance of chlorophyll *a* concentrations above 1.5 mg/m^3 probably hinges upon a seasonal replacement of species, as well as the absence of an overall limitation of growth by a seasonal phase of low incident radiation.

As estimations of chlorophyll *a* are only available for one period of 10 months, it may appear premature to interpret the temporal changes in terms of a recurrent annual cycle. However, the existence of such a cycle is strongly indicated by observations (see Fish 1957, Newell 1960, Talling 1966 a) on other seasonal changes, which include population densities of the principal planktonic diatom, *Melosira nyassensis* var. *victoriae* O. Müll. The major underlying factor for these cyclic events appears to be the wind regime, and specifically the strongest incidence of the south-east trade winds about June. At this season a rapid cooling and final mixing occur in the offshore lake water, and, like the winter radiation minimum at Windermere, can be regarded as the most critical seasonal events for the production of phytoplankton.

A diatom maximum follows the onset of mixing and accounts for the maximum of chlorophyll *a* in August. One probable cause is the redistribution of nutrients previously accumulated and unavailable in the lower, non-illuminated layers during earlier stratification. Another is the reintroduction of a massive inoculum of *Melosira* cells into the upper offshore water, derived from a deep « reservoir » at or near the mud surface and possibly also from more inshore areas. The close of this algal maximum was accompanied, and probably in large part determined, by the sedimentation of apparently healthy *Melosira* cells. Such redistribution of *Melosira* populations, related to water turbulence, were first described in detail by Lund (1954, 1955) for *M. italica* subsp. *subarctica* in several English lakes, including Windermere. However, in Windermere during 1964 they played little part in the seasonal changes of total population density, due to the relatively low concentrations of cells involved. The poor representation is connected with the lateness of the breakdown of seasonal stratification, after which *Melosira* cells are resuspended in the euphotic zone — then under low winter illumination. This restriction of *Melosira* growth in Windermere is discussed by Lund (1954, p. 171) with reference to the South Basin. The timing of destratification also reduces the significance of the event in the supply and utilization of nutrients accumulated in the hypolimnion. In this respect, too, there is a sharp contrast with Lake Victoria.

The second and broader algal maximum in Lake Victoria, centered between November and January, is less easily related to the environmental regime. It was chiefly composed of blue-green algae, and a connection with the relatively shallow depth of the upper mixed layer — then reforming — has been suggested (Talling 1966 a). Thus the sudden deepening of this layer in December-January was accompanied by a very rapid decline of a population of *Anabaena flos-aquae*. The final slow decline of the chlorophyll *a* maximum occurs during the phase of strongest thermal stratification, in which a progressive loss of essential nutrients from the upper productive layers can be expected.

In the preceding and probably over-generalized account, emphasis has been placed upon the effects of seasonally varying incident radiation for Windermere, and of seasonally varying stratification (related to wind regime) for Lake Victoria. It is noteworthy that both the amounts of incident light and the depth of the mixed water column interact in determining the average illumination available per circulating cell. This interaction finds an expression in the earlier onset of the spring growth in the shallower (south) basin of Windermere. A parallel may also exist between the apparent success of diatoms in first utilizing the increasing illumination during spring in the temperate lake, and their ready response to deep mixing in the tropical lake.

The influence of the annual cycles of thermal stratification upon the distribution of algal concentration is illustrated further, by depth-time diagrams, in Fig. 5. During phases of thermal stratifica-

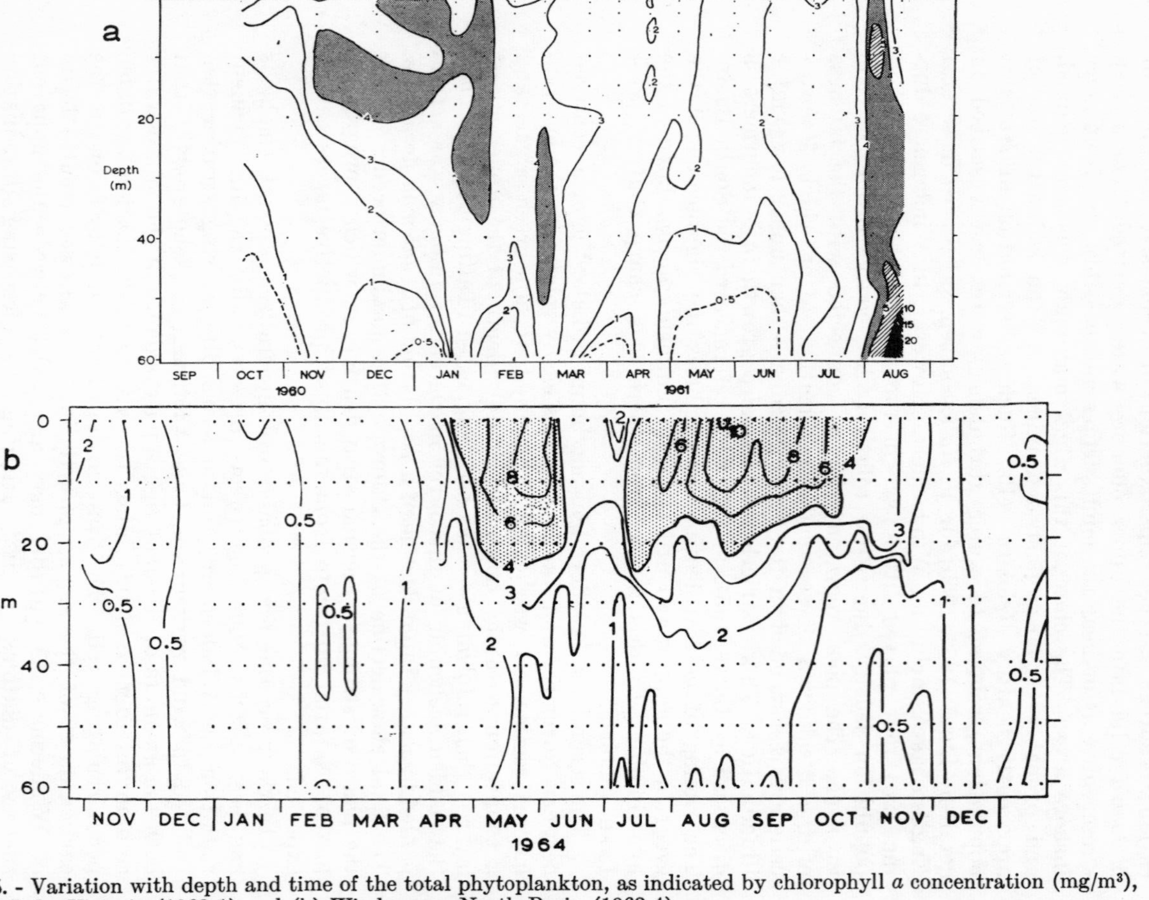

Fig. 5. - Variation with depth and time of the total phytoplankton, as indicated by chlorophyll *a* concentration (mg/m³), in (a) Lake Victoria (1960-1) and (b) Windermere North Basin (1963-4).

tion in both lakes, the thermal discontinuities tend to delimit an upper and more densely populated layer. The relationship in Windermere North Basin — where the vertical density gradients are much steeper — is particularly close, although there are indications of some cell sedimentation through the thermocline. In Lake Victoria the sedimentation of *Melosira* cells, following the principal diatom maximum, accounts for a deep concentration of chlorophyll *a* which is much larger than any recorded in the upper productive layers. During the phase of pronounced thermal stratification in this lake (January to May), the upper mixed layer is considerably deeper than its equivalent (the epilimnion) in Windermere, probably due chiefly to the longer wind fetch on the larger lake. Consequently, in offshore Lake Victoria changes of concentration tend to have a proportionately greater effect in terms of quantities assessed below unit area of water surface. Such estimates for Lake Victoria are discussed by Talling (1965, 1966 a).

PRODUCTION AND PRODUCTIVITY DURING THE PRINCIPAL DIATOM MAXIMA

A closer examination of the chief diatom maxima in the two lakes permits some further comparisons of short-term changes and their regulation. These chiefly concern population dynamics of the principal species — *Asterionella formosa* in Windermere, *Melosira nyassensis* in Lake Victoria — and the interrelations of population increase, nutrient depletion, and rates of organic productivity derived from measurements of gross photosynthetic activity. The *Asterionella* maximum, in Windermere North Basin, is here described from observations made during 1959 (Talling 1966 b). The population changes were broadly similar to those recorded in greater detail for 1947 (Lund *et al.* 1963).

The relevant population dynamics are illustrated, by depth-time diagrams of cell concentrations, in Fig. 6. The obvious differences in these distribution patterns result chiefly from the greater impact of thermal stratification upon the *Asterionella* population, and from the accentuated sedimentation of apparently healthy cells at the close of the *Melosira* maximum. Close relationships exist with the corresponding distribution patterns for chlorophyll *a*; they are modified only slightly by the accompanying growth of some other diatoms in Lake Victoria, and by the increased chlorophyll *a* content of deeper cells recorded in Windermere. The different population patterns in the two lakes are more obviously related to species peculiarities than to climatic characteristics — except that in the fairly deep temperate lake the climate usually determines a subdi-

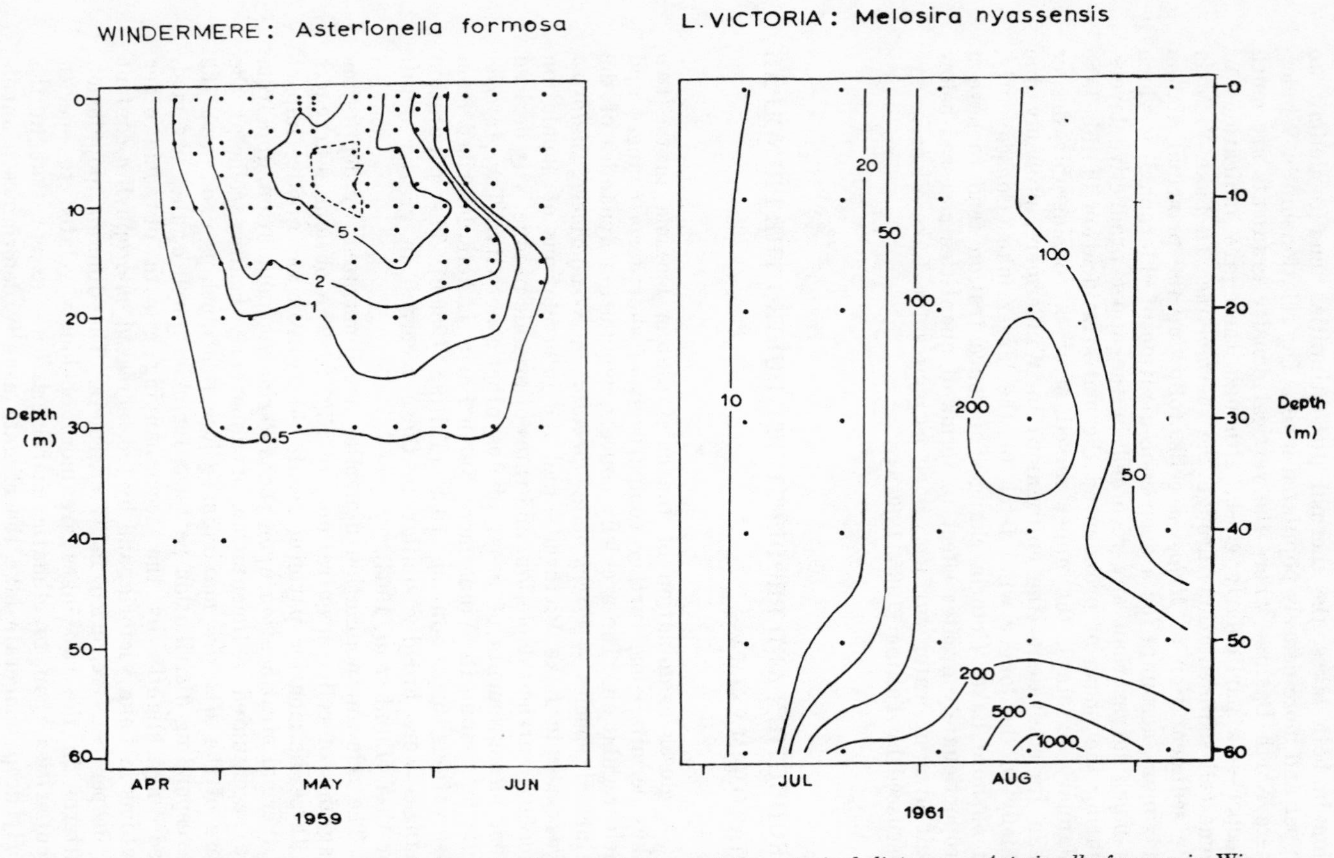

Fig. 6. - Variation with depth and time of cell concentrations of the principal diatoms — *Asterionella formosa* in Windermere North Basin, *Melosira nyassensis* in Lake Victoria — during the seasonal diatom maxima of 1959 (Windermere) and 1961 (Lake Victoria). Isopleths show concentrations in units of 10^9 *Asterionella* cells and 10^6 *Melosira* cells per m^3.

vision of the water-mass by direct thermal stratification before the growth response to increasing vernal illumination is completed.

Some differences in nutrient conditions have already been mentioned which affect, and are affected by, the diatom maxima. The growth in Windermere occurred after the seasonal winter replenishment of phosphate, nitrate, and silicate. In Lake Victoria the mixing which preceded and accompanied diatom growth was associated with some surface increase in phosphate, and less in silicate, but nitrate — here a more probable limiting nutrient — remained at very low levels. The later fall of silica concentration during diatom growth was of similar magnitude (about 1.5 mg/l) to that found in Windermere, but was more deeply felt throughout a less stratified water column. Both initial and final concentrations were much higher, and the latter were very unlikely to have any limiting effect upon diatom growth, despite the much greater estimates per unit area (Talling 1966 a) of crop production and silica removal. Here can be seen a biological consequence of the relatively high concentrations of silicates usually found in tropical freshwaters, possibly related to their mobility in the adjacent soils and muds.

When the comparison of the two diatom maxima is extended to their photosynthetic activity, further significant differences emerge. The evidence, based on periodic measurements of gross rates of oxygen production in short ($\sim$ 3 hour) exposures near midday, is described in detail by Talling (1965, 1966 b). Here the maximum activities recorded, which correspond with the highest population densities, are shown in Fig. 7. They are expressed by depth-profiles of rates per unit volume of water (units: mg $O_2/m^3 \cdot h$) and per unit quantity of chlorophyll *a* (units: mg $O_2/mg \cdot h$), with planimetric integration of the former yielding estimates of photosynthetic productivity per unit area of lake surface (units: mg $O_2/m^2 \cdot h$). The estimate per unit area for Lake Victoria is five times higher than that for Windermere, although the average population densities in the euphotic zones appear quite similar, when estimated in terms of chlorophyll *a* (5-6 mg/m^3). Accurate estimates in terms of cell volume are more difficult, but approximate values are 1.1 mm^3/l for Lake Victoria (Talling 1966 a Fig. 26) and 2.5 mm^3/l for Windermere.

The depth-profiles in Fig. 7 show that two other factors are responsible for the higher photosynthetic productivity in Lake Victoria. One is the greater light penetration, and hence the depth of the euphotic and photosynthetic zones. As was noted earlier, this characteristic is probably related to the preponderance of direct rainfall, as compared to inflows, in the water income of the lake. In Windermere the inflows contribute more non-living organic colored materials, which compete optically with the photosynthetic pigments. They may also be primarily responsible for much higher content (see Table 1) of total iron in Windermere than in Lake Victoria water.

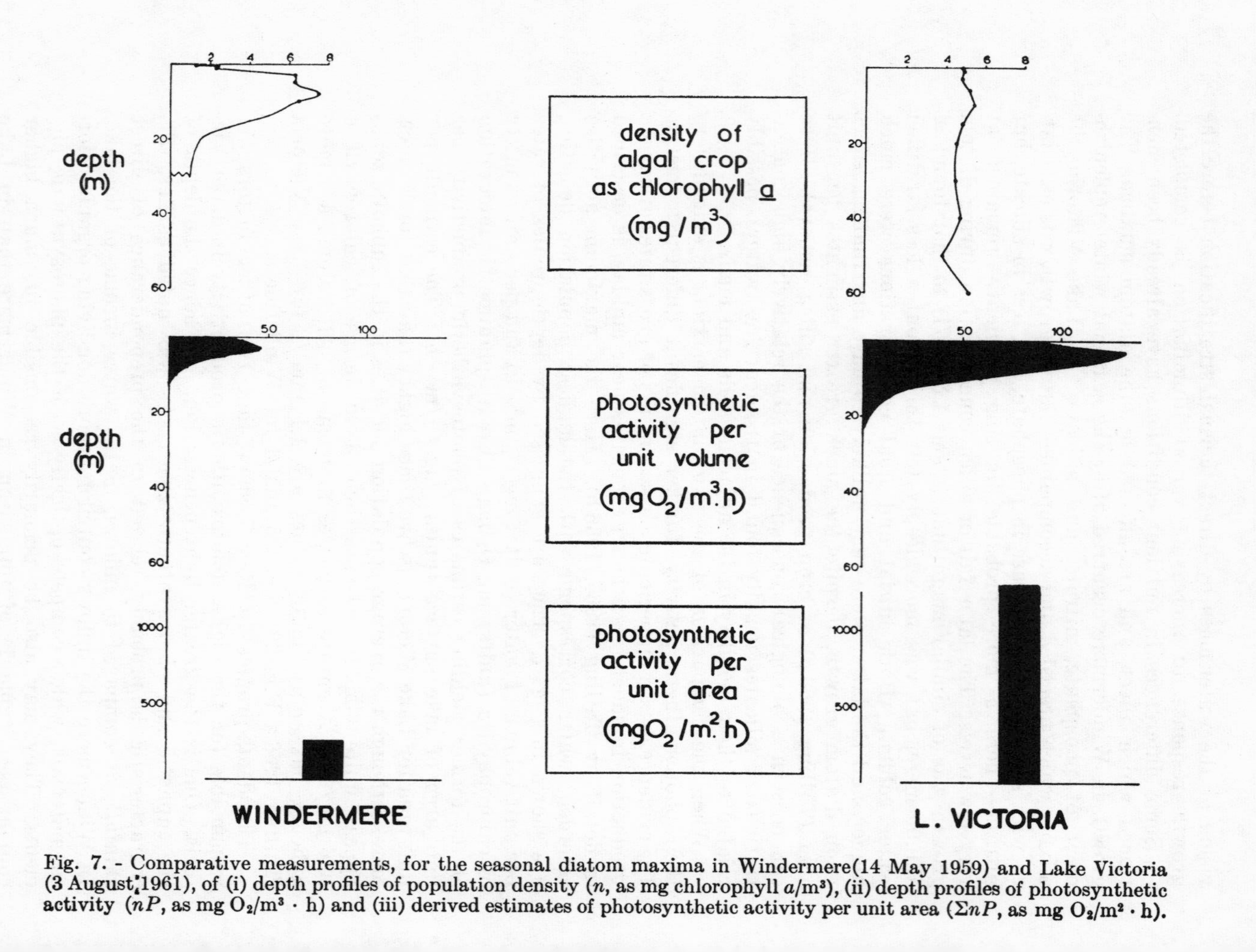

Fig. 7. – Comparative measurements, for the seasonal diatom maxima in Windermere (14 May 1959) and Lake Victoria (3 August, 1961), of (i) depth profiles of population density (n, as mg chlorophyll a/m^3), (ii) depth profiles of photosynthetic activity (nP, as mg O_2/m$^3 \cdot$ h) and (iii) derived estimates of photosynthetic activity per unit area (ΣnP, as mg O_2/m$^2 \cdot$ h).

The second, and more important, factor is the much higher level in Lake Victoria of the light-saturated photosynthetic rate per unit of population (P_{max}). This appears to be approximately four times higher than the level prevalent during the *Asterionella* maximum in Windermere. The latter value is probably depressed to some extent by the limitation of growth in a silica-depleted medium. Such effects of nutrient deficiency are very unlikely to account fully for the large difference under discussion, as only slightly higher values of P_{max} were obtained from an earlier stage of population growth and from *Asterionella* cultures at comparable temperatures (see Talling 1957 a, 1966 b). Nor are species differences likely to be primarily responsible: the high P_{max} value for Lake Victoria is consonant with many other measurements on tropical East African waters (Prowse and Talling 1958, Talling 1957 c, 1965) whereas the *Asterionella* values are representative of various (unpublished) measurements for other phytoplankton in the English lakes. So widespread a trend is most likely to result from the obvious and climatically based difference in temperature between the tropical and temperate areas, which is approximately 10-12 °C for the diatom maxima under discussion (cf. Fig. 4). *Asterionella formosa,* an extra-tropical species, is unable to grow at the temperatures found in Lake Victoria and typical of most tropical lakes (Lund, in press). Nevertheless, its rates of growth and photosynthesis at light saturation are strongly temperature-dependent (average $Q_{10} = 2.1$ to 2.3) in the range from 6 to 16 °C (Talling 1955 b, 1957 a, 1966 b).

The large difference between the two diatom maxima in the hourly rates of gross photosynthetic productivity can be expressed by a formulation which I have previously applied to both temperate and tropical areas (Talling 1955 a, 1957 b, 1957 c, 1965). The basic relationship used can be written

$$\Sigma n P = \frac{n P_{max}}{1.33 k_{min}} ln \frac{I'_o}{0.5 I_k}$$

where $\Sigma n P$ is the integral rate of photosynthesis per unit area (present units mg $O_2/m^2 \cdot h$), n is population density (units mg chlorophyll a/m^3), P_{max} is the light-saturated rate of photosynthesis per unit of population (units mg O_2/mg · h), k_{min} (ln units/m) is the minimum value (over the spectrum) of the vertical extinction coefficient (here applicable to green light), I'_o is the subsurface light intensity (= the surface intensity I_o corrected for surface loss), and I_k the light intensity which measures the onset of light-saturation of photosynthesis. The equation was formulated for populations which are not markedly stratified within the photosynthetic zone, which prevents a simple application to the data (Fig. 7) for Windermere on 14 May 1959. Fortunately, cool and windy weather induced mixing in this zone a few days later (20-22 May), before any appreciable decline in population density. Using measurements from this period, with corresponding values for the Lake Victoria maximum in parentheses, n = 5.7 to 7.5 (4.8), P_{max} = 5.4 to 6.7 (27.5), $n\ P_{max}$ = 38 to 40 (132), k_{min} = 0.43 to 0.47 (0.26), and the term

$$\frac{n\ P_{max}}{k_{min}} = 85 \text{ to } 89\ (508).$$

The last quantity (whose units are here mg $O_2/m^2 \cdot h$) can be regarded as the chief determinant of the large difference in the integral photosynthesis, $\Sigma\, n\, P$, discussed earlier. As pointed out by Talling (1957 c, p. 77), in the calculation the effects of higher tropical values for I'_o are likely to be offset by higher values of I_k, and these quantities are involved as a ratio in a logarithmic term. I hope to give a fuller account of these interrelations elsewhere, but some numerical examples for the two lakes can be found in Talling (1957 a, 1957 b, 1965, 1966 b). Vollenweider (1960) gives further examples from lakes in Sweden, Italy and Egypt.

The ratio $n\,P_{max}/k_{min}$ can also express the consequences of self-shading in phytoplankton, when increase in the population density n produces an increase in the extinction coefficient k_{min}. The increment per unit of population density, k_s, has been calculated as approximately 0.02 (ln units per mg chlorophyll a/m^2) from observations on the *Asterionella* maxima in Windermere North Basin (Talling 1960). Application of this value to the African situation, where values of P_{max} were usually close to 25 mg O_2/mg h, suggested (Talling 1965) that in very dense and self-shading populations the ratio $n\,P_{max}/k_{min}$ would reach a limiting value of about 1250 (units mg $O_2/m^2 \cdot h$). That the actual values from Lake Victoria are well below this supposed ceiling estimate is consistent with the weak, though positive, connection between population changes and the seasonal variation of k_{min} in this lake. For the diatom maxima in Windermere, the corresponding « ceiling » for the ratio — and hence that for the integral photosynthetic productivity — is lowered at least two — or three — fold by the similarly low values of P_{max}. This approach to potential productivity in dense populations emphasises the importance of a possible association between P_{max} values and climatic regime. It also shows the desirability of a distinction between the contributions of the n and P_{max} factors, and the need for more information on the k_s factor and its possible variation.

Two other features, connected with the seasonal timing of population maxima, deserve mention in this comparison of temperate and equatorial situations. At 54°N the daylength varies between 7.4 and 17.2 hours, whereas for Lake Victoria it is practically constant near 12 hours (cf. Smithsonian Meteorological Tables 1951,

Table 171). Since in the temperate lake the principal populations usually occur between the vernal and autumnal equinoxes, the factor of daylength is likely to increase the daily photosynthetic productivity of the temperate lake more than that of the tropical lake. This feature can be illustrated by estimates of the ratio of the daily integral productivity to the hourly rate measured near mid-day. An average value close to 9 was indicated by direct measurements on a Nile reservoir at latitude 15°N (Talling 1957 c, Fig. 12 a and Table 3) and also by indirect calculations for Lake Victoria (Talling 1965). For the *Asterionella* maximum present in Windermere on 22 May 1959, a sunny day of 16 hours duration, calculations based on a field experiment combined with diurnal laboratory measurements (Talling 1966 b) indicate a value of 12. From measurements by Vollenweider and Nauwerck (1961) on a Swedish lake (30 May 1956, latitude 59°N, daylength 18 hours), a still higher value of 14 can be calculated, and an average value of 9.5 appears applicable to estimates by Hepher (1962) from Israel (May to September, latitude 33°N). Besides spurious variation in the ratio due to local weather conditions and algal redistributions with depth, an influence of other variation of photosynthetic capacity may exist. There is some marine evidence that the latter variation is most pronounced at low latitudes (Doty 1959) but it is not clear whether the n or the P_{max} variable is primarily concerned, and corresponding evidence for the trend from freshwaters is lacking.

The influence of daylength introduces a more general question, the differing significance for productivity of the seasonal timing of the diatom maxima. In Lake Victoria this timing appears to be less critical, in view of the small seasonal range of incident daily radiation and temperature. Both these quantities are increasing during the spring diatom maxima in Windermere. Owing to the smaller fraction of non-illuminated water in the shallower South Basin, exponential increase of *Asterionella* usually begins 1-2 months earlier there than in the North Basin, and the maximum populations typically occur in early May before the lake is strongly stratified. The corresponding growth phase in the deeper North Basin is normally delayed (average about 3 weeks) (cf. Lund 1950, Talling 1957 a Fig. 1). Consequently, before mass mortality is induced by the combined action of silica depletion and illumination, the populations experience conditions of incident light and temperature more favorable for photosynthetic productivity than those usually prevalent during the South Basin maxima. A similar differential exists between the earlier *Asterionella* maxima in a small and shallow lake nearby, Blelham Tarn, and those in Windermere North Basin (Talling 1966 b). Though advantages of a late start can thus be traced for photosynthetic productivity per unit area, restrictions upon the total population density per unit area are usually incurred from thermal stratification of the water mass.

DISCUSSION

In these two lakes the seasonal events have been interpreted in terms of an annual cycle, a standpoint familiar for temperate lakes but less so for tropical lakes. In Windermere the annual pattern is ultimately controlled by the varying input of solar radiation; in Lake Victoria a seasonal loss of heat with resulting vertical mixing, dependent upon wind regime, substitutes as the critical and cycle-generating event. Other controlling mechanisms may be such substitutes in other tropical freshwaters. Examples in which primary production has been studied are few; they include the Gebel Aulia reservoir on the White Nile, and the Blue Nile nearby, in which flow-rate is dominant (Rzóska, Brook and Prowse 1955, Prowse and Talling 1958, and unpublished). Further knowledge of temporal patterns, cyclic or otherwise, in tropical lakes under more equable conditions would be of fundamental interest.

Although obviously relevant to the present theme, an adequate comparison of estimated annual photosynthetic productivity in the two lakes is prevented by the incompleteness of seasonal measurements, particularly for Windermere. The extrapolation from hourly to daily estimates is also very approximate. For Lake Victoria 10 of the 12 months are represented; if the average daily estimate ($\sim$ 7 g O_2/m^2: Talling 1965) over this period is applicable to the entire year, an exceptionally high value for gross photosynthesis of $\sim$ 2500 g O_2 (or $\sim$ 950 g C, using a photosynthetic quotient of 1) $/m^2 \cdot$ year is obtained. The only annual estimate available for Windermere (North Basin) refers to net carbon fixation deduced from crop growth and nutrient depletion in 1947, and is 20 g $C/m^2 \cdot$ year (Lund *et al.* 1963). The annual gross photosynthetic productivity is certainly several times higher, but a wide gap still separates the very much higher estimate for Lake Victoria.

For a quantitative and analytical interpretation of these annual estimates, one may consider the value of time integrals of those principal factors discussed in relation to measurements over short periods. The most familiar is the annual integral of incident solar radiation. Records for Windermere during 1964 yield a value of 61750 cal/cm^2, whereas those from a station near Lake Victoria during 1960-61 (quoted by Talling 1965, 1966 a) a value of 154000 cal/cm^2. A comparative « absolute » reference is easily obtained to the corresponding values computed for these latitudes at the top of the atmosphere, which are respectively 199000 and 311500 cal/cm^2 (Smithsonian Meteorological Tables 1951, Table 133). Radiation integrals for the summer half-year, like the maximum daily radiation at midsummer, are relatively insensitive to variation in latitude (though not atmospheric transmission), especially between 0° and 60° (cf. Smithsonian Meteorological Tables 1951, Tables 133, 135, 136; also Fig. 4).

The general significance of light-saturation effects upon photosynthesis makes it difficult to utilize directly such integral estimates (cf. Talling 1957 b), as they cloak differing consequences of variation in daylength and the ratio of I'_o to I_k. I have previously attempted to take these factors into account by a logarithmic function (« light-division-hours », *L. D. H.*), which appears to have a similar average daily value (30-50) for Lake Victoria with some other African waters and for Windermere in spring to summer (Talling 1957 b, 1957 c, 1965).

Crop-duration is another possible time integral, little used in quantitative work on phytoplankton productivity (e. g. Kusnezow 1959 p. 158; Usachev, quoted by Zenkevitch 1963, p. 42) but familiar in terrestrial studies. The most useful single measure of crop quantity, equivalent to the instantaneous vegetation cover, appears to be the content per unit area of the euphotic zone (Σn: present units mg chlorophyll a/m^2). Estimations have been made for Lake Victoria and other East African lakes, and show a close relationship to values obtained for the integral rate of photosynthesis per unit area (Talling 1965, Fig. 8 b). The connecting quantity is the average rate of photosynthesis per unit of euphotic population ($\overline{P}$) as expressed by the relation

$$\Sigma\, n\, P = \Sigma\, n \cdot \overline{P}$$

The magnitude of $\overline{P}$ is primarily determined by the levels of P_{max} and the ln $(I'_o/0.5\ I_k)$ factor (or the latter's derivatives, *L. D.* and *L. D. H.*). For a population homogeneous within the euphotic zone, whose depth is defined in terms of the minimum vertical extinction coefficient as $3.7/k_{min}$ (Talling 1965, p. 23), substitution in equation 4 of Talling (1957 b) yields the relationship

$$\overline{P} = 0.14 \cdot (L.\, D.) \cdot P_{max}$$

Consequently, the hourly midday values of $\overline{P}$ for the Lake Victoria phytoplankton are rather uniform (mostly 15-20 mg O_2/mg · h: Talling 1965, Fig. 8) and relatively high. Lower values obtained from Windermere (to be described elsewhere) reflect lower values of P_{max}, and a strong winter depression is likely in view of the temperature-dependence of P_{max} and the seasonally low values of I_o. The values applicable over longer periods will be further influenced by the variation in daylength, which is expressed in the time-integral *L. D. H.*

The time-integral for the euphotic population content can be measured in units of mg chlorophyll a · days/m². Estimations at the offshore station in Lake Victoria are available (Talling 1965, Fig. 3 d) for 275 days, and yield a value of 12000 mg · days/m². As seasonal changes in the euphotic population density appear not to be large, the probable annual value is $\sim$ 16000 mg · days/m². The corresponding—and more approximate—estimate for Windermere North Basin is 14000 mg · days/m², a very similar figure. This is calculated from the variation of average chlorophyll *a* concentration in the euphotic zone during 1964 (Fig. 3) and a mean value of 10 m — from regular seasonal measurements during 1959 to 1960, and irregular ones in other years — for the depth of this zone. Self-shading effects in dense populations will impose an upper limit to the chlorophyll *a* content of the euphotic zone, and consequently to the annual time-integral. Although a precise estimate is prevented by several factors (which include other pigments and the sizes of algal cells and colonies), the maximum content is probably often reached between 200 and 300 mg/m^2 (Steemann Nielsen 1957, 1962), so that 100,000 mg · $days/m^2$ is an approximate upper limit for the annual time-integral. It is clearly not approached closely in either lake; this result would also be expected from the relatively small influence of the phytoplankton upon light extinction (Talling 1960, 1965, 1966 a).

Reference to Fig. 3 will show the significance for the time-integral of the sparse cover (< 5 mg/m^2) indicated for Windermere North Basin during the three mid-winter months. As discussed earlier, such sparsity can be connected with the seasonally low levels of available illumination for phytoplankton cells in this fairly deep temperate lake. The same environmental condition can be expected to depress the winter values of $\bar{P}$, although direct measurements are not available. The absence of these seasonal effects in Lake Victoria, coupled with higher and probably temperature-dependent levels of P_{max}, suggest a strong climatic element underlying the high estimate for annual photosynthetic productivity in this equatorial lake.

The present interpretation of the photosynthetic productivity per unit area ($\Sigma\, nP$), as the product of $\Sigma\, n$ and $\bar{P}$, has a parallel in a method of growth analysis much used for monocultures of terrestrial vegetation. In this method the product involved is of leaf area index (L; leaf area/ground area) and net assimilation rate (E; rate of dry weight increase per unit leaf area). Although some obvious differences exist, I believe that it would be rewarding to explore the parallels between $\Sigma\, n$ and L (cf. Gessner 1949, Steemann Nielsen 1957, Aruga and Monsi 1963), between daily estimates of $\bar{P}$ and E, and between the two products, in comparative studies of aquatic and terrestrial productivity. Further, in quantitative and analytical comparisons of the annual photosynthetic productivity of phytoplankton under differing climatic regimes, it would appear useful to apply some considerations now familiar in work on terrestrial vegetation (cf. Watson 1956). One may instance the coincident or non-coincident nature of seasonal variation in vegetation cover ($\Sigma\, n$) and average photosynthetic rate ($\bar{P}$), and the utilization of time-integrals of vegetation cover (cf. leaf area duration).

The subject matter of this essay has often illustrated the difficulties in distinguishing effects of climatic-latitudinal factors from those of local limnological features and species peculiarities. The two diatom maxima provide particularly clear examples. However, it would appear that a more explicit recognition of the climatic element is overdue in assessments of primary productivity in freshwaters, although the extensive comparisons implied are unlikely to be securely based without intensive and seasonal studies.

Acknowledgements

I am grateful to Mr. J. Heron for the unpublished determinations of alkalinity, total ionic concentration, and total iron in Windermere; to Mr. A. E. Ramsbottom for the diagram reproduced as Fig. 2 b; to the staff of the Cotton Research Station, Namulonge, Uganda, for unpublished records of solar radiation during 1960-61; and to Dr. J. W. G. Lund, F. R. S., for reading and criticizing the manuscript. My debt to other sources of help has been acknowledged in more detailed accounts of work on the two lakes (Talling 1965, 1966 a, b).

REFERENCES

Aruga, Y., and M. Monsi. 1963. Chlorophyll amount as an indicator of matter productivity in bio-communities. - Plant Cell Physiol. *4*: 29-39.

Cannon, D., J. W. G. Lund, and J. Sieminska. 1961. The growth of *Tabellaria flocculosa* (Roth) Kütz. var. *flocculosa* (Roth) Knuds. under natural conditions of light and temperature. - J. Ecol. *49*: 277-281.

Doty, M. S. 1959. Phytoplankton photosynthetic periodicity as a function of latitude. - J. Mar. Biol. Ass. India *1*: 66-68.

Fish, G. R. 1957. A seiche movement and its effect on the hydrology of Lake Victoria. - Fish. Publ., London *10*: 1-68.

Gessner, F. 1949. Der Chlorophyllgehalt im See und seine photosynthetische Valenz als geophysikalische Problem. - Schweiz. Z. Hydrol. *11*: 378-410.

Hepher, B. 1962. Primary production in fishponds and its application to fertilization experiments. - Limnol. Oceanog. *7*: 131-136.

Heron, J. 1961. The seasonal variation of phosphate, silicate and nitrate in waters of the English Lake District. - Limnol. Oceanog. *6*: 338-346.

Hutchinson, G. E. 1957. A treatise on limnology. I. Geography, physics and chemistry. - Wiley, New York.

Jenkin, P. M. 1942. Seasonal changes in the temperature of Windermere (English Lake District). - J. Anim. Ecol. *11*: 248-269.

Kusnezow, S. I. 1959. Die Rolle der Mikroorganismen in Stoffkreislauf der Seen. - (trans. A. Pochmann). Berlin.

Lund, J. W. G. 1949. Studies on *Asterionella*. I. The origin and nature of the cells producing seasonal maxima. - J. Ecol. *37*: 389-419.

— 1950. Studies on *Asterionella formosa* Hass. II. Nutrient depletion and the spring maximum. - J. Ecol. *38*: 1-35.

— 1954. The seasonal cycle of the plankton diatom *Melosira italica* (Ehr.) Kütz. subsp. *subarctica* O. Müll. - J. Ecol. *42*: 151-179.

— 1955. Further observations on the seasonal cycle of *Melosira italica* (Ehr.) Kütz. subsp. *subarctica* O. Müll. - J. Ecol. *43*: 91-102.

— 1961. The periodicity of μ-algae in three English lakes - Verh. int. Ver. Limnol. *14*: 147-154.

— 1964. Primary production and periodicity of plankton algae. - Verh. int. Ver. Limnol. *15*: 37-56.

— in press. *In* Riley, G. A. (ed.), Marine Biology. II. Second International Interdisciplinary Conference on Marine Biology. - Washington, D. C.

— F. J. H. Mackereth, and C. H. Mortimer. 1963. Changes in depth and time of certain chemical and physical conditions and of the standing crop of

Asterionella formosa Hass. in the North Basin of Windermere in 1947. Phil. Trans. B, *246*: 255-290.

Mackinney, G. 1941. Absorption of light by chlorophyll solutions. - J. Biol. Chem. *140*: 315-322.

Newell, B. S. 1960. The hydrology of Lake Victoria. - Hydrobiologia *15*: 363-383.

Prowse, G. A., and J. F. Talling. 1958. The seasonal growth and succession of plankton algae in the White Nile. - Limnol. Oceanog. *3*: 223-238.

Richards, F. A., and T. G. Thompson, 1952. The estimation and characterization of plankton populations by pigment analysis. II. A spectrophotometric method for the estimation of plankton pigments. - J. Mar. Res. *11*: 156-172.

Rzóska, J., A. J. Brook, and G. A. Prowse. 1955. Seasonal plankton development in the White and Blue Nile near Khartoum. - Verh. int. Ver. Limnol. *12*: 327-334.

Smithsonian Meteorological Tables 1951. Sixth Revised Edition. - Smithsonian Institution, Washington.

Steemann Nielsen, E. 1957. The chlorophyll content and the light-utilization in communities of plankton algae and terrestrial higher plants. - Physiol. Plant. *10*: 1009-1021.

— 1962. On the maximum quantity of plankton chlorophyll per surface unit of a lake or the sea. - Int. Revue ges. Hydrobiol. *47*: 333-338.

Talling, J. F. 1955 a. The light-relations of phytoplankton populations. - Verh. int. Ver. Limnol. *12*: 141-142.

— 1955 b. The relative growth rates of three plankton diatoms in relation to underwater radiation and temperature. - Ann. Bot. *19*: 329-341.

— 1957 a. Photosynthetic characteristics of some freshwater plankton diatoms in relation to underwater radiation. - New Phytol. *56*: 29-50.

— 1957 b. The phytoplankton population as a compound photosynthetic system. New Phytol. *56*: 133-149.

— 1957 c. Diurnal changes of stratification and photosynthesis in some tropical African waters. - Proc. Roy. Soc. B *147*: 57-83.

— 1957 d. Some observations on the stratification of Lake Victoria. - Limnol. Oceanog. *3*: 213-221.

— 1960. Self-shading effects in natural populations of a planctonic diatom. Wetter u. Leben *12*: 235-242.

— 1965. The photosynthetic activity of phytoplankton in East African lakes. Int. Revue ges. Hydrobiol. *50*: 1-32.

— 1966 a. The annual cycle of stratification and phytoplankton growth in Lake Victoria (East Africa). - Int. Revue ges. Hydrobiol. (in press).

— 1966 b. Photosynthetic behaviour in stratified and unstratified lake populations of a planktonic diatom. - J. Ecol. *54* (in press).

— and D. Driver. 1963. Some problems in the estimation of chlorophyll *a* in phytoplankton. Rep. Symp., Primary Productivity in the Pacific. 10th Pacific Science Congress, University of Hawaii, 1961. - U.S. Atomic Energy Commission, Div. Tech. Information, TID-7633. pp. 142-146.

— and I. B. Talling. 1965. The chemical composition of African lake waters. Int. Revue ges. Hydrobiol. *50*: 421-463.

Vollenweider, R. A. 1960. Beiträge zur Kenntnis optischer Eigenschaften der Gewässer und Primärproduktion. - Mem. Ist. Ital. Idrobiol. *12*: 201-244.

— and A. Nauwerck. 1961. Some observations on the C^{14} method for measuring primary production. - Verh. Int. Ver. Limnol. *14*: 134-139.

Watson, D. J. 1956. Leaf growth in relation to crop yield, pp. 178-191. *In* F. L. Milthorpe (ed.), The growth of leaves. - Butterworths Scientific Publications, London.

Wesenberg-Lund, C. 1910. Summary of our knowledge regarding various limnological problems, pp. 374-438. *In* J. Murray and L. Pullar, Bathymetrical survey of the Scottish freshwater lochs, Vol. 1, Edinburgh.

Zenkevitch, L. 1963. Biology of the seas of the U.S.S.R. (Translation by S. Botcharskaya). - Allen and Unwin, London.

CALCULATION MODELS OF PHOTOSYNTHESIS-DEPTH CURVES AND SOME IMPLICATIONS REGARDING DAY RATE ESTIMATES IN PRIMARY PRODUCTION MEASUREMENTS

RICHARD A. VOLLENWEIDER
Istituto Italiano di Idrobiologia
Verbania Pallanza, Italy

For bibliographic citation of this paper, see page 10.

Abstract

Calculation models of photosynthesis-depth curves which result from *in situ* primary production measurements in inland waters and the sea are discussed. Special attention is given to curves showing surface and subsurface light inhibition. In order to satisfy a large variety of experimental conditions, a strict distinction between *instantaneous rate integrals* and *day rate integrals* appeared to be necessary. With regard to the calculation of the latter, the concept of a *reference integral* was developed and symmetrical as well as non-symmetrical trends of *surface rate curves* were considered. Non-symmetrical curves are attributed to some kind of nutrient depletion function.

The mathematical implications following from the theoretical treatment are compared with experimental data; accordingly, reasonable day rate estimates of primary production from short term *in situ* experiments can be obtained by dividing the light day e.g. into 5 equal periods and exposing samples during either the second or third period, or both. From the theory presented one can expect that during either of these periods about 30% (or 55 to 60% during both periods) of the total day rate integral is produced below a unit of surface.

1. INTRODUCTION

The improvement of direct *in situ* measurements of primary production in aquatic environments during recent years has considerably enlarged our view about the basic production properties of many waters. A substantial extension to new waters, as intended by the IBP, however, creates the methodological question of how to restrict individual measurements to that minimum number which gives still a sufficiently reliable estimate of integral photosynthesis, or, in other words, what kind of parameters we need to know and what we have to do with them under different conditions of measurement.

A number of simple, more or less comparable calculation models have been developed by several authors (Steemann Nielsen 1952, Talling 1957, Ryther and Yentsch 1957, Vollenweider 1958, 1960). The following contribution is, in part, a review of such models, and in part an enlargement of the basic concept with the aim of discussing the principles needed in calculating day rates either from a few hour *in situ* exposures or from laboratory or shipboard measurements.

Time and space do not allow consideration of all aspects at the same level. With regard to the complexity of the question, only the most simple, i.e. homogeneous distribution model has been considered. Instantaneous surface rate integrals have been treated for the most part by analytical methods, whereas the discussion of day rate integrals is based primarily on planimetric methods.

2. THE BASIC PROBLEM

Gross production rates beneath a unit surface area depend, among other factors, upon phytoplankton density, species composition, age and physiological stage, as well as on such environmental properties as incident light, light attenuation in the water, temperature, and nutrition conditions.

At present, a complete understanding of the co-action of all these factors is impossible; this is particularly true with regard to the influence exerted by biological parameters on total production.

Some of the external factors, however, are easier to co-ordinate in a simple model which assums that the phytoplankton is homogeneously distributed throughout the photosynthetic layer. A further condition of the model is that some of the basic parameters such as phytoplankton density, light attenuation, temperature, etc. do not change with time.

Let p_z be the production rate at depth z during any small time interval; then the total rate below a unit of surface at time t will be (cf. Fig. 1),

$$\Sigma\, p_{(t)} = \int_{o}^{\infty} p_z \cdot dz \qquad (2.1.)$$

and integrated with regard to the time interval $t_2 - t_1$

$$\Sigma\, p_{(t_2 - t_1)} = \int_{t_1}^{t_2} \int_{o}^{\infty} (p_z \cdot dz)\, dt \qquad (2.2.)$$

As known from earlier works (Talling 1957; Vollenweider 1960) the general solution of both equations will be of the form:

$$\Sigma\, p = F(i) \cdot \frac{p_{opt}}{\varepsilon} \qquad (2.3.)$$

where,

$\Sigma\, p$ = production rate per unit surface (g C/m²),
p_{opt} = production rate per unit volume (g C/m³) at light optimum,
$F(i)$ = a function of the photosynthetic active incident light,
ε = the attenuation coefficient of the light flux in water (1/m).

In order to satisfy equation (2.3.) $F(i)$ must be a dimensionless number.

The basic question involved in the solution of our problem will therefore be to determine *a*) function F(i), *b*) to integrate equation (2.2.) with regard to the time interval t_2-t_1. In the following, the first step is considered in sections 3 to 7; section 8 deals with the second step.

3. MATHEMATICAL EXPRESSION FOR PHOTOSYNTHESIS-LIGHT CURVES.

It is well known that the photosynthesis-light relationship has a relatively simple appearance: at low intensities, rates are primarily limited by light, whereas at higher levels they reach saturation, i.e. they are no longer dependent on light. Reaction curves of this kind can be described by two parameters. These are the initial, almost linear slope and the height of the saturation plateau.

As an analogy to Michaelis' well known enzymatic reaction curve Baly (1935) first used a simple rectangular hyperbola. In fact, some experimental data do fit such expressions, but most findings show that the initial slope is in general steeper than required by the Michaelis hyperbola.

Other expressions have been given later by several authors; the best known is perhaps that of Smith (1936). This expression, which for mathematical reasons will be frequently used in this paper, can be given as follows:

$$p = p_{max} \cdot \frac{ai}{\sqrt{1 + (ai)^2}} \qquad (3.1.)$$

Its mathematical properties are recognized at once; at low *ai*, *p* depends linearily upon *i* up to values of about $1/2a$; at high intensities ($ai >> 1$) *p* becomes identical with p_{max}.

Talling (1957) points out that the term $1/a$, which he designates as I_k, has the dimension of a light intensity and can in fact be interpreted as the intensity at the intersect between the linear slope with the height of the saturation plateau; the onset of light saturation begins at about $I_k/2$.

From a biological point of view, I_k may be considered as a measure of the degree of adaptation endured by the phytocoenosis in question to low light intensities.

Yet, I_k is not an independent parameter. Talling has discussed its relationship to temperature. A further modification is introduced below. Its significance therefore may appear somewhat questionable; nevertheless, I_k has proved to be a practical expedient to bring the function F(i) to a dimensionless number, and this entity is perhaps the best measure to discuss a number of questions on a common ground.

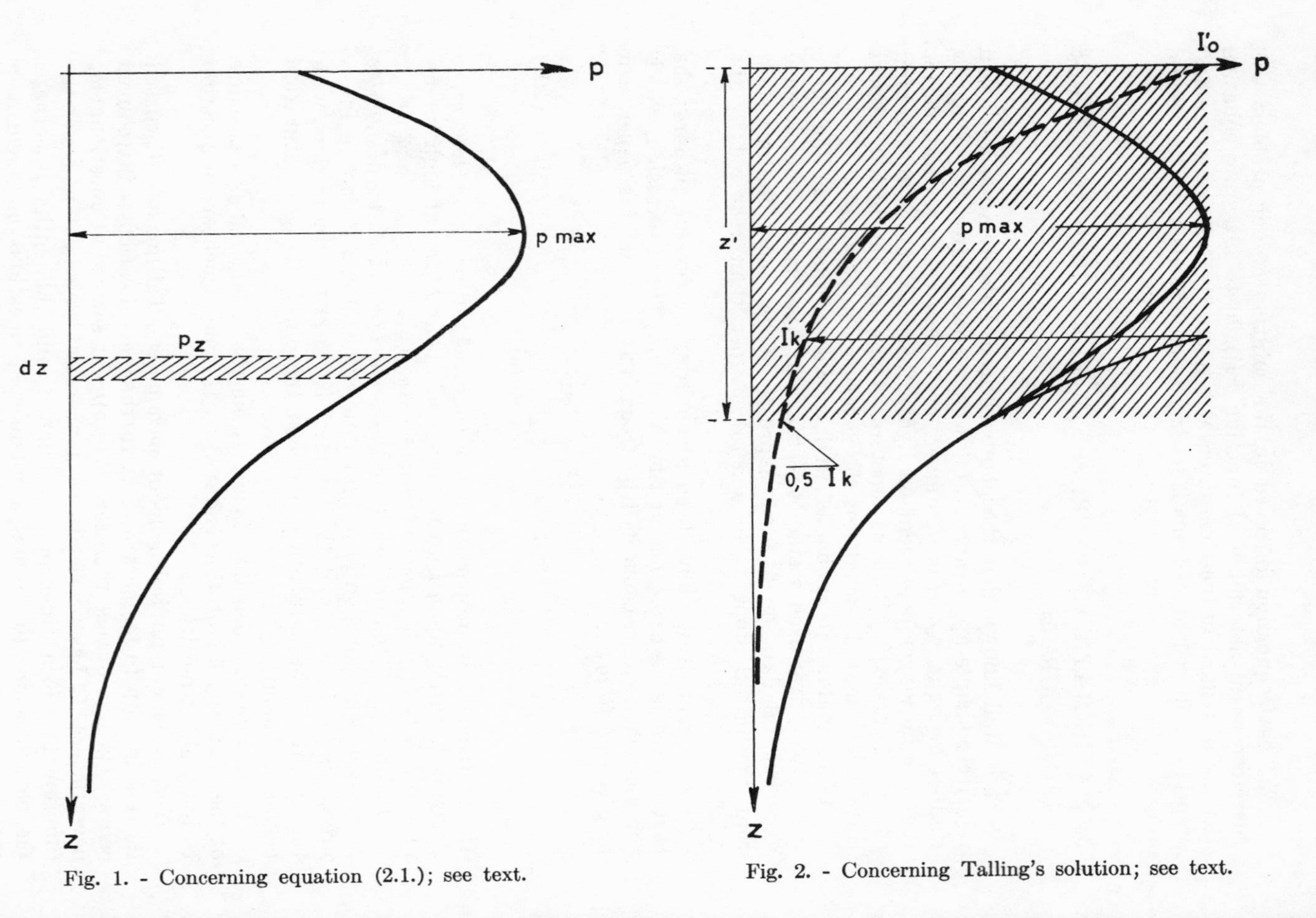

Fig. 1. - Concerning equation (2.1.); see text.

Fig. 2. - Concerning Talling's solution; see text.

Smith's formula in its original form is restricted to ideal cases of photosynthesis-light relations lacking light inhibition phenomena at higher intensity levels; yet, it is well known that *in situ* curves from most fresh and saline waters differ from this. In the following, I have given attention to this by introducing into Smith's formula a correction function $h = \varphi(\alpha i)$, so that

$$p = p_{max} \cdot \frac{ai \cdot \varphi(\alpha i)}{\sqrt{1 + (ai)^2}} \tag{3.2.}$$

Some properties of h can be established at once, i.e. for the limits $i \rightarrow O$, $(h \rightarrow 1)$, and $i \rightarrow \infty$, $(h \rightarrow O)$; furthermore formula (3.2.) must approximate Smith's formula for $\alpha \rightarrow O$.

4. TALLING'S SOLUTION

Talling's solution of equation (2.1.) was primarily derived by planimetric methods. He assumed that integral photosynthesis beneath a unit of surface is equal to a rectangle given by p_{max} and the depth z' at which light intensity is about $0.5\ I_k$ (cf. Fig. 2), so that

$$\Sigma\, p_{(t)} = p_{max} \cdot z'.$$

From Lambert's law follows,

$$0.5\ I_k = I'_o \cdot e^{-\varepsilon z'},$$

or,

$$z' = 1/\varepsilon\,(\ln I'_o - \ln 0.5\ I_k),$$

so that

$$\Sigma\, p_{(t)} = \frac{p_{max}}{\varepsilon} \cdot \ln(2\, I'_o/I_k) \tag{4.1.}$$

This solution has proved to be a satisfactory approximation for many cases (cf. Vollenweider 1960, Rodhe 1965), despite the fact that surface and subsurface light inhibition were assumed to be negligible with regard to the total integral. Beside this limitation, one has to note that the theory breaks down for low subsurface light intensities, i.e. when $I'_o < I_k$.

Assume that p_z at any depth z is sufficiently described by equation (3.1.); then the following analytical integration of Talling's integral can be given.

Let be:

$$p_z = p_{max} \cdot \frac{ai}{\sqrt{1 + (ai)^2}}$$

then:

$$\Sigma\, p_{(t)} = p_{max} \int_{o}^{\infty} \frac{ai}{\sqrt{1 + (ai)^2}}\, dz\,.$$

Introducing from Lambert's law

$$dz = -\frac{di}{\varepsilon\, i}\,,$$

and setting

$$ai = i'\,, \qquad di = \frac{1}{a} \cdot di',$$

we have, with the new integration limits (O and I'_o = subsurface light intensity),

$$\Sigma\, p_{(t)} = -\frac{p_{max}}{\varepsilon} \int_{i=I'_o}^{i=o} \frac{di'}{\sqrt{1 + (i')^2}}$$

and, by considering $a = 1/I_k$,

$$\Sigma\, p_{(t)} = \frac{p_{max}}{\varepsilon} \left[\ln \left(I'_o/I_k + \sqrt{1 + (I'_o/I_k)^2}\,\right)\right] \qquad (4.2.)$$

This equation may also be written as

$$\Sigma\, p_{(t)} = \frac{p_{max}}{\varepsilon} \cdot \mathrm{ArSin}\,(I'_o/I_k) \quad [1] \qquad (4.2a.)$$

ArSin (i) is a tabulated function.

ArSin (I'_o/I_k) is a solution for F(i) of equation (2.3.) satisfying the condition of a dimensionless number; it is noteworthy that F(i) is not a function of the incident light only but a function of the ratio of two light intensities.

The advantage of solution (4.2.) or (4.2a.) over the original equation of Talling is evident since it does not break down at low light intensities; equation (4.2.) turns into equation (4.1.) if $I'_o/I_k >> 1$.

5. OTHER SOLUTIONS FOR F(i)

In order to overcome the difficulties of *ln* ($2\, I'_o/I_k$) breaking down at lower light intensities, one can tentatively test similar functions; so, Talling (1961) proposes *ln* ($1 + I'_o/I_k$), a function which also might be applied to terrestrial photosynthesis estimates.

---

[1] Area Sinus Hyperbolicus (must not be confused with arcsin).

In Table 1 a set of numerical calculations for various ratios of I'_o/I_k and various functions are given. *ln* $(2\ I'_o/I_k)$ corresponds well with *ArSin* (I'_o/I_k) for I'_o/I_k-ratios > 2, but there is no correspondance for values below 1. Confronted with this function, *ln* $(1 + I'_o/I_k)$ is certainly better at low ratios but remains over the whole range of interest about 16-17% behind *ArSin* (I'_o/I_k). In turn, a somewhat better agreement with the latter function would result from *ln* $(1 + 2\ I'_o/I_k)$, but it overestimates the function at low ratios.

Table 1. - Numerical values for various assumptions concerning F(i).

I'_o/I'_k	I'_k/I'_o	ArSin (I'_o/I'_k)	ln $(2\,I'_o/I'_k)$	ln $(1 + I'_o/I'_k)$	ln $(1+2\,I'_o/I'_k)$
0	∞	0	$-\infty$	0	0
0.5	2	0.48	0	0.40	0.69
1	1	0.88	0.69	0.69	1.08
2	0.5	1.44	1.39	1.09	1.61
3	0.33	1.82	1.79	1.38	1.94
4	0.25	2.10	2.08	1.61	2.20
5	0.2	2.31	2.31	1.79	2.40
10	0.1	3.00	3.00	2.40	3.04
15	0.067	3.40	3.40	2.78	3.43
20	0.05	3.68	3.68	3.04	3.72
30	0.033	4.09	4.09	3.43	4.11

The *ln* $(1 + I'_o/I_k)$ may appear appropriate in cases where subsurface light inhibition must be taken into consideration, yet, this function is in reality the solution of F(i) when Baly's equation for p is introduced in (2.1.). Mathematically this means that the underestimation does not refer to the light inhibited part of the basic curve but to its linear range.

Such a kind of compensation is easily produced if one turns from the original argument of an integral to a function of either of its limits (i.e. in our case from p to I'_o). One has therefore to be very carefull in using integrated functions, even if they seem to fit experimental data. It shall be demonstrated below that this critisism must be applied to Talling's solution.

6. GRAPHICAL DETERMINATION OF I_k.

In an earlier paper (Vollenweider 1960) I have shown that I_k values can easily be evaluated from a semi-logarithmic plot of *in situ* C^{14} measurements by determining the intensity of the total light flux (given as % of I'_o) at the intersect of the linear, i.e. light controlled part of the photosynthesis curve, with the height of the photosynthesis plateau. The approximate relative intensity of the photosynthetic active light at any depth is calculated from the relation:

$$i_{total} = \frac{1}{3}\,(i_{630} + i_{530} + i_{430})\ ,$$

all values given as a % (cf. Vollenweider, 1961).

With regard to the choise of ε there is some disagreement between different authors as to whether one should take a mean extinction coefficient (Vollenweider) or the extinction coefficient of the most penetrating wave length (Talling, Rodhe). Experimental data seem, in fact, to justify both ways. However, the disagreement can not be serious if one considers that, with increasing depth, the mean extinction coefficient approximates progressively the extinction coefficient of the most penetrating spectral region, i.e.

$$\lim_{z \to \infty} \varepsilon_v \to \varepsilon_v^{\lambda\, max}$$

A more serious question concerns the exact determination of I_k. Rodhe (1965) uses the intensity resulting from the intersect with the line of the most penetrating wave length; yet, from a physiological point of view there is no reason to assume that only this part of the aquatic light spectrum should determine the onset of the saturation plateau. At light intensities close to the expected I_k the contribution of the remaining spectral regions to the total intensity may be as high as 50 %.

7. INTEGRALS WHICH CONSIDER SUBSURFACE LIGHT INHIBITION

7.1. Steele's integral

In contrast to Talling, Steele (1962) does not use a physiologically justifiable formulation of p with exception of the linear part of the photosynthesis-light reaction. While putting

$$p = a \cdot i \cdot p_{max} \cdot e^{1-ai} \qquad (7.1.1.)$$

he gives more attention to the light inhibited upper part of the photosynthesis curve.

Introducing (7.1.1.) into the basic equation (2.1.) follows:

$$\Sigma\, p_{(t)} = a \cdot p_{max} \int_{o}^{\infty} i \cdot e^{1-ai}\, dz$$

or, by considering Lambert's law for dz and using the new limits,

$$\Sigma\, p_{(t)} = -\,\frac{a \cdot p_{max}}{\varepsilon} \int_{i=I'_o}^{i=o} e^{1-ai}\, di\ ,$$

the solution of which is

$$\Sigma\, p_{(t)} = \frac{p_{max}}{\varepsilon} \cdot e \cdot [1 - e^{-aI'_o}] \qquad (7.1.2.)$$

From differentiation of equation (7.1.1.) one recognizes that $1/a$ is not equal to the former I_k-value but to the light intensity I_m at which p equals p_{max}, so that

$$\Sigma\, p_{(t)} = \frac{p_{max}}{\varepsilon} \cdot 2.72\, [1 - e^{-I'_o/I_m}] \qquad (7.1.2a.)$$

In general I_m is about 1/2 to 1/4 I'_o, ($1 - e^{-I'_o/I_m} = 0.864 - 0.982$), so that equation (7.1.2a.) simplifies to

$$\Sigma\, p_{(t)} = \frac{p_{max}}{\varepsilon}\, (2.4 - 2.7) \qquad (7.1.3b.)$$

This i sthe form given earlier by Vollenweider (cf. Rodhe, Vollenweider and Nauwerck, 1958).

Although (7.1.2a.) does not break down at low light intensities this equation contains two artificial elements: *a*) the maximum conversion factor (2.72); in fact experimental data do not always fit the equation (cf. Saunders, Trama and Bachmann, 1962); *b*) coupling of the linear with the light inhibited part of the curve by the same factor a. However, this handicap is easy to overcome by introducing two distinct values, say a and α ($a \neq \alpha$), so that

$$p = a \cdot i \cdot p_{max} \cdot e^{1-\alpha i} \qquad (7.1.3.)$$

Using this formula in (2.1.), the solution will be

$$\Sigma\, p_{(t)} = \frac{p_{max}}{\varepsilon} \cdot e \cdot \frac{a}{\alpha}\, [1 - e^{-\alpha I'_o}] \qquad (7.1.4.)$$

Measurable expressions for a and α are found as

$$1/ea = I_k, \text{ and } 1/\alpha = I_m,$$

so that the final form may be written as

$$\Sigma\, p_{(t)} = \frac{p_{max}}{\varepsilon} \cdot (I_m/I_k)\, [1 - e^{-I'_o/I_m}] \qquad (7.1.4a.)$$

This is a more flexible formulation of Steele's integral; preliminary calculations have shown that (7.1.4a.) is more appropriate to experimental data than the former equation.

7.2. A generalized approach

As pointed out in section 3, Smith's formula can be generalized by introducing an inhibition function h for which certain boundary conditions have already been established. Taking into account that the basic function represents a hyperbola of higher order, h can be choosen with a corresponding structure, *e.g.*

$$p = p_{max} \cdot \frac{a\,i}{\sqrt{1 + (a\,i)^2}} \cdot \frac{1}{(\sqrt{1 + (\alpha\,i)^2})^n} \qquad (7.2.1.)$$

Varying n and α a family of curves results which may satisfy a great variety of experimental data (cf. Fig. 3).

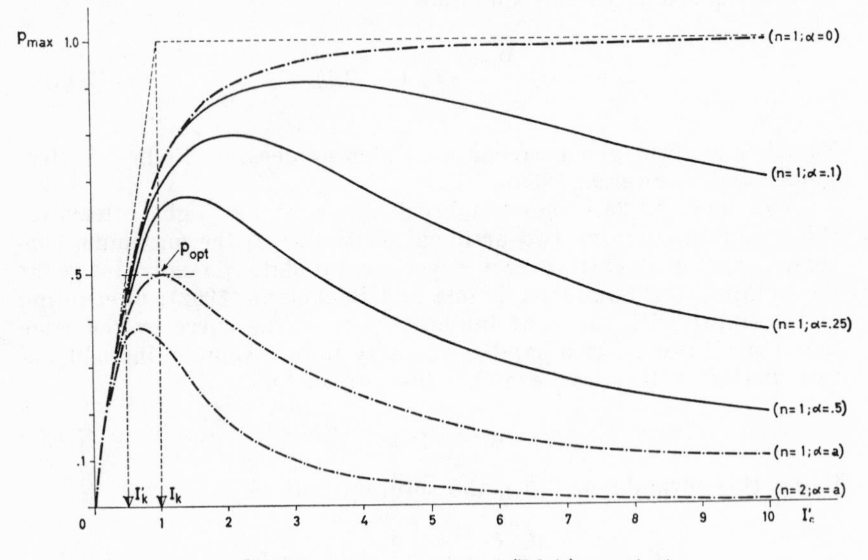

Fig. 3. - Concerning equation (7.2.1.); see text.

The handicap of (7.2.1.) as a general formulation of p lies in difficulties resulting when it is introduced in (2.1.). There is no general solution of such integrals; yet, some specific cases are easier to handle.

Before proceeding consider some of the basic properties of the above equation. In Fig. 3 Smith's equation is represented by curve $n = 1$, $\alpha = O$. Its basic parameters are p_{max} and I_k ($= 1/a$). Any positive value of n and α different from O produces curves which in contrast to the first function have a distinct maximum. It can be recognized at once that this maximum is not identical with p_{max} of Smith's formula: in correspondance with that, Talling's value I_k also assumes a different meaning. For the sake of clarity these basic parameters will be called from here on p_{opt} and I'_k, restricting p_{max} and I_k to the sole case where Smith's equation applies.

A clear distinction of the meaning given to the above parameters is required in view of the following mathematical elaborations; if beyond that there is also a biologically meaningful interpretation of the various terms, this cannot be said without further investigations on the subject.

With this background some of the integrals can now be discussed. Let's first put $\alpha = a$; then the inhibition function reduces to the variation of n; $n = 1, 2, 3...$

$n = 1$. The integration of this case will be simple for the integrand reduces to a well known elementary function;

$$\Sigma\, p_{(t)} = p_{max} \int_o^\infty \frac{a\, i}{1 + (a\, i)^2}\, dz \qquad (7.2.2.)$$

and by changing the variables as previously,

$$\Sigma\, p_{(t)} = \frac{p_{max}}{\varepsilon} \operatorname{arctg} (I'_o/I'_k) \qquad (7.2.3.)$$

Now, as can easily be demonstrated,

$$p_{max}/p_{opt} = I_k/I'_k = 2\,,$$

so that

$$\Sigma\, p_{(t)} = \frac{p_{opt}}{\varepsilon} \cdot 2 \cdot \operatorname{arctg} (I'_o/2\, I'_k) \qquad (7.2.3a.)$$

$n = 2$. This assumption also leads to a relatively simple integration problem,

$$p_{(t)} = p_{max} \int_o^\infty \frac{a\, i}{(\sqrt{1 + (a\, i)^2})^3} \qquad (7.2.4.)$$

the solution of which is:

$$\Sigma\, p_{(t)} = \frac{p_{max}}{\varepsilon} \cdot \frac{I'_o/I_k}{\sqrt{1 + (I'_o/I_k)^2}} \qquad (7.2.5.)$$

From differentiation follows:

$$p_{max}/p_{opt} = I_k/I'_k \sim 2.6\,,$$

so that the final form will be:

$$\Sigma\, p_{(t)} = \frac{p_{opt}}{\varepsilon} \cdot \frac{I'_o/I'_k}{\sqrt{1 + (I'_o/2.6\, I'_k)^2}} \qquad (7.2.5a.)$$

At higher I'_o/I'_k values, the function approximates

$$\Sigma\, p_{(t)} = \frac{p_{opt}}{\varepsilon} \cdot 2.6\,,$$

and is therefore apparently very similar to Vollenweider's simplified relation of Steele's integral.

(4.2.), (7.2.3 a.) and (7.2.5 a.) are given in Fig. 4 as functions of I'_o/I'_k.

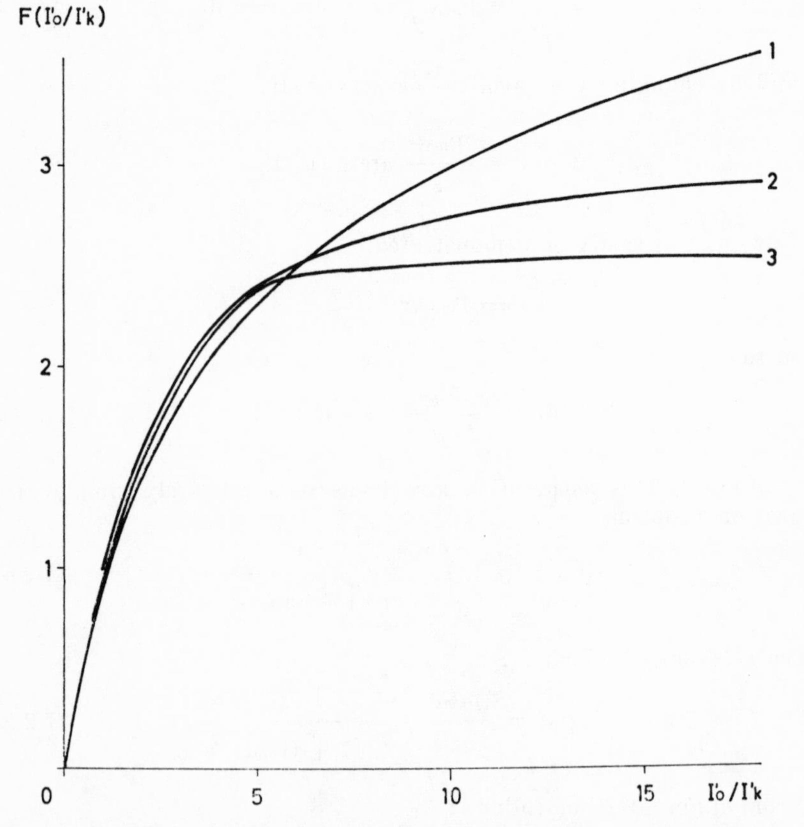

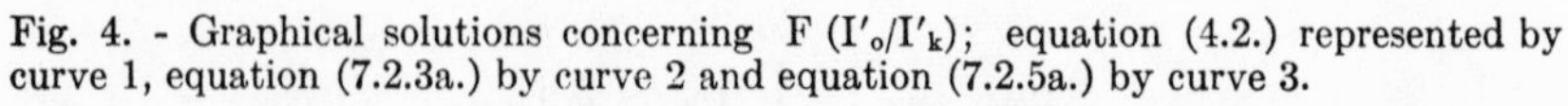

Fig. 4. - Graphical solutions concerning $F\,(I'_o/I'_k)$; equation (4.2.) represented by curve 1, equation (7.2.3a.) by curve 2 and equation (7.2.5a.) by curve 3.

$n > 2$. It does not appear that values higher than $n = 2$ are of specific interest; in any case, the solution of corresponding integrals can always be achieved by the combination of expressions of

$$\sqrt{1 + (I'_o/I'_k)^2}$$

with an *arctg*-function of I'_o/I'_k.

A common feature of all integrals considered here is given by the rapid increase at low I'_o/I'_k values. Up to about $I'_o/I'_k = 5$ they do not differ much from each other; above this value a more or less pronounced turn occurs, and the functions representing basic curves showing surface and subsurface light inhibition tend toward a plateau.

Variations of α. For this case I shall limit the analysis to the family of curves where $n = 1$; cf. Fig. 3. Their principal parameters are related by:

$$p_{opt}/p_{max} = I'_k/I_k = a/(a + \alpha) \ ,$$

a and α can be determined from measurable entities, *i.e.* I'_k and I_m (intensity of the light optimum) by:

$$\alpha = 1/I^2_m \cdot a \ ,$$

and

$$a = 1/2I'_k + \sqrt{(1/2I'_k)^2 - 1/I^2_m} \ .$$

The determination of $\Sigma\, p_{(t)}$ does not lead to a simple solution; for low values of *ai* the solution can be approximated by developing

$$(\sqrt{1 + (a\, i)^2})^{-1/2}$$

in series, yet appears to be of little interest.

For higher values of *ai* one can say that (F(i) in any case lies between *ArSin* (I'_o/I'_k) and *2 arctg* $(I'_o/2I'_k)$; therefore the approximated integral can be given as

$$\Sigma\, p_{(t)} = \frac{p_{opt}}{\varepsilon}\left[\mathrm{ArSin}\,(I'_o/I'_k) - \alpha\left[\mathrm{ArSin}\,(I'_o/I'_k) - 2\,\mathrm{arctg}\,(I'_o/2I'_k)\right]\right] \quad (7.2.6.)$$

This formula turns into (4.2.) for $\alpha = 0$, and into (7.2.3.) for $\alpha = 1$.

7.3. Why does Talling's integral work?

At this point, I would like to introduce some notes about Talling's integral which, as stated above, does not account for subsurface light inhibition. On the other hand, experience shows that applying formula (4.1.) to practical problems, the differences between calculated and planimetric values are rather small even in cases where subsurface light inhibition does not appear to be negligible (cf. Vollenweider, 1960). This phenomenon calls for a more appropriate analysis of what is really measured with Talling's integral.

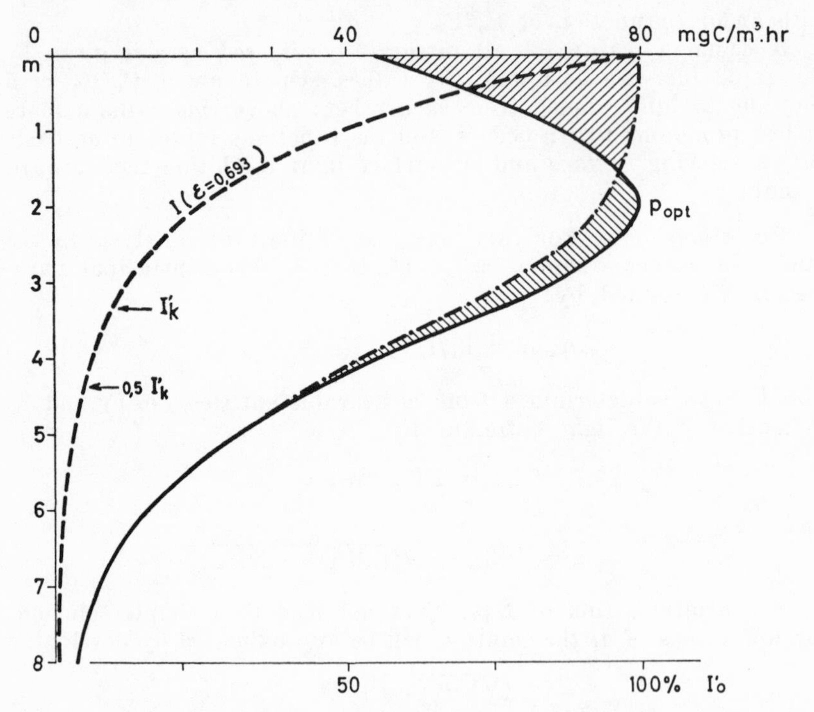

Fig. 5. - Concerning Talling's integral applied to photosynthesis-depth curves showing surface light inhibition; see text.

The idea can be fixed on the following theoretical example. In Fig. 5 the solid line was constructed assuming that with the

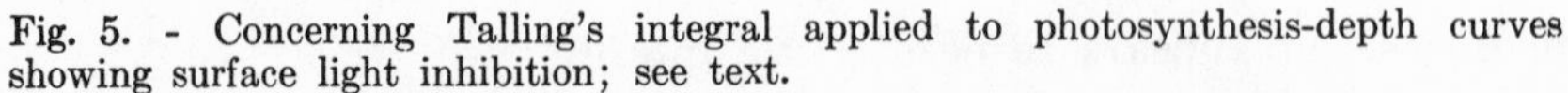

$$p = p_{max} \cdot \frac{a\,i}{\sqrt{1 + (a\,i)^2}} \cdot \frac{1}{\sqrt{1 + (\alpha\,i)^2}},$$

constants: $a = 1$, $\alpha = 0.25$, $p_{max} = 100$ mg C/m^3 · hr, and hence $p_{opt} = 80$ mg C/m^3 · hr, $I'_o/I'_k = 10$, $\varepsilon = 0.693$. By experience the resulting curve looks like many *in situ* curves would look.

By calculation and from planimetric measurements the value of the corresponding integral is found as 341 mg C/m^2 · hr; using Talling's integral for the same assumptions one gets 346 mg C/m^2 · hr, i.e. a value only about 1.2% higher than the former.

Yet, the curve to which Talling's integral applies, *i.e.* Smith's formula, is somewhat different from our simulated *in situ* curve, and is given in Fig. 5 by a pointed-dashed line. This curve intersects the former somewhat above p_{opt}, according to which the integral overestimates the area above and compensates for the underestimation below that point.

From the analytical point of view, Talling's integral can therefore not be considered as the integral of the photosynthesis-depth curve, except, naturally, in that case where Smith's formula applies. With regard to practical calculations, this inadequacy does, however, not appear to be serious, *i.e.* as long as only integral photosynthesis is considered; it could, however, lead to discrepancies if extended to conclusions concerning the properties of the basic photosynthesis-depth curve.

8. CALCULATION OF DAILY RATES OF PHOTOSYNTHESIS

In the preceding sections attention was drawn exclusively to the evaluation of equation (2.1), *i.e.* to the calculation of *instantaneous rate integrals* of photosynthesis as functions of I'_o/I'_k; time was not considered in that connection. Here, we shall proceed to the study of equation (2.2), *i.e.* to the integration of these rates over the time interval t_2—t_1. This problem is directly connected with experimental studies of primary production rates in as far as normal exposures of C^{14}, or similar assays, involve a certain exposure time. Under *in situ* conditions *instantaneous rates* continually change with changing environmental conditions.

In accordance with the model conditions, the only variable of interest will be the change of the incident light changing with time; the change of instantaneous rates can then be described by a time function of the ratio I'_o/I'_k. To simplify, I shall confine the analysis to a *standard light day,* given *e.g.* by a clear, cloudless day, since the course of the incident light may be sufficiently described by the equation:

$$I'_{o\,(t)} = I'_{o\,(max)} \cdot \tfrac{1}{2} (1 + \cos 2\pi/\lambda \cdot t) \qquad (8.1.)$$

Time t is positively or negatively measured with regard to zero time taken at local midday. $I'_{o\,(max)}$ signifies the incident subsurface light at zero time, and λ is a *day length factor* for the light day, and is measured in the same units as t. λ can be found in nautical almanacs for any latitude and for any time of the year.

Day rate integrals can be calculated according to two different methods, the first one based on summation of rates of short intervals, the second one is called here the *reference integral method.*

8.1. Simulation of daily photosynthesis-depth curves

This method is essentially the same as that used by Talling (1957). Selecting anyone of the equations given for p and estimating during a 12 hour standard light day consecutive photosynthesis-depth curves/hour, curves for a whole day or any fraction of it can be simulated by adding the single curves over the desired time interval.

This was done here for two assumptions concerning p:

a) using Smith's function;

b) considering a situation of very strong midday light inhibition at the surface given by the relation used in (7.2.4.).

The results of this hypothetical computations are presented in Table 2 and 3 and Fig. 6 and 7. Morning and afternoon trends are assumed to be symmetrical, so that the summation can be limited to 1/2 day. The following implications are of interest:

a) The depth gradient of the exponential range of the simulated daily curve is practically identical with that of any instantaneous rate curve.

Table 2. - Simulation of a half-day photosynthesis-depth curve assuming p given by Smith's formula; cf. Fig. 6.

p_{max} = 100 mg C/m³ h; ε = 0.693; λ = 12 hrs; $I'_{o\,(max)}/I'_k$ = 10.

Time interval:	0 - 1	1 - 2	2 - 3	3 - 4	4 - 5	5 - 6	Sum (1)
I'_o/I'_k	0.67	2.5	5	7.5	9.33	10.0	
Depth in m							
0	55	93	99	99	99,5	99.5	495
1	32.5	77	93	97.5	99	99	449
2	16.5	52.5	77	85	92	93	370
3	8.2	31	52.5	67	74.5	77	272
4	4.1	15.5	31	43	50	52.5	170
5	2.1	7.8	15.5	23	29	31	94
6	1.1	3.9	7.8	11.6	14.5	15.5	47
7	0.6	2.0	3.9	5.8	7.2	7.8	23
8	0.3	1.0	2.0	2.9	3.6	3.9	12
Sum (1)	93	237	367	385	418	428	(1712-1731)

Calculations:

a) following Talling's formula: $\frac{495}{0.693} \ln 12.5 = 1800$ mg C/m²

b) following reference integral method: $\frac{300}{0.693} \cdot 4.05 = 1760$ mg C/m²

(1) Following Simpson's formula.

b) Although the general shape of the day rate curve is similar to that of the basic curve used, the former is mathematically not identical with the latter. The summation of Smith curves produces a curve slightly less curvated with depth; in the second example, the day curve resembles rather the basic curve used in (7.2.6.) with an α of about 0.3.

c) The I'_o/I'_k resulting for the day rate curve is lower than that pertaining to the midday rate curve; the quantitative effect on the shift depends on the particular kind of basic curve assumed; it is less pronounced in cases representing strong midday inhibition (cf. figures cit.).

As to the possibility of calculating day rate integrals with the aid of analytical formulas, it is clear from the above that in no cases can a simple mathematical expression apply to the problem; yet, with regard to what has been pointed out, one can tentatively try some approximation with any of the formulas previously present-

Table 3. - Simulation of half-day photosynthesis-depth curve, assuming p given by $\frac{a\,i}{(\sqrt{1+(a\,i)^2})^3}$; cf. Fig. 7.

p_{opt} = 100 mg C/m³·h; ε = 0,693; λ = 12 hrs; $I'_{o\,(max)}/I'_k$ = 10.

Time interval:	0 - 1	1 - 2	2 - 3	3 - 4	4 - 5	5 - 6	Sum (1)
I'_o/I'_k	0.67	2.5	5	7.5	9.33	10.0	
Depth in m:							
0	63	91	46	24	16.5	14.7	251
1	34	92	91	64	51	46	353
2	17.5	59	92	99	94	91	407
3	8.9	31	59	79	88	92	313
4	4.5	16.2	32	46	55.5	59	184
5	2.3	8.2	16.2	24	29.5	32	96
6	1.2	4.1	8.2	12.3	15	16.2	49
7	0.6	2.1	4.1	6.2	7.6	8.2	25
8	0.3	1.1	2.1	3.1	3.8	4.1	13
Sum (1)	100	260	327	344	351	354	(1560-1580)

Calculations:

a) following Talling's formula: $\frac{410}{0.693} \ln 15.4 = 1620$ mg C/m²

b) following reference integral method: $\frac{300}{0.693} \cdot 3.6 = 1560$ mg C/m²

(1) Following Simpson's formula

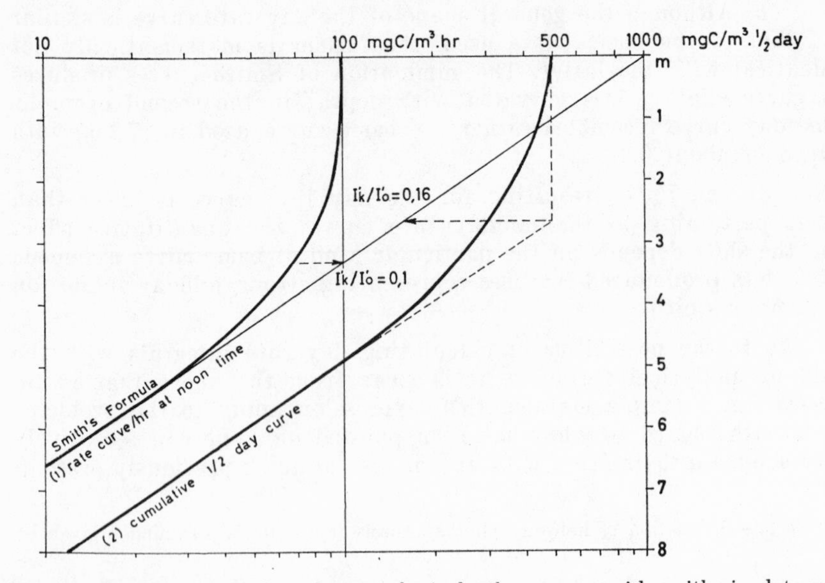

Fig. 6. - Simulated half day photosynthesis-depth curve; semi-logarithmic plot; see text.

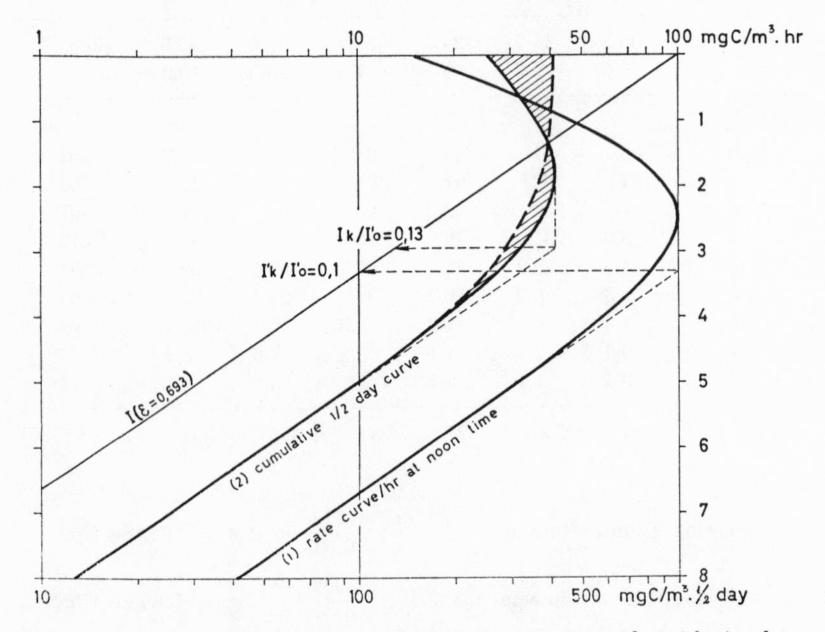

Fig. 7. - Simulated half day photosynthesis-depht curve; semi-logarithmic plot; see text.

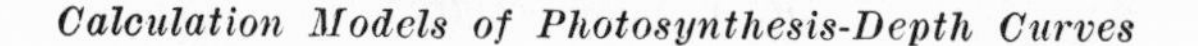

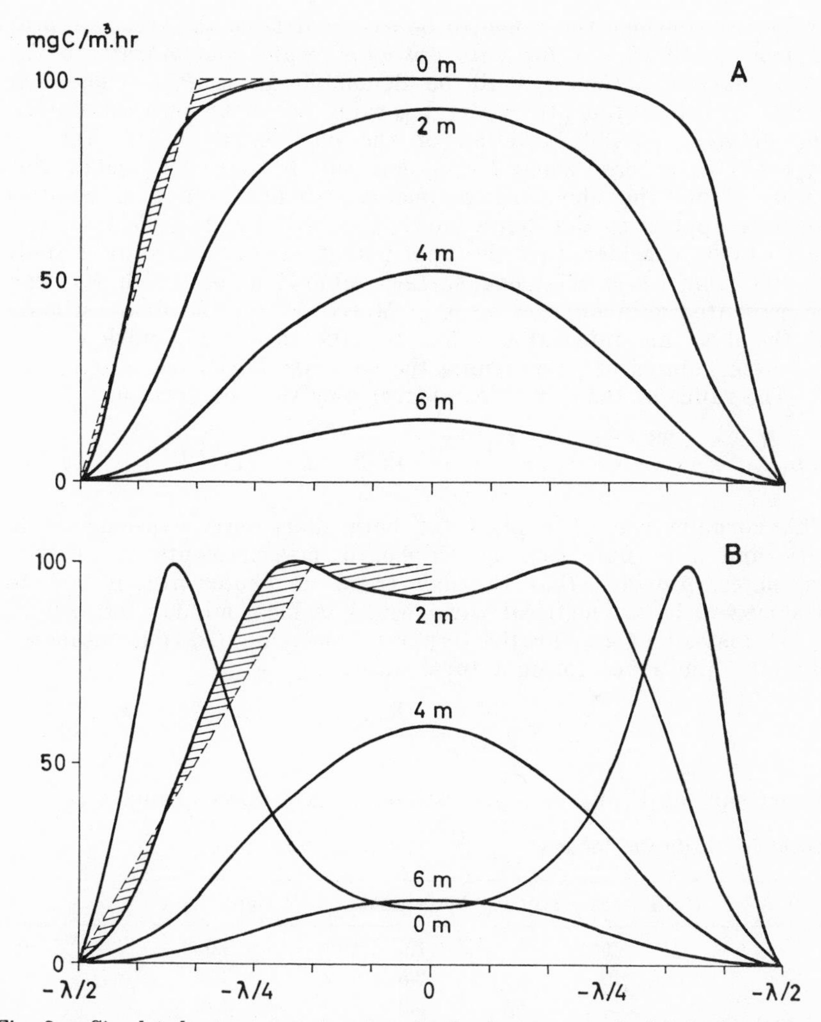

Fig. 8. - Simulated rate curves at various depths during a standard light day for two assumptions: A) Subsurface light inhibition negligible, B) showing strong subsurface light inhibition.

ed for rate integrals. Talling's formula has proved to be satisfactory for this purpose, in as far as also in this case total integrals are only slightly overestimated (cf. Tables 2 and 3).

A slightly different problem arises if one wants to estimate day rate integrals from short term experiments, say one or two hour experiments performed around midday. For this purpose Talling has introduced the concept of *light divisions per hour*. In addition to this method I shall now discuss here a more statistical approach.

Let us consider the range to be expected from the trend of p at various depth of the *day rate optimum,* which in contrats to the *instantaneous optimum* shall be denominated as P_{opt}. The P_{opt} builds up at that depth where p_{opt}-rates per hour proceed during the greatest possible fraction of the day length λ (cf. Fig. 8). P_{opt} will in no case equal $\lambda \cdot p_{opt}$ but will be somewhat below this value. From the above figure one can estimate that, if Smith's formula applies to the depth curve, P_{opt} will be about $5/6\ \lambda \cdot p_{opt}$; this can be considered as the upper limit of P_{opt}. The lower limit results from cases of strong surface inhibition for which P_{opt} can be estimated as about $2/3\ \lambda \cdot p_{opt}$. Moreover, for the final estimate of the day rate integral one has to take into consideration what has been pointed out concerning the *day rate* I'_o/I'_k (cf. 8.1.c.).

The range of the day rate integral may then be given as:

$$\Sigma\, p_{(day)} = \frac{(.83 - .67)\ \lambda \cdot p_{opt}/hr}{\varepsilon} \cdot \ln\left[2\,(.63 - .77)\ I'_o/I'_k\right] \qquad (8.1.1.)$$

This formula can be applied for both short term experiments *in situ* and for conversion of shipboard measurements to *in situ* estimates, provided that the day trend of photosynthesis can be assumed to be symmetrical with regard to local midday (cf. below).

If instead of p_{opt}/hr the day rate optimum (P_{opt}) is measured directly, the above formula turns into:

$$\Sigma\, p_{(day)} = \frac{P_{opt}}{\varepsilon} \cdot \ln r \qquad (8.1.2.)$$

From various I'_o/I'_k *ln* r are estimated as follows (Table 4):

Table 4. - Estimates for ln r

I'_o/I'_k (at midday)	lower limit	upper limit
5	1.9	2.05
10	2.5	2.75
15	3.0	3.15
20	3.3	3.4

Assuming that anyone of these values is equally likely to occur, then the mean value of *ln* r will be something like 2.75; this is in fact close to experience.

8.2. The integration of « surface rate curves ». Reference integral method.

The second method adopted here concerns the integration of *surface rate curves, i.e.* curves which represent the day trend of total photosynthesis below a unit of surface. Curves of this kind can be constructed from (8.1.) and Fig. 4 as a function of I'_o con-

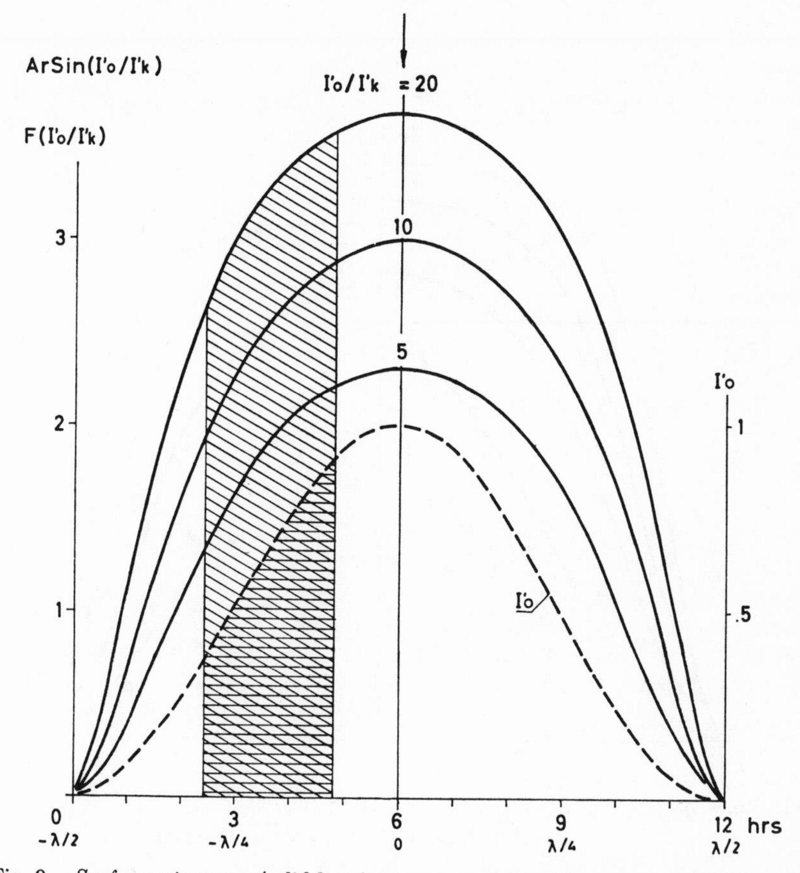

Fig. 9. - *Surface rate curves* (solid lines) for various assumptions concerning $I'_{o(max)}/I'_k$ in relation to the course of the incident light at a *standard light day* (dashed line); symmetrical solutions; the arrow at the top indicates local midday.

$$F\,(I'_o/I'_k) = \text{ArSin}\,(I'_o/I'_k).$$

sidering various assumptions for $I'_{o\,(max)}/I'_k$. For this, our three basic solutions of (2.1.), (*i.e.* 4.2., 7.2.3. and 7.2.5.), were used (cf. Figs. 9 and 10), assuming that the morning and afternoon trends of the curves being symmetrical. This kind of solution shall be called *symmetrical* with regard to local midday; *non symmetrical* solutions are considered later.

Introducing (8.1.) into any of the former equations, the basic equation (2.2.) must be integrated from $-\lambda/2$ to $+\lambda/2$. The solution of the resulting equations is a non-elementary integration problem; yet, with regard to the desired goal, a more rapid, although incomplete, planimetric integration may be as good as a strict analytical one.

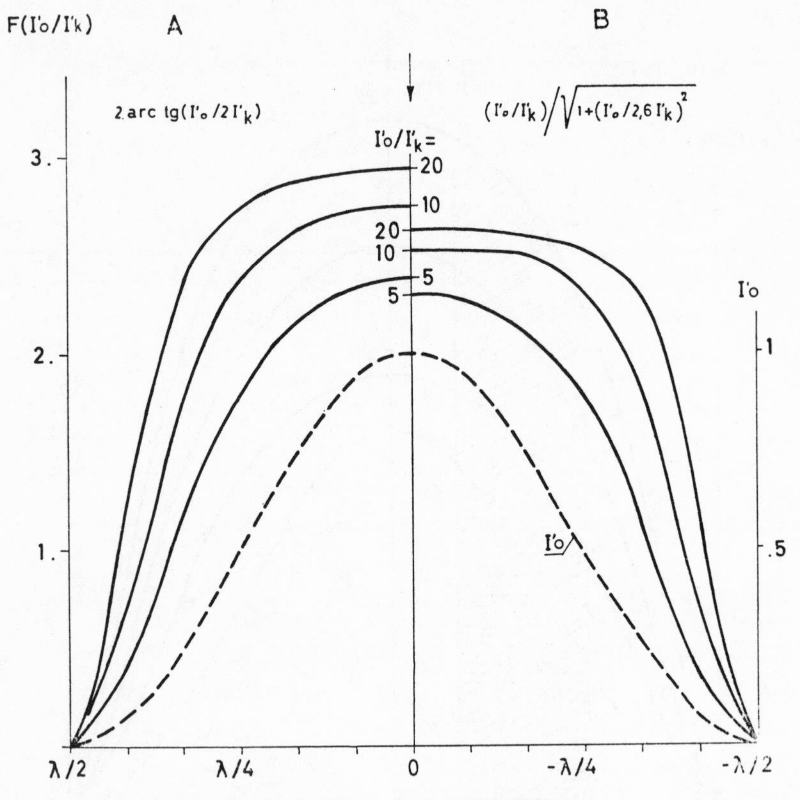

Fig. 10. - As Fig. 9; A) $F(I'_o/I'_k) = 2 \text{ arctg } (I'_o/I'_k)$,
B) $F(I'_o/I'_k) = (I'_o/I'_k) / \sqrt{1 + (I'_o/2.6\, I'_k)^2}$

I shall now introduce what I call the *reference integral method.*

Assume that photosynthesis is perfectly proportional to light intensity at any depth and at any moment during the day, and that the highest rate occuring at midday in the course of a standard day is exactly p_{opt} (mg $C/m^3 \cdot$ hour); then total photosyinthesis beneath a unit of surface would be:

$$(\Sigma\, p)_{ref} = 1/\varepsilon \cdot p_{opt} \cdot \lambda/2 \qquad (8.2.1.)$$

$(\Sigma p)_{ref}$ may be called the *day rate reference integral.* The real day rate integral of photosynthesis is then proportional to the quotient of the area beneath a selected surface rate curve and the area beneath the standard insolation curve (cf. Figs. 9 and 10) which later, in turn, is proportional to equation (8.2.1.). The total integral is therefore:

$$\Sigma\, p_{(day)} = (\Sigma\, p)_{ref} \cdot Q \qquad (8.2.2.)$$

Various values of Q, estimated by planimetric procedure, are given in Table 5 and Fig. 11.

Table 5. - Values of Q (cf. equation 8.2.2.) for various assumptions concerning $F(I'_o/I'_k)$.

	$I'_o/I'_k =$	5	10	20
$F(I'_o/I'_k)$				
1)	$= \mathrm{ArSin}\,(I'_o/I'_k)$	2.85	3.95	5.15
2)	$= 2\,\mathrm{arctg}\,(I'_o/2I'_k)$	3.0	3.85	4.55
3)	$= \dfrac{I'_o/I'_k}{\sqrt{1 + (I'_o/2.6\,I'_k)^2}}$	2.95	3.7	4.2

As indicated in Fig. 9, principally the same procedure can be used to estimate total production of shorter time intervals. A particular example will be discussed in the last section.

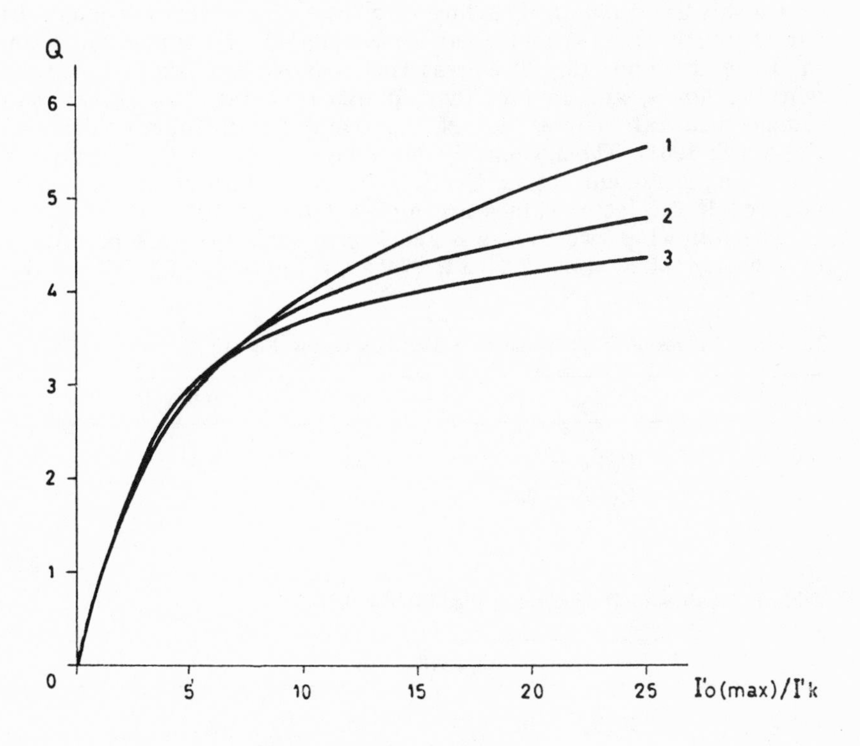

Fig. 11. - Concerning equation (8.2.2.) and table 5.

8.4. Non symmetrical solutions

The assumption made in previous sections, according to which photosynthesis at a given light intensity proceeds at the same constant rate throughout the day, is not generally supported from experimental data. Evening rates at light optimum may drop as low as 1/10 of that during early morning hours (cf. Doty 1959). Evidence exists that this is due, at least in part, to nutrient consumption during intense assimilation.

The drop of rates depending on nutrient consumption can be described by:

$$p_{(t)} = p_{max} \cdot e^{-\nu t} \qquad (8.4.1.)$$

ν shall be called *nutrient depletion coefficient.*

From experimental data, this coefficient has been estimated for two waters in Northern Italy as 0.0745 (Lago Maggiore) and 0.155 (Lago di Mergozzo). There is a perfect correspondance between these coefficients and the trophic level of the lakes concerned. Details shall be given elsewhere (Vollenweider, in prep.).

On this basis, modified surface rate curves for various assumptions concerning F (I'_o/I'_k) and ν can be calculated. Examples are given in Figs. 12 and 13. The resulting curves are skewed toward morning hours, and in fact they fit experimental data better than symmetrical rate curves do (cf. *e.g.* Ohle 1958; Vollenweider and Nauwerck 1961). The asymmetry depends, in addition to the nutrient depletion coefficient, upon I'_o/I'_k too; the skewness is more pronounced if the latter values are higher (cf. Fig. 12).

The following two Tables 6 and 7 give some Q-values pertaining to equation (4.2.) and (7.2.3 a.), (7.2.5 a.) and Figs. 12 and 13 respectively.

Table 6. - Values of Q (cf. equation 8.2.2.) concerning Fig. 12.

ν =	0.0745	0.155
I'_o/I'_k = 10	2.45	1.55
I'_o/I'_k = 20	3.3	2.1

Table 7. - Values of Q concerning Fig. 13, ν = 0.0745

Fig. 13	a)	b)
I'_o/I'_k = 10	2.55	2.45
I'_o/I'_k = 20	3.0	2.75

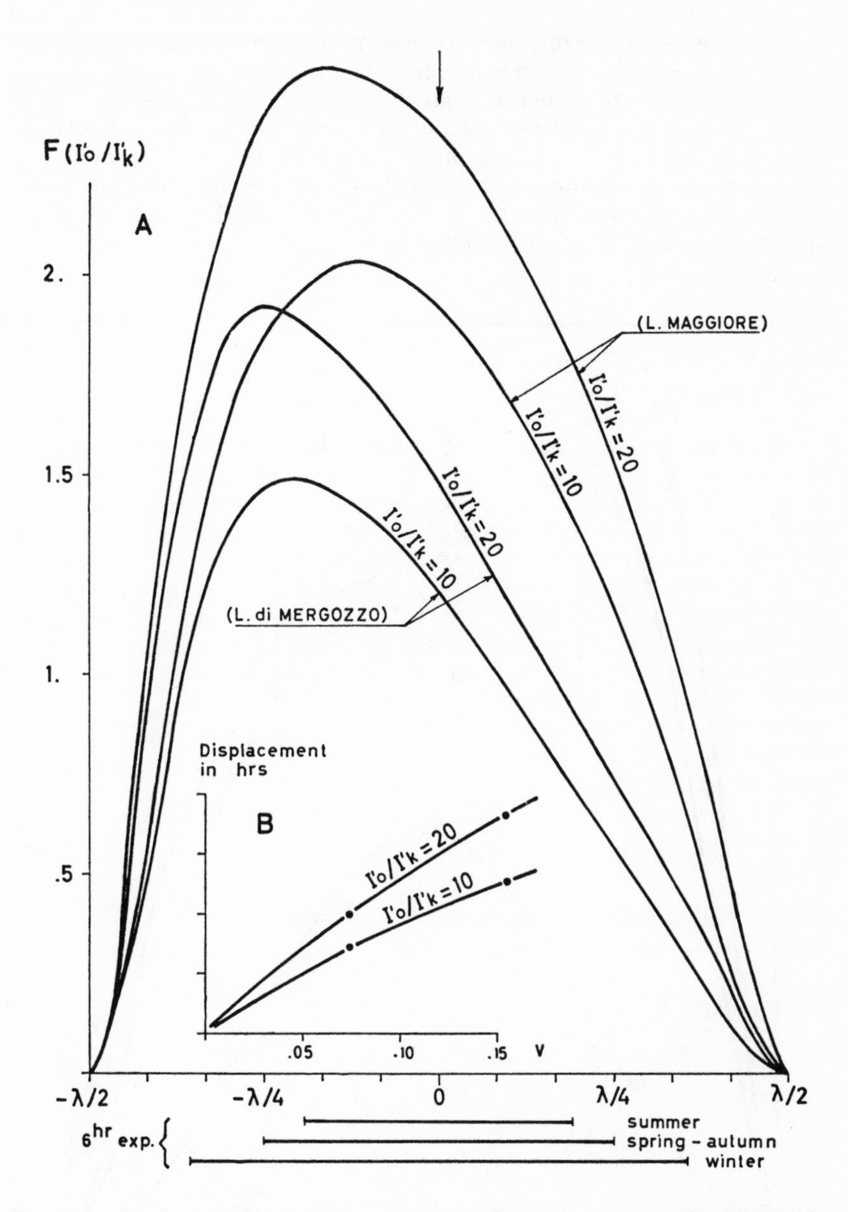

Fig. 12. - A) Asymmetrical surface rate curves for various assumptions concerning $I'_{o\,(max)}/I'_k$ and ν. $F(I'_o/I'_k) = \text{ArSin}\,(I'_o/I'_k)$.

B) Shift of the rate maximum as a function of ν and I'_o/I'_k.

Concerning the horizontal lines at the bottom of the figure cf. appendix.

These values are considerably lower than those resulting from symmetrical solutions. Preliminary calculations on some experimental data are in general somewhat higher than deduced from theory; this is probably due to the fact that the nutrient depletion function is applied here indistinctly to all depths of the photosynthetic layer. The asymmetrical surface rate curves therefore are probably slightly underestimated. However, for a final statement more experimental data are still needed.

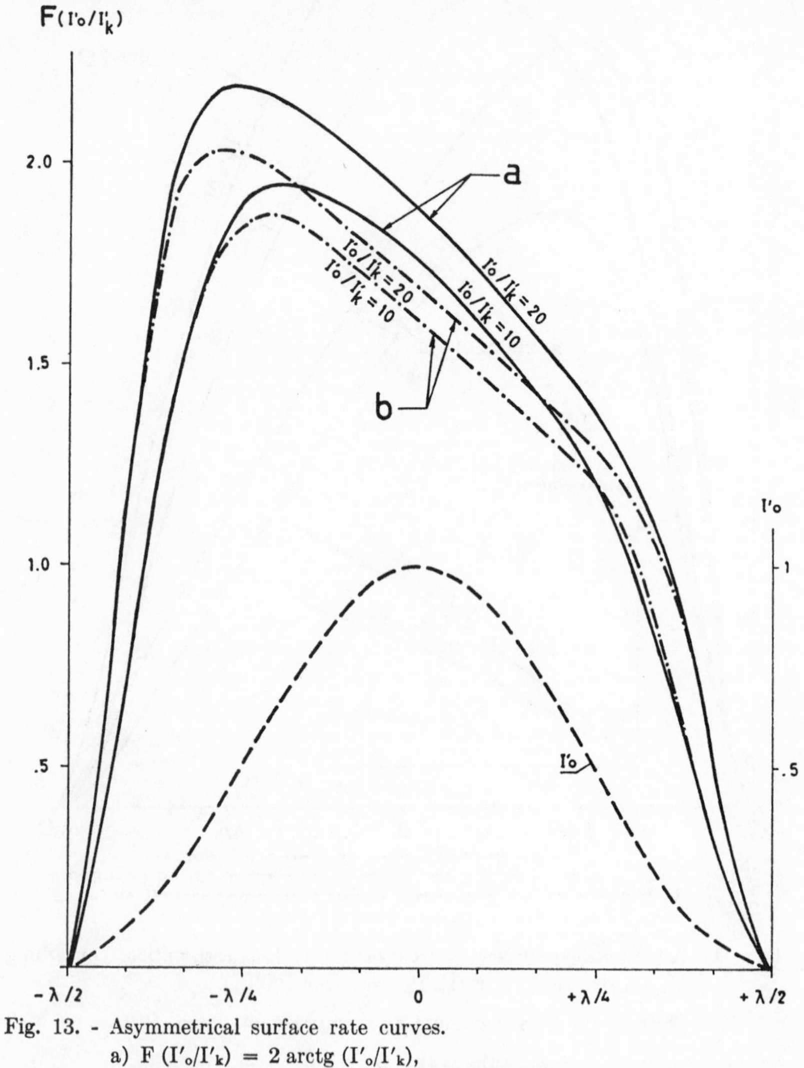

Fig. 13. - Asymmetrical surface rate curves.

a) $F(I'_o/I'_k) = 2 \operatorname{arctg} (I'_o/I'_k)$,

b) $F(I'_o/I'_k) = (I'_o/I'_k) / \sqrt{1 + (2.6\, I'_o/I'_k)^2}$.

DISCUSSION

Both instantaneous rate integrals as well as integrated day-rate curves lead to expressions of the same structure, and, as far as numerical values of various expressions with regard to F(i) are concerned, they appear to be of the same order, ranging from 2 to about 3. For this, most authors who attempted mathematical expressions for photosynthesis integrals did not sufficiently distinguish between *instantaneous rate integrals* and *day rate integrals*; a notable exception is Talling. Yet, this distinction becomes necessary if day rate integrals should be calculated on the basis of exposures lasting only a fraction of the light day.

The evalutation of F(i) can be considered, in a certain sense, also as a statistical problem. Such an approach might be preferable, at least as a starting point, in as far as no hypotheses are introduced as to this function. Yet, this imposes the necessity of referring to a clear classification of the experimental procedures considered, *e. g.* *a*) short experiments not lasting more then two hours; *b*) experiments from 4 to 6 hours symmetrical to local midday; *c*) half day experiments; *d*) 24 hour experiments; etc. So, our formerly used simplified conversion factor 2.4-2.7 was tested on 24 hour experiments (Rodhe, et al., 1958); the interpretation of Rodhe's recent approach leads to a conversion factor of 2.3, referring in part to 24 hour experiments, in part to shorter exposures. Saunders, et al. (1962) claim, on the basis of 24 hour experiments, a higher conversion factor of about 3.3. Talling (1965) finds in African lakes a factor of 2.6 based on exposures of about 2 hours.

From our own work, the best check concerning the validity, or at least the usefulness, of the former considerations, follows from Vollenweider and Nauwerck (1961). There, comparisons were made between experiments lasting a few hours, and cummulative exposures. Three of six experiments were performed in the same way by dividing the light day into five equal periods; the meteorological conditions during those days were very close to what I called a « standard day », as can be seen from the expected incident light sums per period in comparison with the measured ones: cf. Table 8.

Table 8. - Mean values of the incident light energy per period in % during three experiments performed in Lake Erken, 1956.

Period	I	II	III	IV	V
Expected fraction in % of the daily total energy	5.4	25	39.2	25	5.4
measured mean values	5.85	26.1	38.7	25.6	3.75

The day trend of photosynthesis was found to be *non-symmetrical* with regard to local midday. Applying the theoretical treatment discussed in section 8, one should therefore expect that the fractions per period of the total daily photosynthesis are as follows; cf. Table 9 (v was assumed to be 0.075):

Table 9. - Theoretical and measured fractions per period of the total daily photosynthesis per unit of surface.

Period	I	II	III	IV	V
a) using equation (4.2.)					
$I'_o/I'_k = 10$:	11.75	*30.4*	29.8	21.55	6.45
$I'_o/I'_k = 20$:	13.9	*29.75*	28.2	20.8	7.35
b) using equation (7.2.3a.)					
$I'_o/I'_k = 10$:	13.8	*30.8*	27.4	20.9	7.1
$I'_o/I'_k = 20$:	17.0	*29.5*	25.0	19.8	8.7
c) using equation (7.2.5a.)					
$I'_o/I'_k = 10$:	11.7	*31.3*	27.3	21.8	7.9
$I'_o/I'_k = 20$:	17.8	*28.7*	24.3	19.6	9.6
d) *measured mean values*	*9.95*	*30.8*	*30.4*	*21.85*	*6.95*

This table shows that the measured fractions agree excellently with the expected distribution; the best agreement is found in the second and the fourth period whereas the highest deviation between expected and measured values is noted for the first period.

CONCLUSIONS

On the basis of their experiments, Vollenweider and Nauwerck (1961) concluded that, in order to avoid C^{14} losses due to various processus, *in situ* C^{14} experiments should not last more than 4 to 6 hours. On the other hand, restricting the exposure time to only a fraction of the light day creates the not less unsatisfactory conversion problem from short term to day rate estimates.

1) Application of the theoretical treatment presented in this paper shows that these difficulties may be minimized by adopting a procedure slightly different from those normally used. *Day rate estimates would be facilitated if the duration of experiments were chosen*

proportional to the day length factor. So, one could divide the light day into 5 equal periods, as was done arbitrarily in the above discussed Erken experiments, and perform C^{14} exposures during one of the central periods.

2) From the treatment presented can be deduced that the specific shape of the photosynthesis-depth curve is of much less importance with regard to day rate estimates than has generally been thought. The experiments discussed confirm this deduction. The strongest influence will be exerted by the drop of the photosynthetic activity during the course of the day, whereas neither the subsurface light inhibition nor the degree of adaptation to low light intensities (I_k) will produce serious errors.

3) The best period to be selected from a day divided into 5 equal periods will be presumably the second one; as has been shown, the production rate of that period can be expected to be in the order of 30% of the total day rate, provided that nutrient depletion, or its effect on photosynthesis, remains within moderate limits ($v = 0.05$-0.1). With regard to the third period, the uncertainty appears to be somewhat more serious.

4) If instead, no nutrient depletion occurs at all (symmetrical trend of surface rates!), about 25% of the total day rate will be measured during the second and 30% during the third period. One can therefore say that, in any case, during both periods (II + III) about 55%-60% of the total day rate is produced. According to this, the error introduced in estimating day rate integrals of photosynthesis from exposures during the second and third period, will be of the order of ± 10%.

Acknowledgement

This paper is based on a lecture given during the IBP Symposium on Primary Productivity and stimulating discussions with Dr. W. Rodhe, Dr. J. Talling and Dr. C. Goldman.

APPENDIX

CONVERSION OF 6 HR EXPERIMENTS TO DAILY PRODUCTION RATES AT LATITUDE 46°

In our own research we have standardized the exposure time from 9 a.m. to 3 p.m., i.e. to 6 hours irrespective of season. In Upper Italy, such 6 hour experiments cover at summer solstice 38.5%, at winter solstice 71.5% of the geographical light day. The energy fraction of the total theoretical energy of a standard day, falling at a horizontal plane during such exposures, is estimated at summer solstice as about 67%, at winter solstice as about 95%; in between, *i.e.* at spring or autumn equinox it was found as about 81%.

The corresponding fractions of the expected daily photosynthesis per unit of surface area and the corresponding conversion factors are reported in Table 10. In accordance with the theoretical considerations discussed, the variations in the basic assumptions (I'_o/I'_k and asymmetry with regard to local midday) do not appear to have too much importance on the total estimates. With regard to the situations considered here, the uncertainty will in no case be more than ± 10%.

Table 10. - Conversion from 6 hour runs (0900 to 1500) to daily estimates of photosynthesis per unit surface regarding various assumptions concerning I'_o/I'_k and nutrient depletion coefficients, valid for latitude 46°.

a) Day length fractions of 6 hour runs, and fraction of the incident energy per period with regard to total incident energy on a standard day at various seasons.

	winter	spring/autumn	summer
Day length fraction, in %	71.5	50	38.5
Incident energy, in %	95.5	81	67

b) Fractions of total daily photosynthesis to be expected during 6 hour expositions and conversion factors in (parentheses.)

	winter	spring/autumn	summer
ν = 0.075 (e.g. L. Maggiore)			
I'_o/I'_k = 10	91.5	70.5	56
	(1.09)	(1.42)	(1.78)
I'_o/I'_k = 20	89	67.5	53.5
	(1.12)	(1.48)	(1.87)
ν = 0.15 (e.g. L. di Mergozzo)			
I'_o/I'_k = 10	90.5	67.5	52.5
	(1.11)	(1.48)	(1.90)
I'_o/I'_k = 20	88	65	50
	(1.14)	(1.54)	(2.00)

	winter	spring/autumn	summer
c) Conversion range	1.09 - 1.14	1.42 - 1.54	1.78 - 2.00

REFERENCES

Baly, E. C. C. 1935. The kinetics of photosynthesis. - Proc. Roy. Soc. London *117 B* : 218-239.

Doty, M. S. 1959. Phytoplankton photosynthetic periodicity as a function of latitude. J. Mar. Biol. Ass. India, *1*, 66-68.

Ohle, W. 1958. Diurnal production and destruction rates of phytoplankton in lakes. - J. Cons. Explor. Mer. *144* : 129-131.

Rodhe, W. 1965. Standard correlation between pelagic photosynthesis and light. Mep. Ist. Ital. Idrobiol., 18 suppl.: 365-382.

— R. A. Vollenweider and A. Nauwerck. 1958. The primary production and standing crop of phytoplankton. *In* A. A. Buzzati-Traverso, (ed.), Perspectives in Marine Biology. Univ. of California. 299-325.

Ryther, J. H. and C. S. Yentsch, 1957. The estimation of phytoplankton production in the ocean from chlorophyll and light data. Limnol. Oceanogr. *2* : 281-286.

Saunders, G. W., F. B. Trama and R. W. Bachmann. 1962. Evaluation of a modified C^{14} technique for shipboard estimation of photosynthesis in large lakes. - Great Lakes Res. Div., 8, Univ. of Michigan, 1-61.

Smith, E. L. 1936. Photosynthesis in relation to light and carbon dioxide. - Proc. Nat. Acad. Science, Wash., *22* : 504.

Steemann Nielsen, E. 1952. The use of radioactive carbon (C^{14}) for measuring organic production in the sea. - J. Cons. Explor. Mer *18* : 117-140.

Steele, J. H. 1962. Environmental control of photosynthesis in the sea. - Limnol. Oceanogr. 7 : 137-150.

Talling, J. F. 1957. The phytoplankton population as a compound photosynthetic system. - The New Phytologist, *56* : 133-149.

— 1961. Photosynthesis under natural conditions. - Annual Rev. Plant. Physiol. *12* : 133-154.

— 1965. The photosynthetic activity of phytoplankton in East African Lakes. Int. Rev. ges. Hydrobiol. *50* : 1-32.

Vollenweider, R. A. 1958. Sichttiefe und Produktion. - Verh. Int. Ver. Limnol. *13* : 142.

— 1960. Beiträge zur Kenntnis optischer Eigenschaften der Gewässer und Primärproduktion. - Mem. Ist. Ital. Idrobiol. *12* :201-244.

— 1961. Photometric studies in inland waters. Relations existing in the spectral extinction of light in water. - Mem. Ist. Ital. Idrobiol. *13* : 87-113.

— and A. Nauwerck. 1961. Some observations on the C^{14} method for measuring primary production. - Verh. Int. Ver. Limnol. *14* : 134-139.

INDEX

Index of Waters

Index of Organisms

INDEX OF SUBJECTS

INDEX OF SUBJECTS